PROGRESS IN

Nucleic Acid Research and Molecular Biology

Volume 20

PROGRESS IN
Nucleic Acid Research and Molecular Biology

edited by

WALDO E. COHN
Biology Division
Oak Ridge National Laboratory
Oak Ridge, Tennessee

Volume 20

1977

ACADEMIC PRESS
New York San Francisco London
A Subsidiary of Harcourt Brace Jovanovich, Publishers

COPYRIGHT © 1977, BY ACADEMIC PRESS, INC.
ALL RIGHTS RESERVED.
NO PART OF THIS PUBLICATION MAY BE REPRODUCED OR
TRANSMITTED IN ANY FORM OR BY ANY MEANS, ELECTRONIC
OR MECHANICAL, INCLUDING PHOTOCOPY, RECORDING, OR ANY
INFORMATION STORAGE AND RETRIEVAL SYSTEM, WITHOUT
PERMISSION IN WRITING FROM THE PUBLISHER.

ACADEMIC PRESS, INC.
111 Fifth Avenue, New York, New York 10003

United Kingdom Edition published by
ACADEMIC PRESS, INC. (LONDON) LTD.
24/28 Oval Road, London NW1

LIBRARY OF CONGRESS CATALOG CARD NUMBER: 63–15847

ISBN 0–12–540020–9

PRINTED IN THE UNITED STATES OF AMERICA

Contents

LIST OF CONTRIBUTORS	ix
ABBREVIATIONS AND SYMBOLS	xi
SOME ARTICLES PLANNED FOR FUTURE VOLUMES	xv

Correlation of Biological Activities with Structural Features of Transfer RNA
B. F. C. CLARK

I. Biological Activities of tRNAs	1
II. Subclassification and Generalized Primary Structure of tRNAs	4
III. Significance of G · U Base-Pairs	8
IV. Position of Pseudouridine	9
V. Exceptions to the Generalized Structures	10
VI. Three-Dimensional Structure of tRNA^{Phe}	11
VII. Structural Correlations	15
VIII. Correlation of Chemical Reactivity with Three-Dimensional Structure	17
References	18

Bleomycin, an Antibiotic That Removes Thymine from Double-Stranded DNA
WERNER E. G. MÜLLER AND RUDOLF K. ZAHN

I. Introduction	22
II. Reaction of Bleomycin with Native DNA *in Vitro*	23
III. Reaction with Other Polynucleotides *in Vitro*	37
IV. Effect on DNA-, RNA-, and Protein Synthesis *in Vitro*	38
V. Effect on Nucleases *in Vitro*	42
VI. Influence on Nucleic Acids *in Vivo*	43
VII. Influence on Nucleic Acid Metabolism *in Vivo*	46
VIII. Medical Implications	49
IX. Future Directions	51
References	54

Mammalian Nucleolytic Enzymes
HALINA SIERAKOWSKA AND DAVID SHUGAR

I. Introduction	60
II. Endoribonucleases	61

III.	Exoribonucleases	80
IV.	Endodeoxyribonucleases	82
V.	Exodeoxyribonucleases	93
VI.	Endo-Exodeoxyribonuclease with Preference for Poly(dT)	95
VII.	Single-Stranded-Nucleate Endonuclease (Sugar-Nonspecific Endonuclease; 5'-Endonuclease), EC 3.1.4.21	95
VIII.	Sugar-Nonspecific Exonucleases	97
IX.	2′ : 3′-Cyclic-nucleotide 3′-Phosphodiesterase, EC 3.1.4.37	102
X.	Localization and Function of Nucleases	102
XI.	Virus-Associated Nucleases	106
XII.	Nucleases and Cellular Repair Processes	114
	References	120

Transfer RNA in RNA Tumor Viruses
LARRY C. WATERS AND BETH C. MULLIN

I.	Introduction	131
II.	Nucleic Acid Components of RNA Tumor Viruses	134
III.	Transfer RNA in the Free 4 S RNA of the Virus	138
IV.	Transfer RNA Associated with the 70 S RNA of the Virus	141
V.	Function of Transfer RNA in RNA Tumor Viruses	146
VI.	Summary	155
	References	155

Integration versus Degradation of Exogenous DNA in Plants: An Open Question
PAUL F. LURQUIN

I.	Introduction	161
II.	Experimental Systems	163
III.	Attempts to Conclude and Prospectives	196
	References	204

Initiation Mechanisms of Protein Synthesis
MARIANNE GRUNBERG-MANAGO AND FRANÇOIS GROS

I.	Introduction	209
II.	Components Involved in the Initiation Step	210
III.	Total Reaction Sequence	220
IV.	Reversible Association of Ribosomal Subunits	227
V.	Role of Ribosomes and Initiation Factors in Recognizing the Initiation Signals	235
VI.	Pending Questions	268

CONTENTS

VII. Conclusion	274
References	276
SUBJECT INDEX	285
CONTENTS OF PREVIOUS VOLUMES	289

List of Contributors

Numbers in parentheses indicate the pages on which the authors' contributions begin.

B. F. C. CLARK (1), *Department of Chemistry, Aarhus University, Aarhus, Denmark*

FRANÇOIS GROS (209), *Institut Pasteur, Paris, France*

MARIANNE GRUNBERG-MANAGO (209), *Institut de Biologie Physico-Chimique, Paris, France*

PAUL F. LURQUIN (161), *Radiobiology Department, Centre d'Etude de l'Energie Nucléaire, C.E.N./S.C.K., Mol, Belgium*

WERNER E. G. MÜLLER (21), *Institut für Physiologische Chemie, Universität Mainz, Mainz, West Germany*

BETH C. MULLIN* (131), *Carcinogenesis Program, Biology Division, Oak Ridge National Laboratory and the University of Tennessee—Oak Ridge Graduate School of Biomedical Sciences, Oak Ridge, Tennessee*

DAVID SHUGAR (59), *Institute of Biochemistry and Biophysics, Academy of Sciences, Warsaw, Poland*

HALINA SIERAKOWSKA (59), *Institute of Biochemistry and Biophysics, Academy of Sciences, Warsaw, Poland*

LARRY C. WATERS (131), *Carcinogenesis Program, Biology Division, Oak Ridge National Laboratory and the University of Tennessee—Oak Ridge Graduate School of Biomedical Sciences, Oak Ridge, Tennessee*

RUDOLF K. ZAHN (21), *Institut für Physiologische Chemie, Universität Mainz, Mainz, West Germany*

---

* Present address: Biology Department, Wilmington College, Wilmington, Ohio.

Abbreviations and Symbols

All contributors to this Series are asked to use the terminology (abbreviations and symbols) recommended by the IUPAC–IUB Commission on Biochemical Nomenclature (CBN) and approved by IUPAC and IUB, and the Editor endeavors to assure conformity. These Recommendations have been published in many journals (1, 2) and compendia (3) in four languages and are available in reprint form from the NAS–NRC Office of Biochemical Nomenclature (OBN), as stated in each publication, and are therefore considered to be generally known. Those used in nucleic acid work, originally set out in section 5 of the first Recommendations (1) and subsequently revised and expanded (2, 3), are given in condensed form (I–V) below for the convenience of the reader. Authors may use them without definition, when necessary.

I. Bases, Nucleosides, Mononucleotides

1. *Bases* (in tables, figures, equations, or chromatograms) are symbolized by Ade, Gua, Hyp, Xan, Cyt, Thy, Oro, Ura; Pur = any purine, Pyr = any pyrimidine, Base = any base. The prefixes S–, H_2, F–, Br, Me, etc., may be used for modifications of these.

2. *Ribonucleosides* (in tables, figures, equations, or chromatograms) are symbolized, in the same order, by Ado, Guo, Ino, Xao, Cyd, Thd, Ord, Urd (Ψrd), Puo, Pyd, Nuc. Modifications may be expressed as indicated in (1) above. Sugar residues may be specified by the prefixes r (optional), d (=deoxyribo), a, x, l, etc., to these, or by two three-letter symbols, as in Ara-Cyt (for aCyd) or dRib-Ade (for dAdo).

3. *Mono-, di-, and triphosphates of nucleosides* (5') are designated by NMP, NDP, NTP. The N (for "nucleoside") may be replaced by any one of the nucleoside symbols given in II-1 below. 2'-, 3'-, and 5'- are used as prefixes when necessary. The prefix d signifies "deoxy." [Alternatively, nucleotides may be expressed by attaching P to the symbols in (2) above. Thus: P-Ado = AMP; Ado-P = 3'-AMP.] cNMP '-cyclic 3' : 5'-NMP; Bt_2cAMP = dibutyryl cAMP, etc.

II. Oligonucleotides and Polynucleotides

1. Ribonucleoside Residues

(a) Common: A, G, I, X, C, T, O, U, Ψ, R, Y, N (in the order of I-2 above).

(b) Base-modified: sI or M for thioinosine = 6-mercaptopurine ribonucleoside; sU or S for thiouridine; brU or B for 5-bromouridine; hU or D for 5,6-dihydrouridine; i for isopentenyl; f for formyl. Other modifications are similarly indicated by appropriate *lower-case* prefixes (in contrast to I-1 above) (2, 3).

(c) Sugar-modified: prefixes are d, a, x, or l as in I-2 above, alternatively, by *italics* or **boldface** type (with definition) unless the entire chain is specified by an appropriate prefix. The 2'-O-methyl group is indicated by *suffix* m (e.g., -Am- for 2'-O-methyladenosine, but -mA- for N-methyladenosine).

(d) Locants and multipliers, when necessary, are indicated by superscripts and subscripts, respectively, e.g., $-m_2^6A-$ = 6-dimethyladenosine; $-s^4U-$ or $-^4S-$ = 4-thiouridine; $-ac^4Cm-$ = 2'-O-methyl-4-acetylcytidine.

(e) When space is limited, as in two-dimensional arrays or in aligning homologous sequences, the prefixes may be placed *over the capital letter*, the suffixes *over the phosphodiester symbol*.

2. Phosphoric Acid Residues [left side = 5', right side = 3' (or 2')]

(a) Terminal: p; e.g., pppN . . . is a polynucleotide with a 5'-triphosphate at one end; Ap is adenosine 3'-phosphate; C > p is cytidine 2':3'-cyclic phosphate *(1, 2, 3)*; p < A is adenosine 3':5'-cyclic phosphate.

(b) Internal: hyphen (for known sequence), comma (for unknown sequence); unknown sequences are enclosed in parentheses. E.g., pA-G-A-C(C_2,A,U)A-U-G-C > p is a sequence with a (5') phosphate at one end, a 2':3'-cyclic phosphate at the other, and a tetranucleotide of unknown sequence in the middle. **(Only codon triplets are written without some punctuation separating the residues.)**

3. Polarity, or Direction of Chain

The symbol for the phosphodiester group (whether hyphen or comma or parentheses, as in 2b) represents a 3'-5' link (i.e., a 5' . . . 3' chain) unless otherwise indicated by appropriate numbers. "Reverse polarity" (a chain proceeding from a 3' terminus at left to a 5' terminus at right) may be shown by numerals or by right-to-left arrows. Polarity in any direction, as in a two-dimensional array, may be shown by appropriate rotation of the (capital) letters so that 5' is at left, 3' at right when the letter is viewed right-side-up.

4. Synthetic Polymers

The complete name or the appropriate group of symbols (see II-1 above) of the repeating unit, **enclosed in parentheses if complex or a symbol,** is either (a) preceded by "poly," or (b) followed by a subscript "n" or appropriate number. **No space follows "poly"** *(2, 5)*.

The conventions of II-2b are used to specify known or unknown (random) sequence, e.g.,

polyadenylate = poly(A) or $(A)_n$, a simple homopolymer;

poly(3 adenylate, 2 cytidylate) = poly(A_3C_2) or $(A_3,C_2)_n$, an *irregular* copolymer of A and C in 3:2 proportions;

poly(deoxyadenylate-deoxythymidylate) = poly[d(A-T)] or poly (dA-dT) or (dA-dT)$_n$ or d(A-T)$_n$, an *alternating* copolymer of dA and dT;

poly(adenylate,guanylate,cytidylate,uridylate) = poly(A,G,C,U) or $(A,G,C,U)_n$, a random assortment of A, G, C, and U residues, proportions unspecified.

The prefix copoly or oligo may replace poly, if desired. The subscript "n" may be replaced by numerals indicating actual size, e.g., $(A)_n \cdot (dT)_{12-18}$.

III. Association of Polynucleotide Chains

1. *Associated* (e.g., H-bonded) chains, or bases within chains, are indicated by a *center dot* (not a hyphen or a plus sign) separating the *complete* names or symbols, e.g.:

poly(A) · poly(U)   or $(A)_n \cdot (U)_m$
poly(A) · 2 poly(U)   or $(A)_n \cdot 2(U)_m$
poly(dA-dC) · poly(dG-dT)   or $(dA-dC)_n \cdot (dG-dT)_m$.

2. *Nonassociated* chains are separated by the plus sign, e.g.:

2[poly(A) · poly(U)] $\xrightarrow{\Delta}$ poly(A) · 2 poly (U) + poly(A)
or $2[A_n \cdot U_m] \rightarrow A_n \cdot 2U_m + A_n$.

3. Unspecified or unknown association is expressed by a comma (again meaning "unknown") between the completely specified chains.

Note: In all cases, each chain is completely specified in one or the other of the two systems described in II-4 above.

IV. Natural Nucleic Acids

RNA	ribonucleic acid or ribonucleate
DNA	deoxyribonucleic acid or deoxyribonucleate
mRNA; rRNA; nRNA	messenger RNA; ribosomal RNA; nuclear RNA
hnRNA	heterogeneous nuclear RNA
D-RNA; cRNA	"DNA-like" RNA; complementary RNA
mtDNA	mitochondrial DNA
tRNA	transfer (or acceptor or amino-acid-accepting) RNA; replaces sRNA, which is not to be used for any purpose
aminoacyl-tRNA	"charged" tRNA (i.e., tRNA's carrying aminoacyl residues); may be abbreviated to AA-tRNA
alanine tRNA or tRNA^{Ala}, etc.	tRNA normally capable of accepting alanine, to form alanyl-tRNA
alanyl-tRNA or alanyl-tRNA^{Ala}	The same, with alanyl residue covalently attached. [*Note:* fMet = formylmethionyl; hence tRNA^{fMet}, identical with tRNA$_f^{Met}$]

Isoacceptors are indicated by appropriate subscripts, i.e., tRNA$_1^{Ala}$, tRNA$_2^{Ala}$, etc.

V. Miscellaneous Abbreviations

P$_i$, PP$_i$	inorganic orthophosphate, pyrophosphate
RNase, DNase	ribonuclease, deoxyribonuclease
t_m (not T_m)	melting temperature (°C)

Others listed in Table II of Reference 1 may also be used without definition. No others, with or without definition, are used unless, in the opinion of the editor, they increase the ease of reading.

Enzymes

In naming enzymes, the 1972 recommendations of the IUPAC-IUB Commission on Biochemical Nomenclature (CBN)(*4*), are followed as far as possible. At first mention, each enzyme is described *either* by its systematic name *or* by the equation for the reaction catalyzed *or* by the recommended trivial name, followed by its EC number in parentheses. Thereafter, a trivial name may be used. Enzyme names are not to be abbreviated except when the substrate has an approved abbreviation (e.g., ATPase, but not LDH, is acceptable).

REFERENCES*

1. *JBC* **241**, 527 (1966); *Bchem* **5**, 1445 (1966); *BJ* **101**, 1 (1966); *ABB* **115**, 1 (1966), **129**, 1 (1969); and elsewhere.†
2. *EJB* **15**, 203 (1970); *JBC* **245**, 5171 (1970); *JMB* **55**, 299 (1971); and elsewhere.†
3. "Handbook of Biochemistry" (H. A. Sober, ed.), 2nd ed. Chemical Rubber Co., Cleveland, Ohio, 1970, Section A and pp. H130–133.
4. "Enzyme Nomenclature," Elsevier Scientific Publ. Co., Amsterdam, 1973, and Supplement No. 1, *BBA* **429**, (1976).

* Contractions for names of journals follow.

† Reprints of all CBN Recommendations are available from the Office of Biochemical Nomenclature (W. E. Cohn, Director), Biology Division, Oak Ridge National Laboratory, Box Y, Oak Ridge, Tennessee 37830, USA.

5. "Nomenclature of Synthetic Polypeptides," *JBC* **247**, 323 (1972); *Biopolymers* **11**, 321 (1972); and elsewhere.*

Abbreviations of Journal Titles

Journals	Abbreviations used
Annu. Rev. Biochem.	ARB
Arch. Biochem. Biophys.	ABB
Biochem. Biophys. Res. Commun.	BBRC
Biochemistry	Bchem
Biochem. J.	Bj
Biochim. Biophys. Acta	BBA
Cold Spring Harbor Symp. Quant. Biol.	CSHSQB
Eur. J. Biochem.	EJB
Fed. Proc.	FP
J. Amer. Chem. Soc.	JACS
J. Bacteriol.	J. Bact.
J. Biol. Chem.	JBC
J. Chem. Soc.	JCS
J. Mol. Biol.	JMB
Nature, New Biology	Nature NB
Proc. Nat. Acad. Sci. U.S.	PNAS
Proc. Soc. Exp. Biol. Med.	PSEBM
Progr. Nucl. Acid Res. Mol. Biol.	This Series

* Reprints of all CBN Recommendations are available from the NRC Office of Biochemical Nomenclature (W. E. Cohn, Director), Biology Division, Oak Ridge National Laboratory, Box Y, Oak Ridge, Tennessee 37830, USA.

Some Articles Planned for Future Volumes

The Transfer RNAs of Cellular Organelles
    W. E. BARNETT, L. I. HECKER AND S. D. SCHWARTZBACH

Mechanisms in Polypeptide Chain Elongation on Ribosomes
    E. BERMEK

Ribonucleotide Reductase
    F. D. HAMILTON

Regulation of the Synthesis of Aminoacyl-tRNAs and tRNAs
    D. SÖLL

Informosomes and Their Protein Components
    A. S. SPIRIN

Bioenergetics of the Ribosome
    A. S. SPIRIN

Physical Structure, Chemical Modification and Functional Role of the Acceptor Terminus of tRNA
    M. SPRINZL AND F. CRAMER

The Biochemical and Microbiological Action of Platinum Compounds
    A. J. THOMSON AND J. J. ROBERTS

Synthetic Oligodeoxyribonucleotides in the Analysis of DNA Structure and Function
    R. WU

Structure and Functions of Ribosomal RNA
    R. ZIMMERMANN

Correlation of Biological Activities with Structural Features of Transfer RNA

B. F. C. Clark

*Department of Chemistry
Aarhus University
Aarhus, Denmark*

I.	Biological Activities of tRNAs	1
II.	Subclassification and Generalized Primary Structure of tRNAs	4
III.	Significance of G · U Base-Pairs	8
IV.	Position of Pseudouridine	9
V.	Exceptions to the Generalized Structure	10
VI.	Three-Dimensional Structure of tRNA^{Phe}	11
VII.	Structural Correlations	15
VIII.	Correlation of Chemical Reactivity with Three-Dimensional Structure	17
	References	18

I. Biological Activities of tRNAs

Transfer RNA (tRNA), a class of small RNA molecules of molecular weight ratio (M_r) about 25,000, is becoming more and more intriguing as it is found to be implicated in many activities other than those associated with its traditional role in protein biosynthesis. There are of the order of 55 different species of tRNA in a particular cell type, and they appear to arise from longer precursor tRNA molecules. The precursor molecules are primary gene products synthesized by RNA polymerase under the direction of the DNA genome. One of the characteristics of tRNAs is their high content of modified bases compared with other RNA molecules. So far it is not clearly established at which step, and in what order during the maturation of tRNAs, the modified bases are formed.

Now that a three-dimensional structure for one tRNA species is known, there is interest in attempting to consider the functions of tRNA in terms of this structure. A summary of the current state of identification of functions or, more precisely, biological activities of tRNAs in prokaryotes and eukaryotes is given in Table I. Transfer RNA plays a central role in protein biosynthesis (activities 1–9), and much more is known about the biochemistry of the processes involved than in any other function. It is therefore probably more feasible to

TABLE I
ACTIVITIES OF tRNA

Protein biosynthesis
1. Activation of amino acids
2. Recognition by EF-Tu
3. Location in A-site
4. Decoding mRNA
5. Signal for "magic spot"[a]
6. Recognition of initiator tRNA by IF ⎫
7. Location in I-site (part of P-site) ⎬ Special for Initiation
8. Recognition by transformylase ⎭
9. Regulation
   a. Repressor
   b. Feedback inhibitor
   c. Suppression

RNA metabolism
10. As precursor by cleavase and maturation enzymes
11. Enzymes modifying bases
12. C-C-A repair enzyme
13. As peptidyl-tRNA by hydrolase
14. Nuclease degradation
15. Reverse transcriptase primer
16. Selection during viral encapsulation
17. Correlation with 3' end of viral RNA
18. Alteration of *Escherichia coli* ENDO I specificity

Cell wall biosynthesis
19. Transfer of amino acids to wall structure

[a] ppGpp and pppGpp.

relate structure and function in this field. The additional activities or functions are listed for future interest.

In normal protein biosynthesis, each tRNA species is charged with an amino acid (activity 1) by an aminoacyl-tRNA synthetase ("activating enzyme"), and the charged species is then carried to the ribosome in the form of a ternary complex made with elongation-factor-Tu (EF-Tu) and GTP (activity 2). In similar fashion, the unique tRNA species, initiator tRNA, a special class of methionine tRNA, is thought to be carried to the ribosome by an initiation factor and GTP (activity 6). The aminoacyl-tRNA is located by an uncharacterized mechanism in the A-site of the ribosome (activity 3) where it decodes mRNA via its anticodon triplet (activity 4). In contrast, the initiator tRNA, formylmethionyl-tRNA^{fMet} in prokaryotes, and methionyl-tRNA^{fMet} (sometimes called Met-tRNA$_i$) in eukaryotes, is located in the initiation (I)-site on the small ribosomal subunit (activity 7) for decoding the initiator triplet codon. This site becomes part of the ribosomal P-site. Since the prokaryotic Met-tRNA^{fMet} must be formylated, it is also recognized by a special enzyme for this, the transformylase (activity 8).

When prokaryotic cells are starved for amino acids, an unusual role may be detected for uncharged tRNA. The uncharged tRNA is bound to the ribosomal A-site as though in mRNA decoding, but sets off a signal for the formation, by the so-called stringent factor, of unusual guanosine nucleotide derivatives originally called "magic spots" (activity 5), now ppGpp and pppGpp.

A group of somewhat poorly defined (mechanistically) roles for tRNA in the regulation of protein biosynthesis has been collected as "function" 9. These include bacterial roles as a repressor, feedback inhibitor of the aromatic amino-acid pathway, and the well-characterized role as suppressor of nonsense mutations. Additionally in eukaryotes, there are uncharacterized roles relating to the binding to tryptophan pyrrolase (in *Drosophila*) and the inhibition of protein synthesis in virally infected animal cells by the degradation of one or more essential tRNA species, a process that seems to accompany interferon production.

In addition to their role in protein synthesis, tRNAs have activities concerned with a number of reactions conveniently classified as being concerned with RNA metabolism (activities 10–18). The tRNA is trimmed to size and matured from the precursor molecule by a series of as yet poorly characterized enzymes presumably linked to other metabolic roles (activities 10–12). The C-C-A-repair enzyme (activity 12) certainly repairs tRNAs with incomplete 3' ends and probably is concerned with maturation as well. If peptidyl-tRNA should fall off the ribosome, there is a peptidyl-tRNA hydrolase (activity 13) that can remove the peptide, thus permitting the tRNA to be recycled for use in protein biosynthesis. Little is known about tRNA turnover, but specific nucleases (activity 14) must be involved.

Recently some interesting properties of eukaryotic tRNAs have been identified (activities 15–17) with regard to virus metabolism. Reverse transcriptase from RNA tumor viruses uses a specific tRNA^{Trp} as a primer during synthesis of virally coded DNA. Furthermore, a certain number of selected tRNA species (perhaps ten to fifteen) are incorporated noncovalently into RNA tumor virus particles during encapsulation from the cell membrane.[1] It is also well established that many viruses, especially plant viruses, have elements of tRNA structure that permit their 3' ends to be charged specifically with an amino

---

[1] See article by Waters and Mullin in this volume.

acid (activity 17), e.g., turnip yellow mosaic viral RNA with valine. Activity 18 is not well-defined, but the specificity of bacterial endonuclease I for double-stranded cuts in DNA is altered to a "nicking" property (single-strand cleavage) when it binds tRNA.

Finally, there is a special class of tRNAs that is chargeable and that transfers amino acids into cell-wall structures (activity 19). These tRNAs, best characterized for a series of staphylococcal tRNA^{Gly} species, do not contain all the constant features of the general cloverleaf structure, presumably including those for ribosome binding. However, their primary structures can be arranged in normal cloverleaf structures (1).

II. Subclassification and Generalized Primary Structure of tRNAs

The information from 77 different tRNA sequences known in July 1976 and listed in Table II (Barrell and Clark (1), Clark and Klug (2), plus 14 new structures) has been conveniently incorporated into standard "cloverleaf" forms as shown in Fig. 1. This remarkable feature of all the primary structures was first proposed by Holley et al. (3) and is based on Watson–Crick base-pairing. The simple classification shown in Fig. 1 is based on size (see Table II for species).

Thus we have small and large tRNAs dependent upon the size of the extra arm (see also Fig. 2 and Table II). The fourteen new structures are those for Ec Arg$_2$ y Arg$_2$, Ec Cys, T4 Gln, Ec Gly$_2$, Ec Lys, Rbl Lys$_{2A(2B)}$, Rbl Lys$_3$, An fMet, Bsu fMet, Sf fMet, Rl Ser$_3$, Bs Val$_{2A}$, y Val$_{2A}$.

In Fig. 2 (p. 6), I have incorporated the information from the small class 1 sequences into a standard generalized cloverleaf. For this information, 56 of the 63 class 1 species indicated in Table II have been

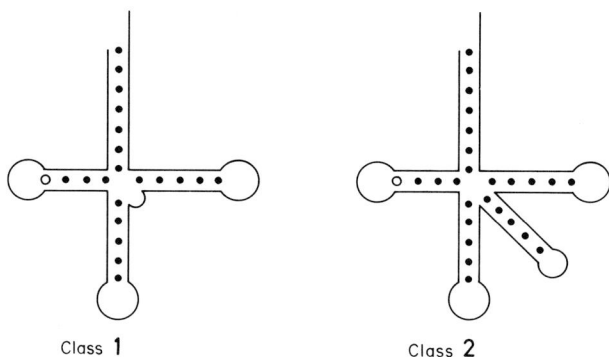

FIG. 1. Simplified classes of tRNA. The open circle indicates that a Watson–Crick base-pair is not always found in this position.

TABLE II
CLASSES OF tRNA ACCORDING TO ARM SIZES AND STRUCTURE CORRELATIONS[a]

Class 1 (63)	
A(36)	4 base-pairs in b-stem (D stem)
	5 bases in extra loop III and containing m^7G
(and with A9)	Ec Ala$_1$, Ec Arg$_1$, Ec Arg$_2$, Ec Asp$_1$, Ec Gly$_3$, Ec + Sal His$_1$, Ec Ile$_1$, Ec Lys, y Lys, Rbl Lys$_{2A(2B)}$, Rbl Lys$_3$, Svt Lys$_4$, An fMet, Ec Met, y Met$_3$, Mye + Rbl Met$_4$, Ec Phe, Mp Phe, Bs Phe, y Phe, Wg + Ps Phe, Rbl Phe, T4 Pro, Ec Trp, Ec Val$_1$, Ec Val$_{2A}$, Ec Val$_{2B}$, Bs Val$_{2A}$, Mye Val$_1$
(and with m^1G9 or G9)	y Cys, Ec fMet, Bsu fMet, y fMet, Mye + Rbl + St + Xl + Smg + Hup fMet, y Trp, Chi + Bl Trp
B(6)	4 base-pairs in b-stem (D stem)
	5 bases in extra loop III without m^7G
	y Ala$_1$, Tu Ala$_1$, y Arg$_3$, Hay Lys, Sf fMet, Ec Thr
C(7)	4 base-pairs in b-stem (D stem)
	4 bases in extra loop
	y Asp$_1$, Ec Glu$_1$, Ec Glu$_2$, Ec + Sal Gly$_1$, Sta Gly, Wg Gly, y Gly
D(14)	3 base-pairs in b-stem (D stem)
	Small extra arm with several bases (3–5)
	y Arg$_2$ (5), Ec Cys (4), Ec Gln$_1$ (5), Ec Gln$_2$ (5), T4 Gln (5), y Glu$_3$ (4), Ec Gly$_2$ (4), T4 Gly (4), Tu Ile (5), y Tyr (5), Tu Tyr (5), y Val$_1$ (5), y Val$_{2A}$ (5), Tu Val (3)
Class 2 (14)	3 base-pairs in b-stem (D stem)
	Large extra arm with several bases (13–21)
	Ec + Sal Leu$_1$ (15), Ec Leu$_2$ (15), T4 Leu (14), y Leu$_3$ (13), y Leu$_4$ (13), Ec Ser$_1$ (16), Ec Ser$_3$ (21), T4 Ser (18), y Ser$_1$ (14), y Ser$_2$ (14), Rl Ser$_1$ (14), Rl Ser$_3$ (14), Ec Tyr$_1$ (13), Ec Tyr$_2$ (13)

[a] Abbreviations: An, *Anacystis nidulans;* Bl, Beef liver; Bs, *Bacillus stearothermophilus;* Bsu, *Bacillus subtilis;* Chi, chicken; Ec, *Escherichia coli;* Hay, haploid yeast; Hup, human placenta; Mp, *Mycoplasma;* Mye, myeloma; Ps, *Pisum sativum;* Rbl, rabbit liver; Rl, rat liver; Sal, *Salmonella typhimurium;* Sf, *Streptococcus faecalis;* Smg, Sheep mammary gland; Sta, *Staphylococcus;* St, Salmon testis; Svt, Svt 2 cells; Tu, *Torulopsis utilis;* Wg, wheat germ; Xl, *Xenopus laevis;* y, yeast.

used: clear exceptions, based on functions such as fMet species and Sta Gly, to the generalized form are omitted.

As shown in Fig. 2, the Watson–Crick base-pairs give rise to four double-helical stem regions a, b, c, and e, three of which are closed by non-base-paired loop regions I, II and IV. Another point of nomenclature illustrated in Fig. 2 is that a *stem* plus a *loop* is also called an *arm*. Most of the tRNA cloverleaf forms have remarkably constant regions. There is a phosphate at the 5′ end, whereas at the 3′ end, where the amino acid is attached, there is a common sequence C-C-A. In addi-

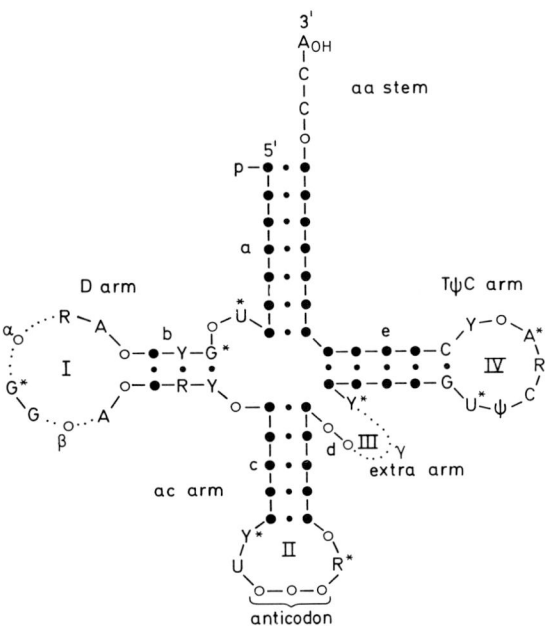

FIG. 2. Class 1 generalized cloverleaf. Solid circles (● · ●) indicate bases involved in helical stems containing Watson–Crick base-pairs. Open circles (○) signify non-Watson–Crick base-paired bases in cloverleaf arrangement. Starred nucleosides are positions where modifications of that nucleoside can possibly occur. R = purine nucleoside, Y = pyrimidine nucleoside. The variable regions (dotted stretches) are indicated by α, β and γ.

tion, stems a, c, and e contain 7, 5 and 5 base-pairs, respectively, and loops II and IV each contain 7 non-base-paired nucleotides.

Stem b contains 3 or 4 Watson–Crick base-pairs (Fig. 2, open circles in stem b; and see Table II), but our knowledge of the three-dimensional structure of tRNA permits us to propose that the various non-Watson–Crick base-pairs occurring in the fourth position of the stem b can be accommodated without distortion of the tertiary structure (2). There is then no point in differentiating between tRNAs based on whether or not this position is a Watson–Crick base-pair (2). The variable regions are confined to loops I and III and stem d. The arms can be also referred to for convenience by trivial historical names as shown in Fig. 2, e.g., loop I + stem b as the D arm, since it usually contains some dihydrouridine; loop II + stem c as the anticodon (ac) arm, since the loop contains the anticodon; loop III + stem d as the "variable finger" or "extra" arm; and loop IV + stem e as the T-Ψ-C arm. Stem a is also called the amino-acid (aa) stem, since this is where the amino acid is attached.

There are also many invariant and semi-invariant nucleotide positions in the generalized structure. The invariant positions are shown by the appropriate nucleoside symbols whereas the semi-invariant positions are shown by R (A or G) or Y (C or U). The dotted stretches in the diagram are regions of variable length. For example, the extra arm can vary in length from 3 to 21 nucleotides. In contrast, the D loop (I) is much less variable in length—from only 7 to 10 nucleotides long when the base-pair b4, even though not of the Watson–Crick type, is considered as part of the D stem. (These variables include class 2.)

It was an interesting but open question until recently what the invariant and semi-invariant nucleosides represented. It was not possible to decide whether these positions in the primary structure were conserved for metabolic or for structural reasons. Now that a three-dimensional structure has been determined (4, 5), the primary structural information can be viewed in a new light, and we can see a structural role for most of these special nucleosides. In this context, it should be pointed out that a semi-invariant base-pair in the D stem at position b2 (see Fig. 2) was noted only after a knowledge of additional H-bonding interactions between bases and backbone in the tertiary structure (2, 6).

As shown in Table II, the class of small tRNAs (class 1) can be conveniently subdivided according to structural characteristics. Class 1 (63 structures) contains 4 base-pairs in the D stem (stem b) and a small extra loop (3–5 nucleotides). Class 2 (14 structures) has only 3 standard base-pairs in the D stem and a large extra loop of 13 to 21 nucleotides. On the basis of the sequences in the extra loop and standard base-pairs in the D stem, class 1 can be conveniently subdivided into subclasses A, B, C and D. Subclass 1A contains 5 nucleotides, one of which is m^7G, in the extra loop (III), 1B also contains 5 nucleotides in the extra loop, but there is now no m^7G, subclass 1C is irregular in that its extra loop contains only 4 nucleotides, and subclass 1D contains only 3 Watson–Crick base-pairs in the D stem. The tertiary structure has been determined for yeast tRNA^{Phe}, which belongs to the subclass 1A and will probably also accommodate class 1B, 1C and class 1D structures (4). For example, a short extra arm, only 4 nucleotides in length, but making all the tertiary interactions, can be built with the excision of the residue U47, which is not involved in the tertiary bonding of yeast tRNA^{Phe} (4, 5).

The list of different structures shown in Table II is somewhat arbitrary since some closely similar structures, such as the minor glutamic tRNA$_1$ (Ec Glu$_1$)(5a) and major glutamic tRNA$_2$ (Ec Glu$_2$), are both listed while uncertain sequences and spontaneous mutants showing evidence of gene duplication are not. Furthermore, it is likely that the sequences determined for Ec Arg$_1$ and Ec Arg$_2$ are the same, Ec Arg$_2$ being correct but containing the modification reported for Ec Arg$_1$ (K. Chakraburtty and S. Nishimura, personal communication).

III. Significance of G · U Base-Pairs

The wealth of tRNA primary structural information now available permits several interesting analyses of the secondary structure to be made. In particular we have examined the occurrence of non-Watson–Crick base-pairs such as G · U, U · U, C · A or G · A in stem regions. There tends to be only one of these in a given molecule, and where there is more than one, no two are adjacent. It is therefore likely that the RNA double-helix can internally accommodate a single non-Watson–Crick base-pair without serious distortion.

The most common non-Watson–Crick base-pair occurring in tRNA is G · U. Figure 3 shows the positions and frequency of occurrence of this base-pair in the known class 1 primary structures. The frequency of occurrence of the probably equivalent G · Ψ pair is shown in parenthesis. So far, G · U base pairs have not been found in stem position a1, b3, c2, c3 and c5, nor in e4 and e5. Furthermore, only a single occurrence has been noted for c4. Although G · U base-pairs occur frequently and, according to our knowledge of the tertiary structure (4, 5), without gross helix distortion, there appears to be a particularly strong conservation of pure Watson–Crick base-pairs in the anticodon stem (c) and the loop end of the TΨC stem (e). It is probable that there is a structural reason for this conservation.

It may be asked why G · U base-pairs and other non-Watson–Crick base-pairs occur at all in the stem regions. Clearly a G · U base pair in the middle of a piece of helix requires a certain accommodation of the helix backbone if two H bonds are to be made. It is therefore a potential point of "weakness" in such a stem, but it could possibly be a special point for enzyme recognition. This may apply particularly to the G · U base-pairs in the aa stem (a).

G · U base-pairs at the end of a helix may play a different role. Thus it has been found in the 0.25 nm model of yeast tRNA^{Phe} [Ladner et al. (6)] that the phosphate of nucleotide 49 in yeast tRNA^{Phe} (i.e., at the break in the long double-helix formed by the aa stem stacked on the TΨC stem; cf. Fig. 7) is moved from its regular double-helical position in order to allow the adjacent nucleotide in the extra loop to make a

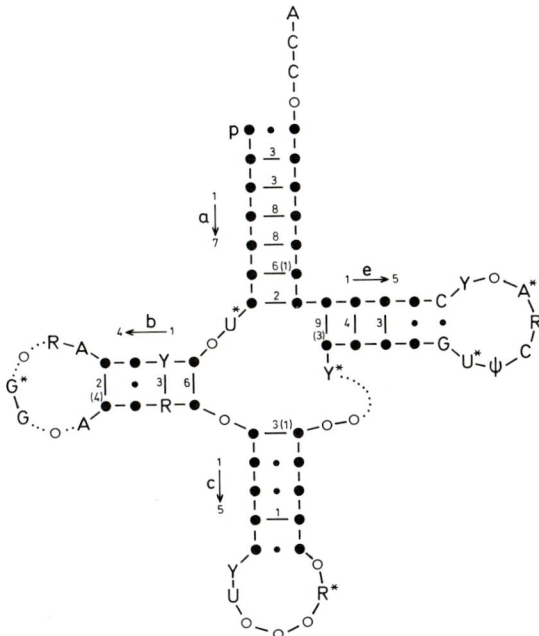

FIG. 3. G · U positions in tRNA (Class 1). The number of occurrences of G · U base-pairs at each position in the cloverleaf (shown by dashes and numerals, followed by number of G · Ψ occurrences in parentheses). The arrows give the direction of the standard numbering of base-pair positions.

nearly right-angle bend with it. A G · U base-pair in this position could be very helpful in relieving this tight corner. It is perhaps significant that the highest frequency of G · U pairs observed is at position e1, and moreover that, in eleven out of twelve cases, the G is always on one strand and the U(or Ψ) on the other. Similar remarks apply to G · U pairs in position b1, which is close to another sharp bend in the backbone (cf. Fig. 7).

IV. Position of Pseudouridine

The positions of Ψ occurring in class 1 tRNA structures are shown in Fig. 4 on a standard generalized cloverleaf. Except for the possible structural involvement of the Ψ contained in the common loop-IV sequence G-U*-Ψ-C, it seems unlikely that these Ψ positions have significant structural roles. More likely they are concerned with metabolic roles largely as yet unidentified, with the exception that the Ψ in the position after R in the anticodon loop (II) appears to confer

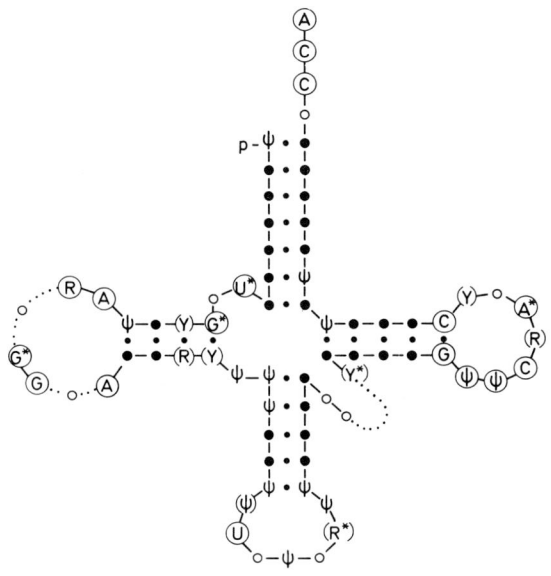

FIG. 4. Position of pseudouridine (Ψ) in class 1 tRNAs.

repressor activity on bacterial tRNA^{His1} in connection with the histidine biosynthetic pathway.

V. Exceptions to the Generalized Structure

There are several exceptions to the generalized tRNA structure shown in Fig. 2. Apart from the non-Watson–Crick base-pairs in helical stems listed below, the main exceptions are confined to tRNAs that have special functional roles. The most important are the prokaryotic and eukaryotic initiator tRNAs. The glycine tRNAs involved in cell-wall metabolism also cannot be fitted into the generalized structure because they lack the pair of Gs in the D loop and have a G instead of Ψ in the TΨC loop. An interesting exception to the generalized structure for a noninitiator tRNA is the myeloma tRNA^{Val_1} which has an A instead of the semi-invariant Y near the 3′ end of the TΨC loop IV and the sequence G-U-Ψ-C for G-U*-Ψ-C.

The positions of G · U base-pairs in double-helical stems have already been discussed. In these stems, G · Ψ is considered to be the same as G · U, and A · Ψ the same as A · U. When the simplified classification of tRNA structures into two classes is used (Fig. 1), then the other non-Watson–Crick base-pairs that occur in stems are U · U, C · C, A · A, G · A, C · A, Ψ · U and Ψ · Ψ. No more than one of these unusual base-pairs is found in a stem, and there are altogether only a few instances. Two interesting examples of such exceptions are found for position c5 in the ac stem: the tRNAs for both y Met_3 and Mye + Rbl Met_4 contain $\Psi \cdot \Psi$ here.

A further type of exception should be noted. Ec $tRNA^{His_1}$ contains eight base-pairs in the aa stem. It is thus longer by one nucleotide at the 5'-end and the single-stranded region at the end of the aa stem contains only the C-C-A. Recently (G. Mazzara and W. McClain, personal communication) an exception to position Y in the extra arm (Fig. 2) has been noted for Ec Cys, and a number of variants in modification of the constant sequence G-U*-Ψ-C (loop IV) has been determined. Especially interesting is the variant G-Ψ-Ψ-C of the reverse transcriptase primer tRNA, Chi Trp.

VI. Three-Dimensional Structure of $tRNA^{Phe}$ [2]

After tRNA was first crystallized (7), we searched for a species that would give crystals suitable for an X-ray crystallographic analysis to high resolution (8). A systematic study of over ten species from E. coli and yeast yielded suitable crystals of a monoclinic form (9) ($P2_1$) of yeast $tRNA^{Phe}$ with a smaller unit cell than the related orthorhombic form of the same species reported by Kim et al. (10), and better ordered than a different orthorhombic form obtained earlier by Cramer and his colleagues (11). The three-dimensional structure of the yeast $tRNA^{Phe}$ in the monoclinic form was solved to 0.3 nm resolution in Cambridge by the method of isomorphous replacement (4), using five heavy-atom derivatives. One of these, Pt, was located by chemical methods with respect to the nucleotide sequence and served to confirm the assignment of residues to the electron density map (12). A photograph and silhouette of the Kendrew skeletal model built to fit the electron density are shown in Fig. 5. A similar structure has also been proposed for $tRNA^{Phe}$ crystallized in the related orthorhombic form (5). At this resolution, a small part of the structure involving TΨC loop interactions with the D loop was not unambiguous (4). However, the second stage of the X-ray crystallographic analysis to 0.25 nm resolution (6) resolved the ambiguity. The chain-tracing of the ribose-phosphate backbone was completed, and many more features of the molecular stereochemistry have been revealed. In addition to the base-pairs of the cloverleaf arrangement of yeast $tRNA^{Phe}$ (Fig. 6), many other interactions have been deduced (see Fig. 7) that fix the

[2] See discussion of NMR contributions to three-dimensional structure of tRNAs by Kearns in Vol. 18 [Ed.].

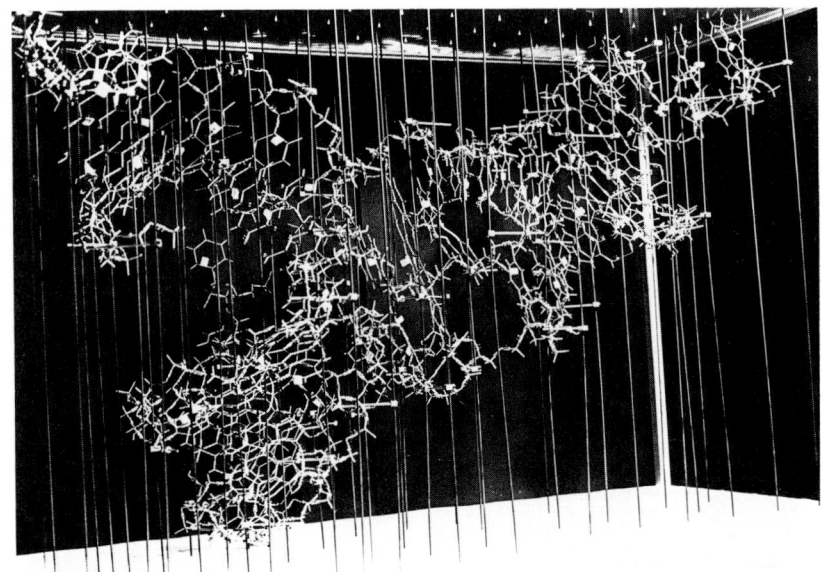

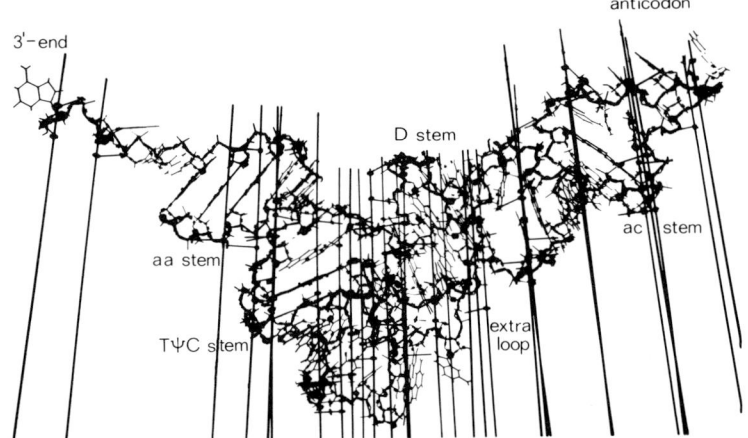

FIG. 5. Photograph and silhouette of 0.3 nm model of yeast tRNA^{Phe}.

three-dimensional structure of yeast tRNA^{Phe} showing how the cloverleaf of Fig. 6 is folded. The 0.3 nm and 0.25 nm tRNA models have already been described in some detail (*4, 6*). Here I summarize some of the structural features that appear to have functional significance.

The molecule depicted in the model (Fig. 5 and schematically in Fig. 7) can be described in terms of three major skeletal substructures:

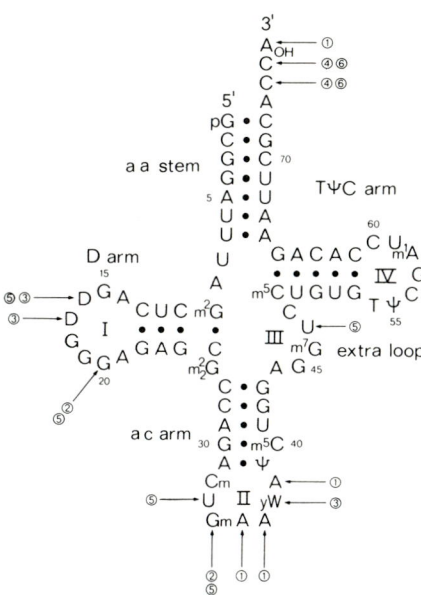

FIG. 6. Chemical reactivity of yeast tRNA^{Phe}. The arrows indicate points of chemical reaction in the structure. Reagents used *(2)* were: (1) perphthalic acid, specific for A residues; (2) Kethoxal, for exposed G residues; (3) NaBH$_4$ reduction; (4) methoxyamine; (5) carbodiimide; (6) I$_2$/TlCl$_3$:

(1) the long double-helix formed by the amino-acid stem and TΨC stem stacking on top of each other; (2) the augmented D helix forming the central part or "thorax" of the molecule, consisting of the D stem, augmented laterally by interactions with the short "stretcher" region 8-9, with the extra loop III and with a part of the D loop; (3) the anticodon stem tilted off by about 20° to the D stem and apparently hinged to it by hydrogen bonds between nucleotides A44 and m$_2^2$G26 and also G45 and m^{2}G10. The molecule has been described as L-shaped *(5)*, but the two major long double-helical stretches, i.e. (1) and (2) + (3) above, are arranged in the shape of the letter T *(4)*.

To these skeletal substructures are joined five additional regions with established, or potential, functional roles. (4) At one end of the long double helix there is the 3' C-C-A end, where the amino acid is attached. (5) At the other end, the invariant TΨC loop is tightly folded and interacts with a part (constant G-G of the D-loop) in a complex set of interactions that have been interpreted in detail in the electron density map at 0.25 nm resolution *(6)*. Briefly, U59 stacks on the base-pair G15 · C48, T54 is base-paired with m^{1}A58 in a reverse Hoogsten type, G57 intercalates between G18 and G19 giving rise to a stack of 4

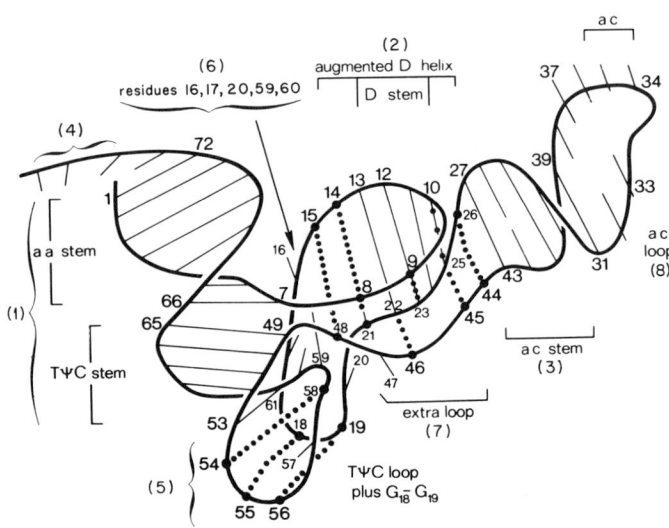

FIG. 7. A schematic diagram *(6)* of the folding and the tertiary interactions between bases of yeast tRNA^{Phe}. The ribose-phosphate backbone is represented by a continuous line. Base-pairs in the double-helical stems are represented by long light lines, and nonpaired bases by shorter lines. Base-pairs additional to those in the cloverleaf formula are indicated by dotted lines.

purines (58, 18, 57 and 19) and interacts with the backbone of the D loop, C56 forms a Watson–Crick base-pair with G19, and G18 appears to interact with Ψ55 and also the TΨC backbone on either side of it *(6)*. The interaction of the TΨC loop with the invariant G18 and G19 probably masks the potential of the T-Ψ-C-G sequence for its proposed role of binding to the ribosomal A-site during peptide bond formation *(13)*. (6) The variable regions in the D loop perhaps form a discrimination site (or part of one) for enzymes to distinguish between tRNAs (not necessarily the aminoacyl-tRNA synthetase recognition). The two sections α and β of D are near each other and form a surface patch on the molecule; they each contain a variable number (1 to 3) of nucleotides and, either on their own or together with the nearby non-base-paired residues 59 and 60, could well provide a general enzyme discrimination site. The variable extra loop (or, at least, a part of it, e.g., U47 in tRNA^{Phe}), region (7), may also play such a role in some cases. Finally, the decoding function of tRNA is provided for by region (8), the anticodon loop, containing the anticodon. This loop contains a stack of five bases on the 3′ side, but the other two bases C32 and U33 are also stacked separately on the end of the anticodon stem *(4)* but with a sharp bend between U33 and Gm34 *(6)*. It is possible that the loop conformation changes during the translocation of the tRNA.

From the X-ray studies, we now have a picture of the ordered complexity of a folded RNA molecule, a complexity as great as that of a protein. An interesting by-product is a detailed picture of the stereochemistry of a G · U base-pair in a double-helical stem, where the pairing is that predicted by the "wobble" hypothesis.

In the account of the 0.3 nm tertiary structure (4), attention was focused on the interactions between bases. We mentioned only a few of the other H-bonding interactions, deferring discussion until they could be checked at 0.25 nm. We have now traced an extensive network of H bonds, including ribose and phosphate groups as well as bases, in which almost half the bonds involve 2'-OH groups of riboses as acceptors, donors, or both (6). A prominent set of H bonds stabilizes the T-form made in the structure by the two long double-helices which meet approximately at right-angles. Another set reinforces the links between bases at the bends of the D and T-Ψ-C loops.

In the 0.3 nm model, all sugar conformations were assumed to be C3'-endo, as found in RNA double-helices (6). In the more detailed model building occasioned by fitting the 0.25 nm map, we changed the pucker of ten of the sugars to the C2'-endo conformation (6). Although the resolution of the map is still not sufficient to show the shape of a sugar unequivocally, the conformation can be deduced from the restrictions imposed on it by the relative dispositions of the two phosphate groups on either side and of the base that emanates from it. A good example is that of nucleotide m^1A58 whose sugar is C2'-endo in the 0.25 nm model. In the 0.3 nm model, the phosphate and base densities were well fitted, but the C3'-endo ribose was not. Fitting the ribose density moved the base of 58 too close to its base-paired partner, T54, which was well fixed. After the change, the C2'-endo sugar resided in its proper density, and the base-pair was comfortably made.

It should also be pointed out that the topography of the model deduced from the crystallographic study agrees with a companion study of the chemical reactivity of yeast tRNA^{Phe} in solution (see later). There is thus no reason to presume that the structure in the crystal departs significantly from that in solution.

VII. Structural Correlations

The structural integrity of the molecule as a whole depends upon the central part, where elements of four chains come close together and a number of base triples or pairs are found. These additional base-pairs or -triples are stacked or intercalated so as to make their H bonds inaccessible to water molecules. The tertiary structure is further stabilized by the stacking of nonpaired bases on each other or on base-pairs, as for example in the anticodon loop. Indeed, about 90% of the bases can be said to be internal in a similar way to hydrophobic groups in proteins.

The elucidation of the tertiary structure of yeast tRNA^{Phe} naturally led us to ask whether other tRNA primary structures could be folded in the same way. In particular, could other bases be substituted into the base-pair or base-triple positions of yeast tRNA^{Phe} so as to give equivalent interactions? We found (4, 14) that most changes of sequences in class 1 can be compensated by concomitant changes elsewhere, so that local geometry in the three-dimensional structure remains essentially unchanged. These changes are listed in Table III.

Levitt (15) first noticed a coordinated base change at positions 15 and 48 (tRNA^{Phe} numbering, see Fig. 6) by examining tRNA primary structures. However, the pair is not of the Watson–Crick type (4) and, moreover, is asymmetric, since only purine R is found in position 15. Structural considerations (14) require such an asymmetry forming a reverse Watson–Crick base pair. G · C can be replaced by A · U to give the same disposition of H bonds in this nonstandard base-pair. Both y tRNA^{Glu3} and Ec tRNA^{Gly1} provide interesting exceptions to this structural correlation; they contain A15 · C48 instead of the usual semi-invariant A15 · U45 or G15 · C48.

Subdivisions of class 1 into 1A and 1B in Table II are based on the two nucleoside positions 46 and 9 (in the numbering for tRNA^{Phe}).

TABLE III
CORRELATED BASE CHANGES IN CLASS 1[a]

Base structure	Direction of chains
Base-pair at positions 15–48	
G15 : C48	Parallel
A15 : U48	Parallel
Base triple at positions 13–22–46	
C13 : G22 : m^{7}G46	Antiparallel/antiparallel
U13 : A22 : A46	Antiparallel/antiparallel
U13 : G22 : G46	Antiparallel/antiparallel
Base triple at positions 12–23–9	
U12 : A23 : A9	Antiparallel/parallel
G12 : C23 · G9	Antiparallel/parallel
U12 : A23 · m^{1}G9	Antiparallel/parallel
Base pair at 26–44	
A26 · G44	Antiparallel
G*26 : A44	Antiparallel

[a] Each dot represents one hydrogen bond.

Position 46 has either m⁷G or not (G or A), and this is the basis for the subdivision into the two groups 1A and 1B. However the base-triple 46 · 22 · 13 can be made equivalent in these subclasses, with nearly the same disposition of glycosyl bonds *(4, 14)*. Nucleoside 9 is G, m¹G or A and here again a base-triple 9 · 23 · 12 can be made for all members of subclasses 1A and 1B, though the number of hydrogen bonds is not the same in all cases *(4, 14)*. The coordinated base changes are summarized in Table III.

The model is perhaps also directly relevant to sequences of class 1C. In the model, it is possible to remove the exposed nucleotide U47 and still bridge the gap between 46 and 48 by the phosphate-sugar linkage, maintaining all the tertiary interactions. Class 1C could have this type of structure: any base-triples equivalent to 13 · 22 · 46 would now have to be made in a somewhat different way, but still preserving the relative disposition of the pieces of backbone being bridged by the base-base interactions. By an extension of the argument, the model is also relevant for class 1D. If, as discussed earlier, a base-pair, albeit nonstandard, can be fitted into position b4, an equivalent augmented D helix might be made without disruption of the pattern of the four chains that come together to make it.

VIII. Correlation of Chemical Reactivity with Three-Dimensional Structure

The bases of yeast tRNA^{Phe} that react with a carbodiimide reagent and with methoxyamine in solution have been identified *(16)*. This information has been combined with other results from the literature to produce a composite picture *(2, 17)* of base accessibility in yeast tRNA^{Phe} shown in Fig. 6. The bases that react chemically can be correlated with exposed positions in the three-dimensional structure of tRNA. Those that do not react are either in the double-helical regions or are involved in maintaining the tertiary structure. These results confirm the usual structural assumptions made about reactivity in chemical studies and hence give confidence for extending such work to other tRNA structures.[3] Furthermore since the chemical studies are carried out on tRNAs in solution rather than in the crystalline state, they also support the contention that there is no significant change in the conformation of yeast tRNA^{Phe} upon crystallization.

This picture for chemical reactivity is very similar to that obtained earlier for the bacterial initiator tRNA, tRNA^{fMet} *(18)*, which also be-

---

[3] See chapter by Hayatsu in Vol. 16 of this series; also R. P. Singhal in *Biochemistry* **13**, 2924 (1974) [Ed.].

longs to class 1. It is therefore reasonable to propose that the overall tertiary structure of bacterial initiator tRNA is similar to that described for yeast tRNA^{Phe}. The special function of *E. coli* tRNA^{fMet} is probably explained by a subtle recognition of an initiation factor for locating it in the correct ribosomal site rather than a different folding of the tRNA.

Recent physical-chemical studies, especially those using NMR, identify the H bonds involved in the tertiary interactions between bases (B. R. Reid, personal communication; also *19, 20*). These studies are now also interpreted in terms of the same structure existing in solution as in the crystal form.

Although we now have strong evidence confirming that the tRNA conformation is the same in solution as in the crystal form, it is unclear whether the structure changes its conformation for its role in protein biosynthesis. Indeed, it is likely that the conformation does change, and this is effected by the many different interactions that the tRNA undergoes with other protein and nucleic acid components. To gain more insight into these putative conformational changes, we need information on the different tRNA · protein complexes formed during the steps of protein biosynthesis from chemical, X-ray diffraction and other physical chemical studies.

A detailed reference list for the tRNA activities discussed in this chapter can be found in a recent review by Rich and RajBhandary *(21)*.

Acknowledgments

The structural work described herein was carried out at the Medical Research Council's Laboratory of Molecular Biology in Cambridge, in collaboration with Dr. Aaron Klug's group. For recent developments, I am grateful for financial support by the University of Aarhus, Danish Natural Science Research Council Grant No. 511-3820 and NATO Research Grant No. 893.

References

1. B. G. Barrell and B. F. C. Clark, "Handbook of Nucleic Acid Sequences." Joynson Bruvvers, Oxford, 1974.
2. B. F. C. Clark and A. Klug, *Proc. FEBS Meet., 10th* **39**, 183–206 (1975).
3. R. W. Holley, J. Apgar, G. A. Everett, J. T. Madison, M. Marquisee, S. H. Merrill, J. R. Penswick and A. Zamir, *Science* **147**, 1462–1465 (1965).
4. J. D. Robertus, J. E. Ladner, J. T. Finch, D. Rhodes, R. S. Brown, B. F. C. Clark and A. Klug, *Nature* **250**, 546–551 (1974).
5. S. H. Kim, F. L. Suddath, G. J. Quigley, A. McPherson, J. L. Sussman, A. H. J. Wang, N. C. Seeman and A. Rich, *Science* **185**, 435–440 (1974).
5a. M. Uziel and A. J. Weinberger, *NARes* **2**, 464 (1975).
6. J. E. Ladner, A. Jack, J. D. Robertus, R. Brown, D. Rhodes, B. F. C. Clark and A. Klug, *PNAS* **72**, 4414–4418 (1975).
7. B. F. C. Clark, B. P. Doctor, K. C. Holmes, A. Klug, K. A. Marcker, S. J. Morris and H. H. Paradies, *Nature* **219**, 1222–1224 (1968).

8. R. S. Brown, B. F. C. Clark, R. R. Coulson, J. T. Finch, A. Klug and D. Rhodes, *EJB* **32**, 130–134 (1972).
9. J. E. Ladner, J. T. Finch, A. Klug and B. F. C. Clark, *JMB* **72**, 99–101 (1972).
10. S. H. Kim, G. Quigley, F. L. Suddath, A. McPherson, D. Sneden, J. J. Kim, J. Weinzierl, P. Blattmann and A. Rich, *PNAS* **69**, 3746–3750 (1972). (See also Kim in Vol. 17 of this series.)
11. F. Cramer, F. von der Haar, K. C. Holmes, W. Saenger, E. Schlimme and G. E. Schulz, *JMB* **51**, 523–530 (1970). (See also Cramer *et al.* in Vol. 11 of this series.)
12. D. Rhodes, P. W. Piper and B. F. C. Clark, *JMB* **89**, 469–475 (1974).
13. V. A. Erdmann, M. Sprinzl and O. Pongs, *BBRC* **54**, 942–948 (1973).
14. A. Klug, J. E. Ladner and J. D. Robertus, *JMB* **89**, 511–516 (1974).
15. M. Levitt, *Nature* **224**, 759–763 (1969).
16. D. Rhodes, *JMB* **94**, 449–460 (1975).
17. J. D. Robertus, J. E. Ladner, J. T. Finch, D. Rhodes, R. S. Brown, B. F. C. Clark and A. Klug, *NARes.* **1**, 927–932 (1974).
18. S. E. Chang, *JMB* **75**, 533–547 (1973).
19. G. T. Robillard, C. E. Tarr, F. Vosman and H. J. C. Berendsen, *Nature* **276**, 363–9 (1976).
20. D. R. Kearns, This Series. **18**.
21. A. Rich and U. L. RajBhandary, *Annu. Rev. Biochem.* **45**, 805–860 (1976).

Bleomycin, an Antibiotic That Removes Thymine from Double-Stranded DNA[1]

> WERNER E. G. MÜLLER
> AND RUDOLF K. ZAHN
>
> *Institut für Physiologische Chemie*
> *Universität Mainz*
> *Mainz, West Germany*

I. Introduction	22
II. Reaction of Bleomycin with Native DNA *in Vitro*	23
A. Production of Strand Scissions	24
B. Reaction Conditions	30
C. Base Specificity	33
D. Reaction Mechanism	34
III. Reaction with Other Polynucleotides *in Vitro*	37
A. Single-Stranded DNA	37
B. DNA · RNA Hybrid	38
C. RNA	38
D. Poly(ADP-ribose)	38
IV. Effect on DNA-, RNA- and Protein Synthesis *in Vitro*	38
A. DNA-Dependent DNA Polymerases	39
B. DNA-Dependent RNA Polymerases	42
C. RNA-Dependent DNA Polymerase	42
D. DNA Ligase	42
E. Protein-Synthesizing System	42
V. Effect on Nucleases *in Vitro*	42
A. Deoxyribonucleases	42
B. Ribonucleases and Phosphodiesterases	43
VI. Influence on Nucleic Acids *in Vivo*	43
A. Effect on DNA	43
B. Effect on Chromatin	44
C. Effect on RNA	46
VII. Influence on Nucleic Acid Metabolism *in Vivo*	46
A. Influence on DNA Synthesis	46
B. DNA Repair	48
C. Influence on Gene Expression	48
VIII. Medical Implications	49
A. Cell Kinetics	50
B. Tumor Specificity	50
C. Antiviral Activity	51

[1] Dedicated to Professor Dr. Hamao Umezawa (Institute of Microbial Chemistry, National Institute of Health, Tokyo, Japan), whose antibiotic studies have contributed to the rational development of antitumor and antibacterial agents.

IX. Future Directions... 51
    A. Tool for Studying Nucleotide Sequences in DNAs............. 51
    B. Tool for Studying DNA Synthesis 53
    C. DNA Repair Mechanisms 53
    References 54

I. Introduction

In the last 20 years, several proteins with molecular-weight ratios (M_r) in the range between 1000 and 10,000 have been isolated from different strains of Actinomycetales from the genus *Streptomyces* that produce single-strand breaks in DNA. They can be subdivided into a group of acidic proteins [e.g., neocarzinostatin *(1)*] and into a group of basic proteins. That agent belonging to the latter group, the most thoroughly studied one, is the antitumor agent bleomycin, isolated by Umezawa *et al.* *(2)* from cultures of *Streptomyces verticillus*. It is a potent antibiotic against a variety of microorganisms *(3)* and an active cytostatic agent *in vitro* as well as *in vivo* (survey: *4*). More than 200 species of bleomycin have been isolated and characterized as complex basic glycopeptides, and their structures have been determined *(5)*. They resemble each other with respect to their physicochemical properties and their structures. The chemical structures of the main moieties are identical and consist of five amino acids, L-gulose, and 3-*O*-carbamoyl-D-mannose, and a terminal cation (Fig. 1). The various bleomycins differ from each other only in the terminal cation moiety, which consists of an amine (or a polyamine). The M_r of bleomycin-A_2, for example, amounts to 1484 *(7)*.

From extensive studies by several groups, the mode of action of the bleomycins has been elucidated to a great extent. The data available show them to be members of a new class of DNA-modifying agents, the quasi-enzymes.

In *in vitro* systems, bleomycin first removes thymines from native DNA by hydrolysis of the *N*-glycosidic bonds without modifying the deoxyribose moiety. In a second step, single-strand scissions occur at the sites of the nonglycosidic deoxyribose moieties, resulting in the formation of 3'-OH and 5'-P termini. It is suggested that bleomycin is bound to DNA by interaction of the positively charged terminal amine moiety with the negatively charged phosphate group in DNA; intercalation seems to be involved in binding. Bleomycin is inactivated by copper and zinc ions, probably by chelate formation, which might cause a contortion of the antibiotic. The action is enhanced by a coincubation with intercalating agents. A reaction mechanism is proposed by which the quasi-enzymic character of bleomycin as a nucleosidase

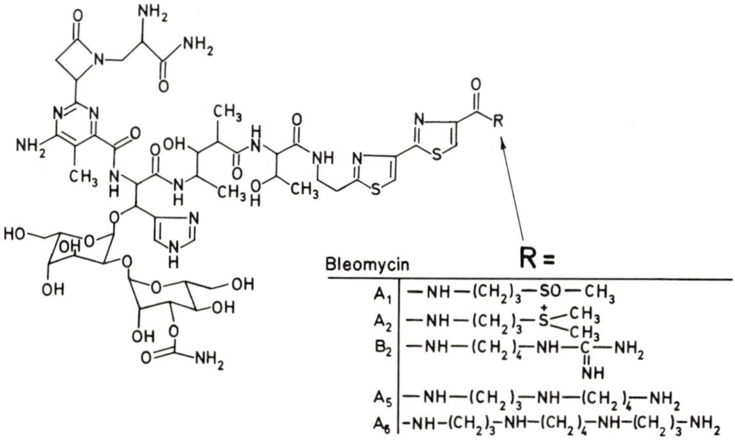

FIG. 1. Chemical structure of bleomycin. The terminal amine moieties of bleomycin-A_1, -A_2, -B_2, A_5 and -A_6 are given (6).

is described, assuming that the carboxyl amide group of β-aminoalanine is the active site of the molecule. Single-stranded DNA, RNA and poly (ADP-ribose) are not modified by bleomycin. In isolated enzyme systems, it affects the activities of DNA-dependent DNA polymerases and DNA-dependent RNA polymerases as well as DNases in an indirect way; it first modifies DNA, which then affects the polymerases through its aldehyde groups and the DNases by reduction of their substrate affinity.

In intact cell systems, bleomycin reduces DNA synthesis selectively by the induction of single-strand breakages; RNA synthesis, as well as protein synthesis, is not affected. Repair of DNA chain breaks occurs rapidly, possibly by activation of a DNA-repair DNase. Cell progression is inhibited by bleomycin at the end of the S-phase and the early half of the G_2-phase, thus showing the drug to be a possible synchronizing agent. Bleomycin is detoxified by the bleomycin-inactivating enzyme.

Antitumor activity of bleomycin is dependent upon uptake, inactivation and activation, leading to tissue- and organ-specific actions. The observed antiviral activity of bleomycin does not seem to be specific, but to be due to an overall inhibition of DNA synthesis.

II. Reaction of Bleomycin with Native DNA *in Vitro*

There is a large body of evidence concerning the interaction of bleomycin with native DNA. The findings can be summarized as follows.

A. Production of Strand Scissions

Umezawa and his colleagues first reported (8, 9) that bleomycin causes single-strand scissions in native DNA, as detected by alkaline sucrose sedimentation velocity. This observation was confirmed by other groups not only by the same method (10, 11), but also by gel filtration (12, 13) and the nitrocellulose filter technique (14) of Geiduschek et al. (15). After modification through chain breakages, the DNA shows a marked decrease of the t_m (16). The effect of bleomycin on the thermal denaturation of DNA is dependent not only on the concentration of the antibiotic and the incubation conditions but also on the t_m of the respective DNA; e.g., in the presence of 50 µg of bleomycin per milliliter, the t_m value of calf thymus DNA (71°C) decreases by 23°C, that of salmon-sperm DNA (85°C) by 22°C, and that of *Mycobacterium smegmatis* DNA (89°C) by 12°C (16). The renaturation behavior of DNA is not affected after incubation with 40 µg/ml (16).

1. Binding

Bleomycin binds to DNA as shown by experiments with labeled bleomycin (17) by UV and circular dichroic (CD) difference spectroscopy (12, 18), by equilibrium dialysis (17, 19) and by fluorescence spectroscopy (19). The results show a binding ratio of one molecule per 350 deoxyribonucleotide residues, as determined by running the complex through a Sephadex G-100 gel column; the binding ratio determined by equilibrium dialysis is much higher and amounts to 10–15 : 1 (17, 19). Almost no binding is observed to rRNA (17). The UV difference spectra with DNA-bleomycin mixtures reveal a shoulder at 260 nm and a maximum at 280 nm (increase of absorbance) (12, 18). The CD difference spectrum obtained for the mixture shows a negative maximum at 290 nm and a broad, shallow positive deflection with a maximum near 258 nm (18). These data also indicate a physical interaction between isolated DNA and bleomycin and the fact that the drug definitely binds to DNA.

Two moieties in the bleomycin molecule assumed to be involved in its binding to DNA are the bithiazole moiety and the terminal amine moiety. From a theoretical point of view, Murakami et al. (20, 21) claimed that bleomycin interacts with helical DNA by partial intercalation of one of the thiazole rings in the bithiazole part. In Fig. 2, the relative orientation of the planar tripeptide part of the molecule [γ-aminopropyldimethylsulfonium; 2'-(2-aminoethyl)-2,4'-bithiazole-4-carboxylic acid; L-threonine] with the double-helix DNA is given. The concept of partial intercalation of one thiazole ring in the vicinity of the thymine base of DNA is outlined in Fig. 3. The theoreti-

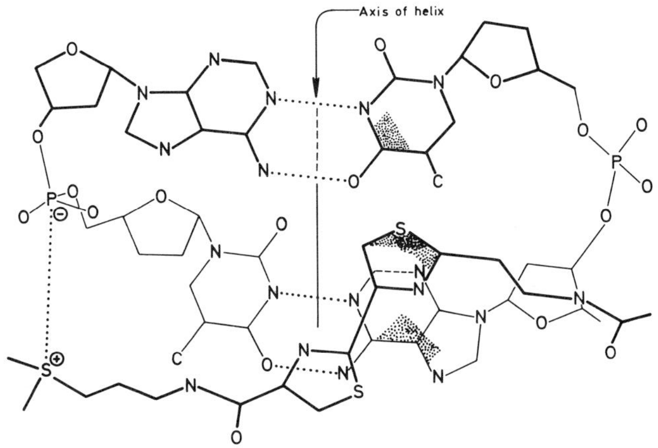

FIG. 2. Proposed relative conformation of the planar part of bleomycin with poly(dA-dT). The base-pair Ade · Thy, shown by thick lines, represents the upper layer and the other base-pair Ade · Thy, shown by thin lines, represents the lower layer. The planar part of bleomycin, including the bithiazole moiety (very thick line), is situated between the two layers of the paired bases. The stippled areas show the overlapping parts to clarify the concept of partial intercalation. Only one thiazole ring in bleomycin is in contact with the neighboring bases on one strand of DNA *(21)*.

cal analysis of the electronic interaction between the bases of DNA and the bithiazole moiety of the drug revealed *(21)* that this interaction can be expressed as the intermolecular electron transfer from the $2p_n$ orbital of the thymine N-3 to the $3d\pi$ and/or 4s orbital(s) of the thiazole sulfur. This localized intermolecular electron transfer results in the formation of a covalent bond between these two atoms: the nitrogen is an electron donor and the sulfur is an electron acceptor *(21)*. Among

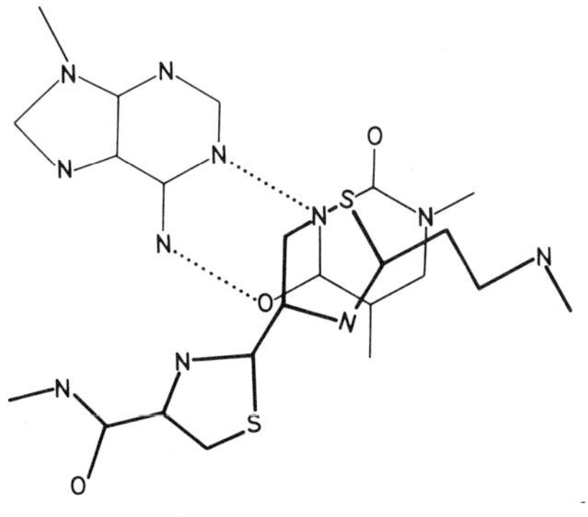

FIG. 3. View of the proposed partial intercalation *(21)*. The scheme represents the overlapping of thymine with the partially intercalated thiazole ring. The thymine (thin line) is lying under the thiazole ring (thick line).

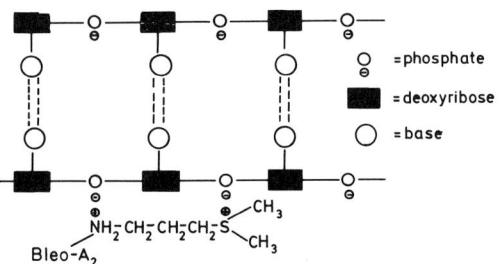

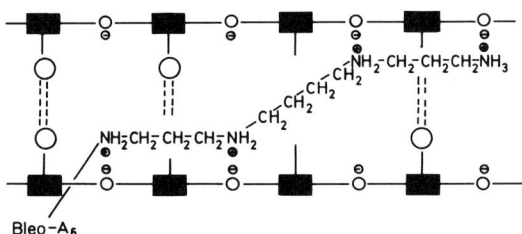

FIG. 4. Possible electrostatic interaction of the terminal positively charged groups of bleomycin with the negatively charged phosphate groups in DNA. *Above:* bleomycin-A_2; *below:* bleomycin-A_6.

the four bases, only thymine can form a comparatively stable localized bond with the partially intercalated thiazole ring through the localized intermolecular electron transfer. The strong binding of the thiazole group with DNA has also been confirmed by Chien *et al.* (19).

The complex formed between bleomycin and DNA by partial intercalation can be further stabilized by heteropolar interactions (22). Owing to the physical structure of bleomycin, the terminal amine moieties can bind ionically with the negatively charged phosphate groups in the DNA (21) (Fig. 4). In previous studies, some evidence was presented (23, 24) that the binding of polyamines to DNA shows a relative preference for the (dA + dT)-rich regions of DNA. Later, Liquori *et al.* (25) proposed that the protonated amino groups form hydrogen bonds with phosphate oxygens in the narrow groove of the DNA molecule. From these reports and the data of Tsuboi (26), we suppose (Fig. 4) that most of the bleomycin species, e.g., A_2, are bound to the phosphate groups only on one strand of the DNA; some, like the spermine-carrying A_6, can form a bridge between the two polynucleotide strands.

2. REMOVAL OF BASES

During the reaction of bleomycin with DNA *in vitro*, free bases are released *(12, 27, 13, 28)*. At low concentrations, (up to 50 µg/ml) and in the presence of up to 50 mM dithiothreitol, only thymine is released *(10, 12, 13, 22, 28)*. At concentrations above 50 µg/ml, all four bases are liberated from the DNA *(27, 22)*. The amount of thymine released from a particular DNA is dependent on its content of adenine and thymine *(12)*. The results from experiments with DNAs of different base compositions can be fit into a semilogarithmic scheme, in which the base-splitting activity (expressed by thymine liberation) depends in a linear fashion on the mole fraction of dAdo + dThd in double-stranded DNAs (Fig. 5). Besides the base composition of the DNA, the molecular topology of the DNA seems to play a determining role in this activity of bleomycin; comparing poly(dA-dT) and poly(dA · dT),

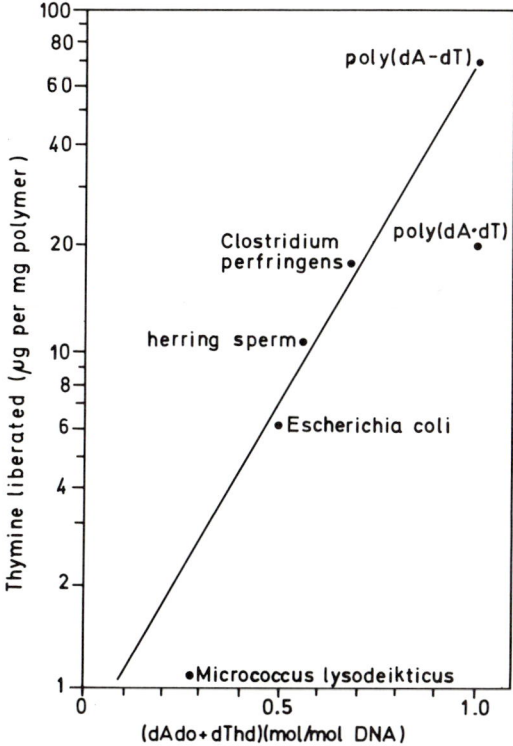

FIG. 5. Dependence of bleomycin activity (expressed by thymine liberation) on DNA composition. The incubation mixture (0.5 ml) consisted of 50 µg of bleomycin, 4 mM dithiothreitol and 20 µg of the different double-stranded nucleic acids *(12)*.

the homopolymer is 30% less susceptible to the drug than the heteropolymer (Fig. 5).

After release of free bases from DNA, free aldehyde groups can be detected by the bisulfide-iodine procedure (12). In an assay (1 ml) with 100 µg of bleomycin, 100 mM 2-mercaptoethanol and 125 µg of poly(dA-dT), 0.14 mol of aldehyde groups per mole of phosphate could be titrated. In a parallel experiment, 52% of all the thymine was released as the free base. This means that 58% of the deoxyriboses that lost their N-glycosidic thymine gave a positive aldehyde reaction. Using the 2-thiobarbituric acid method of Saslaw et al. (29), we could show (22) that in the absence of thiol compounds, 2-thiobarbituric acid does not react with bleomycin-treated DNA. From these experiments, we suppose that the bases are released from nucleic acid by hydrolysis of N-glycosidic bonds, resulting in free bases and some nonglycosidic sugar moieties, present in an equilibrium between the aldehyde and the furanose forms (Fig. 6).

Two attempts were made to prove the existence of nonglycosidic deoxyribose moieties in bleomycin-treated DNA, first by testing the alkali sensitivity and second by observing the temperature sensitivity of the nonglycosidic sugar moieties (22). To determine DNA chain

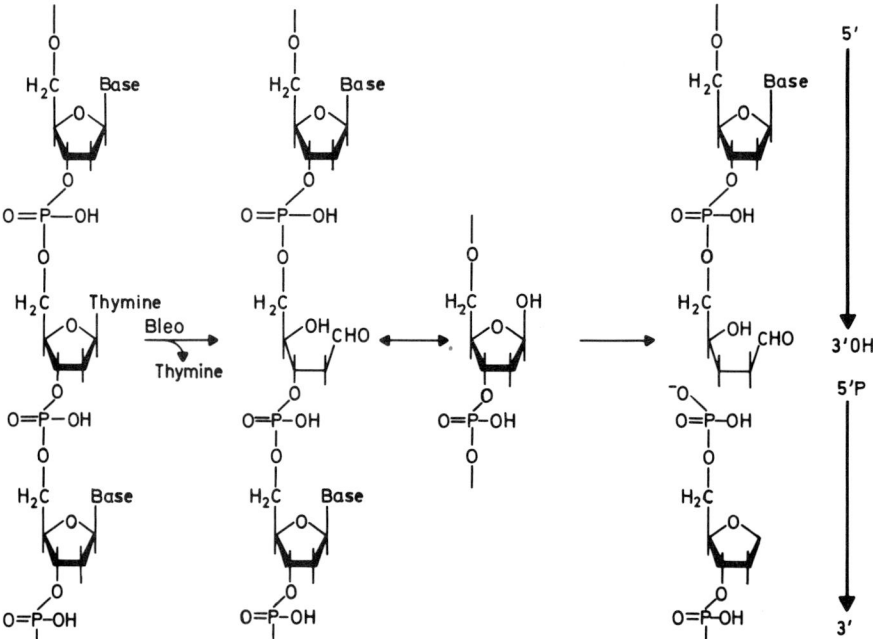

FIG. 6. Schematic representation of bleomycin-caused degradation of DNA. The mechanism of strand scission is taken from Brown and Todd (30).

length, the nitrocellulose filter assay technique (15) was applied. This method rests on the observation that only relatively large polynucleotide chains are retained on filters. The results showed (22) that under the incubation conditions of 5 µg of bleomycin per milliliter and 1.2 µg/ml of DNA (from E. coli), there is no significant decrease in the amount of DNA retained by the filter compared to the control (bleomycin, 5 µg/ml). However, after treatment with either 1 M KOH or high temperature (70°C), the retention of DNA decreases considerably. From these data, we have to conclude that bleomycin treatment does not change the molecular weight of the DNA. The modified DNA carries nonglycosidic sugar moieties that occur randomly in internucleotide linkages (Fig. 6). The alkali sensitivity and temperature sensitivity of these nonglycosidic deoxyribose moieties in bleomycin-treated DNA have previously been observed in other modified DNAs, such as apurinic acid and apyrimidinic acid (31–33).

3. Strand Scission

After the release of free bases, strand scissions of bleomycin-treated DNA occur at the nonglycosidic sugar sites with formation of 5'-phosphate and 3'-hydroxyl termini (34). Such termini are also observed at the heat-induced chain cleavages at apurinic sites (33); however, alkali-catalyzed chain cleavage at apurinic sites results in the formation of 3'-phosphate and 5'-phosphate termini (35). During the bleomycin-induced breakage, no release of deoxyribose, deoxyribose phosphate or inorganic phosphate is observed (27). Consequently, this modification most probably proceeds according to the following sequence (Fig. 6): first, removal of bases (primarily thymine) as free bases from DNA by cleavage of the N-glycosidic bonds, and second, cleavage at the 3' site of the nonglycosidic deoxyribose group between this sugar and the phosphate residue. In preliminary studies determining the relation between thymine liberation and strand scission, it appeared that, per one single-strand scission, five to six thymine moieties are released (36).

From studies by several groups (8, 10, 12) it is known that bleomycin at low concentrations causes only single-strand breaks in DNA. In this respect, it acts as an endonuclease. At higher concentrations, double-strand breaks are observed, due to the occurrence of single-strand breaks on opposite positions in native DNA. These results indicate that the bleomycin molecule has only one active site. After exhaustive treatment, single-stranded DNA fragments with an M_r of about 4000 remain (37). From these results, it seems highly unlikely that this resistance is due to some specific sequence of bases since the base composition of the resistant DNA is similar to that of the native DNA from which it was derived *(37)*.

B. Reaction Conditions

The conditions for the bleomycin reaction are known to a large extent.

1. Concentration

Both the initial rate of the reaction and the yield of DNA fragmentation are dependent on the concentration of the drug *(38, 28, 39)*; at the beginning of the reaction, the rate of strand scission is approximately proportional to reaction time (Fig. 7A). The dependence of the action on concentrations is known *(38)*.

2. pH Dependence

The action of bleomycin is strongly dependent on the pH value *(38, 39)*; the optimum is at pH 9.1. Two other distinct maxima occur at pH 8.7 and 10.1; no activity is observed below 6 or above 13 (Fig. 8). Bleomycin has one pK_a' at 7.3 which is due to the β-amino group in the terminal β-amino-L-alanine moiety *(38, 40)*. Since its activity increases

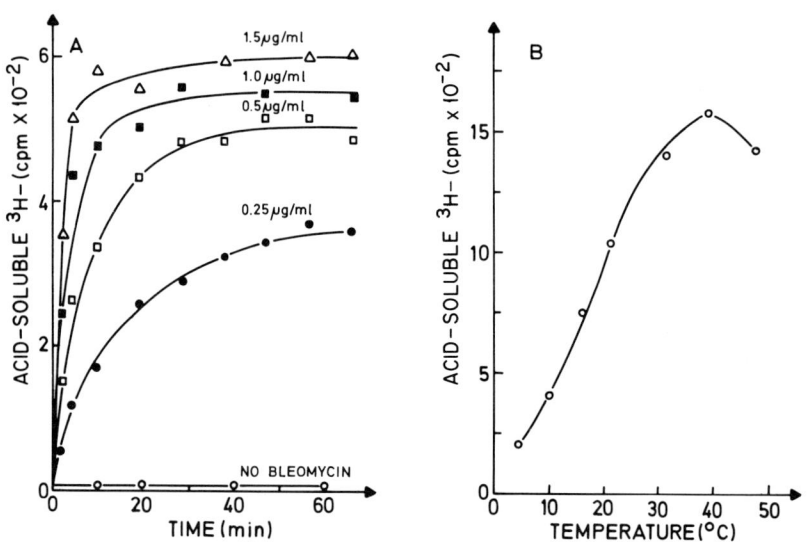

FIG. 7. Effect of time (A) and temperature (B) on degradation of DNA. The reaction mixture (volume in A: 0.2 ml; in B: 0.1 ml) consisted of 0.57 μg DNA (40 × 10³ cpm) per milliliter, 1 mM 2-mercaptoethanol, 1 mM ATP, 1 mM $MgCl_2$, 0.5 μg of bleomycin per milliliter (in B) and a pH of 8.5. As a measure of the bleomycin-caused degradation, the (perchloric) acid-soluble radioactivity is given *(28)*.

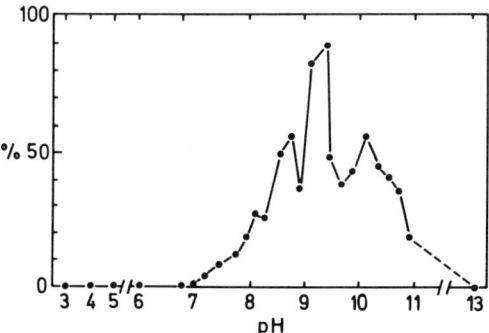

FIG. 8. Effect of hydrogen ion concentration on bleomycin action *(38)*. The activity of bleomycin is expressed as percentage of maximal activity.

markedly in the pH range from 7 to 8, and is negligible below pH 7, it has been concluded that this amino group is essential to the activity *(38)*.

3. TEMPERATURE

With bleomycin-B_2, an optimum in the reaction with DNA occurs at 20°C *(38)*, and with A_2, at 37°C *(28)*; see Fig. 7B. The temperature coefficient Q_{10} *(40)* is 3 *(38)* and 1.2 respectively *(28)*. From the published data *(38)*, the following activation parameters can be calculated *(41)*: $\Delta H\ddagger = 21.8$ kcal/mol, and $\Delta G\ddagger = 24.4$ kcal/mol.

The stability of bleomycin is considerable; it loses no activity during an incubation for 15 min at 37°C in the absence of dithiothreitol or 2-mercaptoethanol *(40a)*.

4. EFFECTS OF IONS

The degradation activity of bleomycin is dependent on the presence of Mg^{2+} *(28)*. Maximal acitivity is observed with 4–10 mM Mg^{2+} *(28, 12)*, and omission of Mg^{2+} causes a 23% reduction of the activity. Fe ions (0.1 mM–0.01 mM) also stimulate the reaction *(13)*; 1 mM $CuSO_4$ or 0.1 mM $ZnSO_4$ abolishes *in vitro* activity of bleomycin *(16, 38)*, probably as a consequence of a chelate formation resulting in a contortion of the bleomycin molecule and a reduction in its affinity for DNA. Chelation by zinc ion could occur in the middle part of the molecule between the hydroxyl-oxygen atom of threonine, the imidazole nitrogen atom of histidine, the nitrogen atom in one thiazole ring, and the sulfur atom in the other *(20)*. Further, zinc and copper ions, which are bound to the phosphate of the phosphodiester bond of the DNA by ionic linkages *(42)*, could interfere with the proposed electrostatic binding of bleomycin to DNA. Addition of 5 mM EDTA to the reaction mixture, containing none of these ions, abolishes all activity *(43, 28)*.

5. NUCLEOTIDES

The purine nucleotides, ATP and GTP, accelerate the bleomycin-induced degradation of DNA *(28)*. β,γ-Methylene-ATP, α,β-methylene-ATP and ADP are less effective. The enhancing effect is seen only when $MgCl_2$ is added in equimolar concentrations to the reaction mixture. AMP and pyrophosphate have no effect on the reaction.

6. OTHER FACTORS

The DNA-splitting activity of bleomycin is greatly enhanced by 2-mercaptoethanol *(8, 10, 12)* and by dithiothreitol *(12)*. The concentrations usually used were below 50 mM and thus had no effect on the integrity of the DNA *(38, 12)*. The reducing agents L-ascorbic acid (2mM) and sodium borohydride (10mM) stimulate the reaction *(44)*. Hydrogen peroxide stimulates the bleomycin reaction *(9)*; the concentrations needed to obtain stimulation are up to 100 mM *(44)*. In the absence of both reducing and oxidizing agents, bleomycin is still active, although less so *(43)*.

We have tested the effect of different radical scavengers, such as 2-aminoethyl-isothiouronium bromide *(45)*, Fe^{3+} *(46)* and iodine *(46)* on the activity of bleomycin toward poly(dA-dT); all were without influence *(12)*. The degradation of DNA seems to be largely prevented by the removal of oxygen *(44)*.

Actinomycin D in a concentration of 1 μg/ml stimulates the ability of bleomycin (0.2 μg/ml) to degrade SV40 DNA (1 μg/ml) while ethidium bromide (1–5 μg/ml) under the same conditions inhibits the reaction *(47)*. However, both ethidium bromide and actinomycin D enhance the bleomycin-induced breakage of native DNA *(48)*. In Fig. 9 it is shown that also steffimycin B, an (A · T)-specific DNA-binding agent with potential intercalating capacity *(49)*, stimulates the bleomycin activity on DNA *in vitro*. While bleomycin (40 μM) alone released only 7100 cpm of thymine, in the presence of 104 μM steffimycin B it split off 14,200 cpm. None of these enhancing compounds at the concentrations used caused fragmentation of DNA in the absence of bleomycin. At present the only explanation for the observed stimulatory effect of these DNA-binding ligands is that they cause a conformational distortion in double-stranded DNA such that bleomycin acts more readily *(47)*. Interesting in this connection is the finding *(47, 50)* that bleomycin fragments that do not contain the above-mentioned binding sites [the bithiazole moieties (causing partial intercalation) and the terminal amine groups (causing electrostatic in-

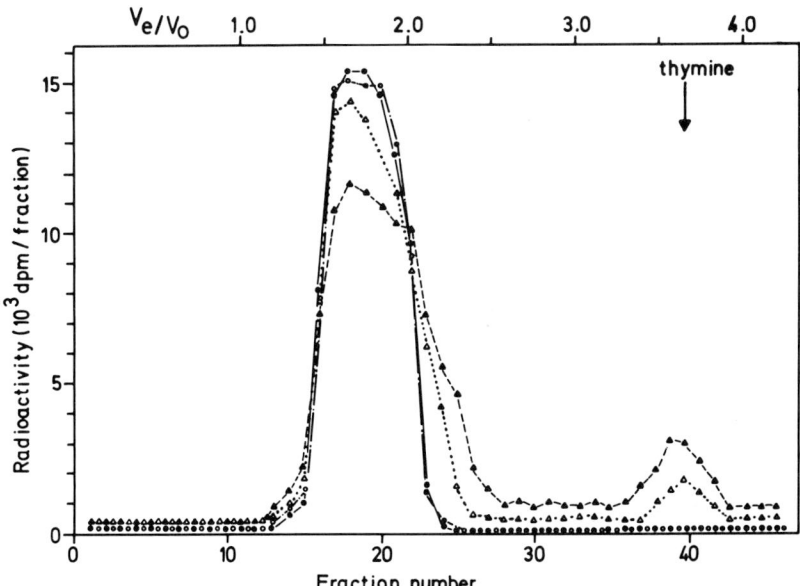

FIG. 9. Influence of steffimycin on the DNA-splitting activity of bleomycin. The drug, at concentrations of 0 and 40 μM, was incubated for 10 minutes at 37°C in a Tris buffer (40 mM, Tris · HCl, 60 mM KCl, 6 mM MgCl$_2$ 1 mM dithiothreitol pH 8.5) together with [^{14}C]thymidine-DNA (E. coli; 23 μCi/mg) in a concentration of 12.5 μM nucleotides (= 4.1 μg DNA/ml) in the absence or in the presence of 104 μM steffimycin B. An aliquot of 0.5 ml was chromatographed on a Sephadex G-200 column (12). Arrow, V_e/V_o value of thymine. ●——●, Control (sample without bleomycin and steffimycin B); incubation mixture with 40 μM bleomycin (△······△), with 104 μM steffimycin B (○–·–○) and with both 40 μM bleomycin and 104 μM steffimycin B (▲--▲).

teractions)] are inactive. In view of the findings that DNA-binding ligands that cause a conformation distortion of the DNA enhance the activity, it seems very likely that the binding moieties in the bleomycin molecule may also cause conformational changes in the DNA double helix.

C. Base Specificity

From the early experiments (12), it is known that, at lower concentrations (100 μg/ml), bleomycin selectively releases thymine from double-stranded DNA whereas at higher concentrations (12,000 μg/ml) (27) all four bases are liberated from the DNA backbone. In more recent reports (13, 22), this thymine-splitting specificity has been confirmed. The concentration-dependence of the reaction is shown in Fig. 10.

In lac operator DNA, 70% of the bleomycin targets are thymine

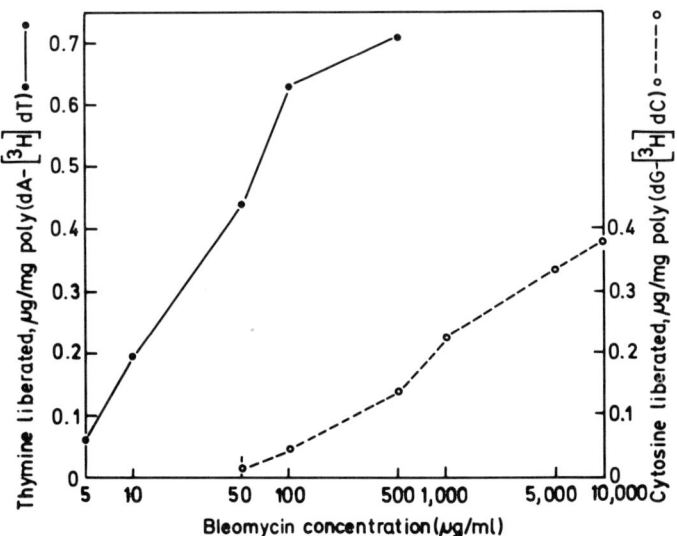

FIG. 10. Influence of bleomycin on poly(dA-dT) and poly(dG · dC) (22). The assay mixture contained 8 μg/ml of the polymer, 20 mM dithiothreitol and different concentrations of bleomycin. The ordinates show the amounts of Thy or Cyt, respectively, liberated from the polymers used.

bases, 22% are cytosine, and 8% are adenine bases; guanine is not affected at all (W. Gilbert, private communication). Interesting is the finding that 92% of the thymines attacked are adjacent to guanine.

D. Reaction Mechanism

The base-splitting reaction of bleomycin is characterized (Section II, B, C) by a pronounced time-dependence, a pH optimum around 8.5–9.5, a marked temperature dependence, an activation energy of 22 kcal/mol, a dependence upon an activating ion (Mg^{2+}), and a high specificity for thymine. The reaction products are the free, unaltered base and nonglycosidic sugar moieties. Oxidizing agents, inactivate bleomycin during the reaction (39) whereas reducing agents do not (22). One molecule of bleomycin in the presence of 2-mercaptoethanol liberates one molecule of thymine from poly(dA-dT) without losing its activity (22); the reaction stops because of limiting amounts of the poly(dA-dT).

The search for the catalytic site of bleomycin was not without success. An enzyme that inactivates bleomycin can be isolated from various organs (51–53). This enzyme acts like an aminopeptidase and hydrolyzes the amide group of the terminal β-amino-L-alanine moiety (5, 54, 55) (Fig. 11). Carboxyl amide groups can act as tautomeric catalysts for the aminolysis of 4-nitrophenyl acetate (56). The carboxyl amide

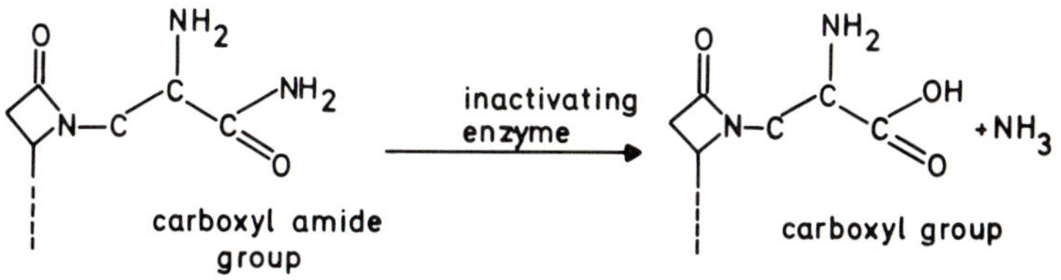

FIG. 11. Action of bleomycin-inactivating enzyme.

group in the β-amino-L-alanine moiety of bleomycin can exist in the two tautomeric states of amide and imide, depending on the hydrogen ion concentration. From this, we suggested that bleomycin acts in a way similar to nucleosidases and termed it a "quasi-enzyme" (22).

A scheme of the quasi-enzyme reaction of bleomycin is shown in Fig. 12. In the first step of the proposed mechanism, one hydrogen atom of the amide group in the carboxyl amide moiety forms a hydro-

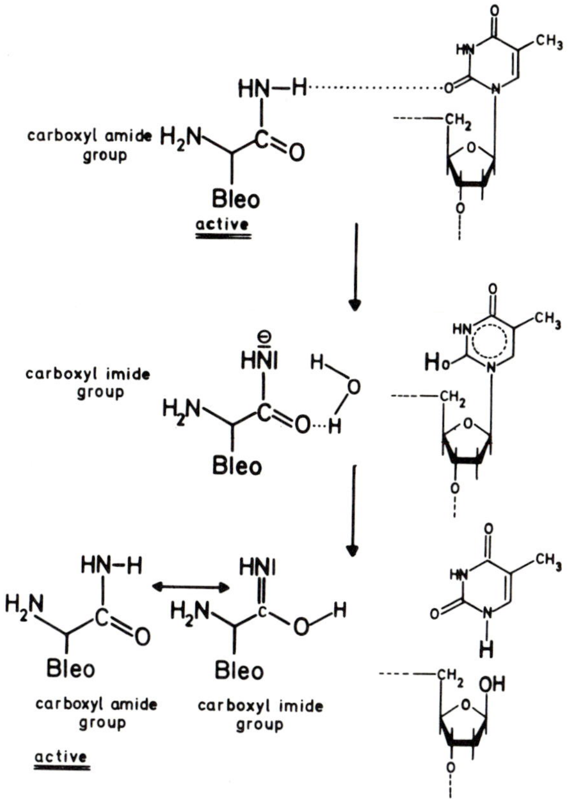

FIG. 12. Proposed mechanism of hydrolysis of N-glycosidic bond between base and deoxyribose.

gen bond with the 2-keto group of the thymine. This keto group of thymine is the only one among the DNA bases that is exposed to the narrow groove of the DNA double helix. In the second step, the hydrogen atom is transferred to a keto group with formation of an alcohol group. The carboxyl amide group in the bleomycin molecule is converted into a carboxyl imide group carrying a negative charge. The oxygen atom of this group forms a hydrogen bond with one hydrogen atom of a water molecule. In the third step, this hydrogen atom is linked to the oxygen atom. The resulting hydroxyl group, derived from the water molecule, is transferred to the carbon-1 atom in the deoxyribose moiety with elimination of the sugar-base link. The end products are free thymine, a nonglycosidic deoxyribose moiety, and a carboxyl imide group in the bleomycin molecule, which is in equilibrium between the imide and the amide forms. During this sequence of reactions, the drug is not inactivated.

The following experimental evidence supports this mechanism of hydrolysis of the N-glycosidic bond. First, bleomycin is inactivated by removal of ammonia from the amide group of the β-amino-L-alanine moiety in the bleomycin molecule, indicating that this group is involved in the reaction (51). Second, bleomycin is not inactivated during the base-splitting process in the absence of thiol-containing compounds. Third, bleomycin is inactivated at pH values above 13. At a pH of 13, the amide group in the carboxyl amide moiety should exist solely in the imide form and thus carry a negative charge; a transfer of a hydrogen atom from the imide group to the keto group is impossible. Fourth, it is known that tautomeric catalysis is temperature- and pH-dependent and shows an activation parameter of around $\Delta H = 15$ kcal/mol (56). Future results will show whether thiol groups, which stimulate the bleomycin-DNA reaction, are involved in this reaction as e.g. hydrogen donors (57). It is also theoretically possible that OH-radicals derived from the enhancing agent H_2O_2 (9) attack the carbon-1 atom of deoxyribose (58) and thus facilitate the cleavage at the N-glycosidic bond.

From theoretical considerations, Murakami et al. (21) suggest a different reaction mechanism for the thymine liberation. The idea is the direct involvement of the planar bithiazole ring not only in the binding of the drug to DNA (by partial intercalation), but also in the splitting reaction. In the first step, an increase of dipositive nature in the glycosidic bond occurs as the result of intermolecular electron transfer from the N-3 of thymine to the adjoining thiazole ring (mainly to the vacant $3d\pi$ and $4s$ atomic orbitals of the sulfur atom). In the second phase, the transformation of electronic hybridization of N-1 from the trigonal plane sp^2 state to another state, the tetrahedral sp^3 state, occurs. The final step includes the hydrogen atom transfer from a thiol compound to the N-1 of the thymine. This process might be facilitated through the increase of free valency of the directional hybridized orbital, including lone-pair electrons in the N-1 of the thymine. The cleavage of the N-glycosidic bond is completed with consumption of a water molecule. The reaction products again are free thymine and nonglycosidic deoxyribose moieties.

These two models are in good agreement with the experimental data, e.g., that bleomycin is not inactivated during the base-splitting reaction and that thymine is split off from DNA with high selectivity.

Besides these two models, which describe bleomycin as a quasi-enzyme, two other mechanisms could be discussed. First, we consider a radical mechanism. However, as described above, none of the studies using radical scavengers *(12, 13, 44)* convincingly show an effect on the activity. Second, one may consider an alkylation of bases. This is unlikely, owing to findings that bleomycin is not inactivated during the cleavage reaction and that no alkylated bases or sugar moieties are found as reaction products.

Considerable information is available about single-strand scissions of nucleic acids at the P-O and C-O bonds *(58a)*. From experiments with apurinic acid, it is known that the rate of chain breakage at the nonglycosidic deoxyribose groups at neutral or alkaline pH is enhanced by primary amines *(59, 59a)*. It has been suggested that this effect is due to an interaction of the aldehyde groups of the deoxyribose residues with the amines *(59)*, probably involving β-elimination. The strength of the amine-promoted splitting reaction is lower with monoamines than with diamines *(33)*. In the bleomycin-B_2 main molecule, for instance, four primary and five secondary amine groups are present in addition to the terminal chain of the molecule consisting of oligoamine agmatine (Fig. 1). Thus, in the bleomycin molecule, many amine groups could combine with the free aldehyde group of the deoxyribose and thus weaken the internucleotide bonds of nonglycosidic sugar moieties.

III. Reaction with Other Polynucleotides *in Vitro*

A. Single-Stranded DNA

Single-stranded, intact, and circular DNA from ϕX174 phage is degraded by bleomycin to an apparently lesser extent than double-stranded DNA *(43)*. Denatured DNA, doubtless containing some double-stranded regions, is degraded by bleomycin less effectively *(12, 44)*. Single-stranded synthetic deoxypolymers, like poly(dT) and poly(dI-dT), are not affected at all *(12)*. This finding is supported by the experiments of Asakura *et al.* *(47)* showing that poly(dT) and poly(dA) do not protect SV40 DNA from bleomycin action while poly(dA-dT) efficiently protects the viral DNA from cleavage.

B. DNA · RNA Hybrid

The hybrid, poly(dT · rA), has been tested *(60)* for its sensitivity to bleomycin. The poly(rA) moiety of the hybrid was not degraded by the drug, while the poly(dT) portion was degraded to some extent. However, the poly(dA) moiety in poly(rU · dA) is not degraded by bleomycin (unpublished results).

C. RNA

RNA binds only one molecule of bleomycin per 8.8×10^3 nucleotides *(17)*. Thus it is not surprising that it causes also no fragmentation of single-stranded RNA [ribosomal RNA, poly(C), poly(U), Qβ-RNA] or of the double-stranded poly(A · U) *(12)*. The reasons for the selectivity of bleomycin to degrade only DNA and not RNA may be (a) a thymine specificity, (b) a difference in the geometry of the two polynucleotides [B-conformation of native DNA and A-configuration of double-helical RNA *(61)*] or (c) a deoxyribose specificity. The first possibility can be ruled out *(62)* because bacteriophage PBS-1 DNA, in which thymine is replaced by uracil, reacts with bleomycin as much as does thymine-containing DNA. The second possibility is also unlikely because in DNA · RNA hybrids, which are presumably present in the A-helical conformation *(61)*, the DNA part is degraded by the drug *(60)*. No experimental data to prove the third possibility are available.

D. Poly(ADP-ribose)

The third polynucleotide besides DNA and RNA that exists in eukaryotic nuclei, poly(ADP-ribose) is not affected by bleomycin even at high concentrations *(22)*.

IV. Effect on DNA-, RNA-, and Protein Synthesis *in Vitro*

Bleomycin affects the activity of several enzyme systems, especially those involved in programmed synthesis. Those enzymes that use DNA as substrate or template are particularly influenced. Table I shows that bleomycin sensitively affects DNA-dependent DNA polymerases and DNA-dependent RNA polymerases, while the RNA-dependent DNA polymerase is not influenced.

TABLE I
QUANTITATIVE COMPARISON OF THE ACTION OF
BLEOMYCIN ON DIFFERENT POLYMERASES (63)[a]

Enzyme	Form of the enzyme	Source of the enzyme	Bleomycin concentration causing 50% inhibition (ED_{50}) (μg/ml; ±SD)
DNA-dependent DNA polymerase	—	RML virus	1.2 ± 0.3
	α	Oviduct	5.7 ± 0.5
	β	Oviduct	3.4 ± 0.4
	Pol I	Escherichia coli	4.9 ± 0.5
DNA-dependent RNA-polymerase	I	Oviduct	6.9 ± 1.2
	II	Oviduct	7.8 ± 1.2
		E. coli	4.8 ± 0.6
RNA-dependent DNA polymerase	—	RML virus	>100

[a] The experiments were performed with 2 mM dithiothreitol and native DNA, with the exception of the tests with Rauscher murine leukemia virus RNA-dependent DNA polymerase (RML virus), where the internal viral RNA was used. The isolation and the assay conditions of DNA and RNA polymerases from quail oviducts were as previously described (64).

A. DNA-Dependent DNA Polymerases

The first report on the influence of bleomycin on a DNA-dependent DNA polymerase showed that, at 40 μg/ml and with very low concentrations (1 μM) of the stimulating agent 2-mercaptoethanol, it stimulates the E. coli DNA polymerase (pol I) activity for a short period (10 minutes) and thereafter causes a marked inhibition (65). The effect on DNA polymerases was studied later in more detail (63, 66–69). As shown in Table I, viral (Rauscher murine leukemia virus), bacterial (E. coli) and eukaryotic DNA-dependent DNA polymerases are sensitively inhibited by bleomycin at 1–6 μg/ml. The most sensitively inhibited DNA polymerase is that isolated from the oncogenic virus; this finding was later confirmed using the enzyme isolated from the Moloney sarcoma virus (70). The inhibition of the enzyme is of the noncompetitive type in viral (66), bacterial (67), and eukaryotic systems (63). The K_i for the viral enzyme is 0.9 μg of bleomycin per milliliter (66) and for the bacterial enzyme 4.9 μg/ml (67) (see Fig. 13, left). The degree of inhibition of the enzyme activity is dependent on the (dA + dT)-content of the native template DNA used in the assay (67). The concentration of bleomycin needed for a 50% inhibition of the enzyme activity is the lower the higher the (dA + dT)-content of the DNA template (Fig. 14); in other words, the drug causes the strongest inhibition in those enzyme assays that contain a DNA

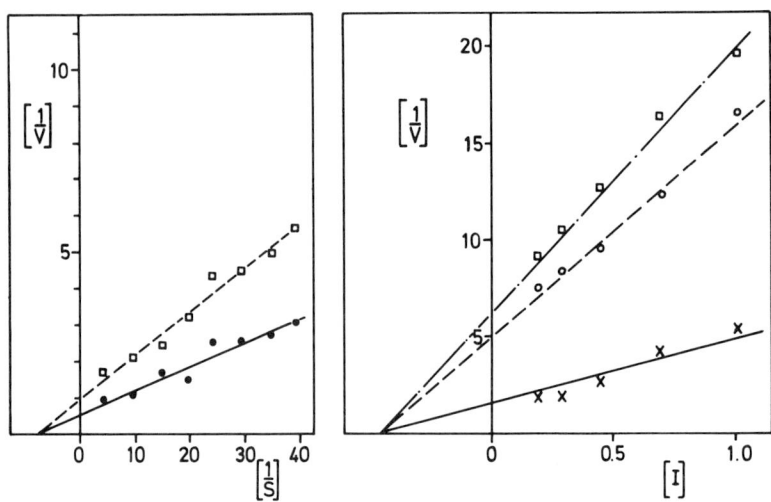

FIG. 13. Inhibition of *Escherichia coli* DNA-dependent DNA polymerase (pol I) by bleomycin (67). *Left:* Plot according to Lineweaver and Burk. The reactions were performed with different concentrations of native herring sperm DNA (S) and in the presence of 0 (●——●) and 7 μg of bleomycin (□——□) per milliliter. Ordinate: [1/V], reciprocal values of the initial reaction velocity in $10^9 \times 20$ min · mg protein per mole of dTTP incorporated; abscissa: [1/S], reciprocal values of the template concentration (ml/mg DNA). *Right:* Plot according to Dixon. The reactions were performed with 90 μg (×——×), 20 μg (○--○) and 13 μg (□-·-□) of poly(dA-dT) per milliliter as template and different concentrations (μg/ml) of bleomycin-treated DNA (= I).

template with a high (dA + dT). The finding that the enzyme system containing poly(dA-dT) is much more sensitive than the one containing poly(dA · dT) is striking, and is a hint that the antibiotic does not affect the enzyme directly but rather indirectly via a modification of the DNA template. This assumption has been proved experimentally (67, 63). Native DNA was incubated with bleomycin; subsequently the modified DNA was isolated and separated from the drug, and added to the DNA polymerase assays. This DNA caused an inhibition of the enzyme activity (Fig. 13, right). Also, in this experiment, a noncompetitive inhibition is observed. The inhibitor constant for the modified DNA is low and amounts to 0.45 μg/ml (67). Therefore the reaction sequence leading to an inhibition of DNA polymerase in the *in vitro* system can be formulated as follows: (a) modification of native DNA by bleomycin; and (b) inhibition of DNA polymerase by modified DNA.

At low concentrations of bleomycin (40 μg/ml and only 1μM 2-mercaptoethanol), the incorporation rate in an assay using DNA polymerase (pol I from *E. coli*) and native DNA increases for a short period of time after the start of the incubation (65). The explanation lies in a generation of primer sites in the native DNA template for the DNA polymerase; as reported above, bleomycin causes single-strand breaks with formation of 3'-OH termini (34), which serve as binding and primer sites for DNA polymerase I of *E. coli* (71).

The sites responsible for inhibition on the bleomycin-modified DNA are the aldehyde groups; after the reduction of these groups, the inhibitory potency is almost completely lost (71a). It is well known that apurinic acid sensitively inhibits the DNA polymerase isolated from mouse lymphoma cells noncompetitively; apurinic acid in which aldehyde groups have been reduced shows a much lower inhibitory potency (72).

In view of these observations, it is not surprising that the inhibitory potency of bleomycin in the DNA polymerase assay is enhanced by thiol-containing compounds (67). Dithiothreitol is a much stronger stimulating agent than 2-mercaptoethanol (67). The inhibitory potency is greatly enhanced by DNA ligands specific for dA and dT (73). The inhibition of *E. coli* DNA polymerase I caused by bleomycin is potentiated by coincubation with Hoechst 33258, a compound that attaches to the outside of the double DNA helix in the major groove (74), or with steffimycin B, an agent with potential intercalating capacity (49).

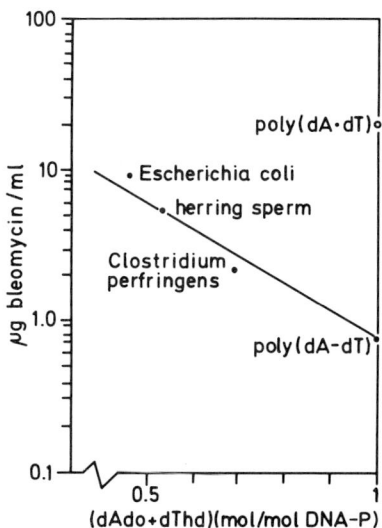

FIG. 14. Influence of bleomycin on DNA-dependent DNA polymerase (*Escherichia coli*) reaction using DNA templates with different molar (dAdo + dThd)-composition (67). The amount of bleomycin causing a 50% reduction of the enzyme activity is plotted against the mole fraction of (dAdo + dThd) in the template.

B. DNA-Dependent RNA Polymerases

The activity of DNA-dependent RNA polymerases from prokaryotes *(67)* and eukaryotes *(75, 4)* is also impaired by bleomycin when native DNA is used as template (see Table I). The most sensitively inhibited step in the RNA polymerase reaction sequence is the initiation of the reaction *(67)*. The inhibitory effect on DNA-dependent RNA polymerase is slightly diminished by high concentrations of RNA *(4)*. This is probably due to the binding of bleomycin to RNA, even though it is low *(17)*.

C. RNA-Dependent DNA Polymerase

The RNA-dependent DNA polymerase isolated from Rauscher murine leukemia virus is only very slightly affected by bleomycin (see Table I). At very high concentrations (100 µg/ml, in 2 mM dithiothreitol), the enzyme activity is reduced by only 25% *(66, 70)*.

D. DNA Ligase

One report *(65)* deals with an inhibition of DNA ligase isolated from T4 phage-infected *E. coli* B. However, this finding does not seem to be very conclusive because bleomycin was not removed from the DNA before the ligase reaction; therefore, it is possible that it is not the polynucleotide ligase reaction per se that is inhibited by the drug but that during the enzyme reaction new cleavages have been formed by the action of bleomycin.

E. Protein-Synthesizing System

Protein biosynthesis in a cell-free system isolated from the sea urchin *Sphaerechinus granularis* is not affected by bleomycin even at concentrations as high as 100 µg/ml *(67)*.

V. Effect on Nucleases *in Vitro*

A. Deoxyribonucleases

Nucleases using native DNA as substrate are strongly affected by bleomycin *(76)*. Both deoxyribonuclease I (bovine pancreas) and deoxyribonuclease II (porcine spleen) are competitively inhibited; deoxyribonuclease I activity is reduced 10 times more than deoxyribonuclease II activity *(76)*. In these cases also, bleomycin does not affect the enzyme activities directly, but does so indirectly via the modified DNA. Also bleomycin-free, modified DNA causes an inhibition of the two nucleases; the K_i for deoxyribonuclease I is 0.17 µg/ml, and for deoxyribonuclease II, it is 2.5 µg/ml. The exact mechanism of action is not known, but it can be assumed that the enzymes bind to, but do not hydrolyze, the modified DNA. The evidence presented rules out the possibility that the inhibition is caused by competition between the drug and the nuclease for DNA, as in the cases of actinomycin D and ethidium bromide *(77)*.

The inhibition of the deoxyribonucleases is strongest using poly(dA-dT) and 3.4 times weaker with poly(dA · dT); no inhibition is observed with poly(dG · dC) and the single-stranded poly(dA) and poly(dT) *(76)*.

B. Ribonucleases and Phosphodiesterases

The activity of ribonuclease A and B (beef pancreas), ribonuclease T_1 *(Aspergillus oryzae)*, phosphodiesterase I *(Crotalus adamanteus)* and phosphodiesterase II (beef spleen) are not influenced by bleomycin *(76)*.

VI. Influence on Nucleic Acids *In Vivo*

The predominant target macromolecules for bleomycin in the intact cell are the nucleic acids.

A. Effect on DNA

After being taken up by the cell, bleomycin enters the cell nucleus and binds to DNA *(78–80)*. These studies have been performed with mercury- or ^{14}C-labeled material. The binding ratio for bleomycin *in vivo* has been estimated to be 1 mol per 3.2×10^8 daltons DNA *(80)*. Most is located on the nuclear membrane and none is found in the nucleoplasm *(78, 79)*. It was not possible in these studies to determine to which structures on the nuclear membrane the antibiotic is bound; it could be that either the nuclear membrane per se, the membrane-associated DNA, or membrane-bound enzymes are the target sites. The chromatin complex does not seem to bind the drug *(78)*.

Umezawa *et al.* *(80)* clearly demonstrated that strand scission of DNA occurs after intracellular binding of bleomycin to DNA. They found that almost each molecule bound to DNA causes one strand scission. The drug causes strand scission of DNA in viruses (vaccinia virus) *(11)*, bacteria *(E. coli)* *(81, 82)*, and eukaryotic cell systems (L cells, mouse fibroblasts, HeLa cells, rat liver cells, rat ascites hepatoma cells) *(80, 83–87)*. Fragmentation of the DNA of cells in culture occurs as early as 30 minutes after incubation with bleomycin *(84)*. The M_r's of the DNA fragments thus produced in rat ascites hepatoma cells have been determined to be 5×10^8 (160 S) to 2×10^9 (280 S) *(80)*, 100 S for mouse fibroblasts *(84)* and 3×10^7 for *E. coli* *(82)*. The number of scissions per strand of DNA in hepatoma cells has been estimated to be 8 *(80)*. With lower concentrations of the drug, only single-strand breaks were produced, while higher concentrations are needed to yield double-strand breaks *(84)*; in *E. coli*, a considerable amount of intracellular DNA becomes acid-soluble after prolonged incubation with bleomycin at concentrations higher than 1 µg/ml *(81, 82)*. The nature of the termini of the DNA at these strand scissions is not known, but they seem to differ from those caused by X-rays *(81)*.

The sensitivity of intracellular DNA of *E. coli* toward bleomycin seems to depend on the physiological condition of the bacteria *(81)*; after incubation for 30 minutes at 100 µg/ml, only 10% of the DNA in growing bacteria remain acid-insoluble compared to 80% in bacteria in the stationary phase.

There are two observations concerning the site of bleomycin action in the nuclei of mammalian cells. The results of Ono *et al.* *(88)* indicate that after exposure of rat hepatoma cells to the antibiotic, a liberation of DNA from a membrane DNA-complex occurs. However, it could not be determined whether this arose from a reaction of bleomycin with DNA or not. In the second report *(89)*, it was found that nucleoplasmic DNA is 20–30 times more sensitive to bleomycin than is nucleolar DNA.

In lysogenic bacteria (KMBL-λ, *E. coli* K-12 D-10 and *B. subtilis* PBSH), bleomycin induces bacteriophage production *(81, 90)*. This induction is much higher in growing bacteria (33%) than in stationary bacteria (2%) *(81)*.

B. Effect on Chromatin

The ability of bleomycin to affect DNA in deoxyribonucleoprotein is dependent on its functional state *(91)*. For this investigation, the DNA · protein was extracted from tissues containing a large amount of proliferating cells (oviducts from immature quails pretreated on three successive days with estrogen hormone) or of nonproliferating cells (oviducts from immature quails treated on 12 successive days with estrogen hormone). The accessibility of the complex to bleomycin was studied by determination of the amount of thymine liberated from [³H]thymine-labeled DNA in the chromatin complex after treatment with the drug. After incubation with 200 µg/ml and in the presence of 5 mM dithiothreitol, 2.4 µg of thymine were released from 500 µg of DNA in the complex isolated from proliferating cells; only 1.5 µg of thymine were released under the same conditions from the DNA · protein isolated from nonproliferating cells. This finding indicates that DNA in the complex from proliferating cells is 1.6-fold more accessible to the drug than DNA in the complex from nonproliferating cells. After pretreatment with KCl, the accessibility of DNA in the complex increases; the higher the KCl concentration used for removal of protein from DNA · protein, the higher the splitting activity of the drug toward the DNA in the complex. For example, a pretreatment of DNA · protein isolated from nonproliferating cells with 0.6 M KCl results in a 5.8-fold increase of the amount of thymine released upon incubation with 200 µg of bleomycin/ml in 5 mM dithiothreitol compared with the release from untreated material. At this concentration of KCl, primarily the very lysine-rich histones and the slightly lysine-rich histones are removed from DNA · protein *(92)*.

It is now of interest that the accessibility of the DNA in deoxyribonucleoprotein from nonproliferating cells to bleomycin can be increased by incubation with 0.8 M KCl to an extent almost as great as that of untreated DNA · protein from proliferating cells *(91)* (see Fig. 15). From this finding, one can attribute a DNA-protecting function against bleomycin action to the F1 histone. From the model of Van Holde *et al. (93)*, this histone species does not seem to be located with the other histone fractions in the major groove of the double helix; it is more

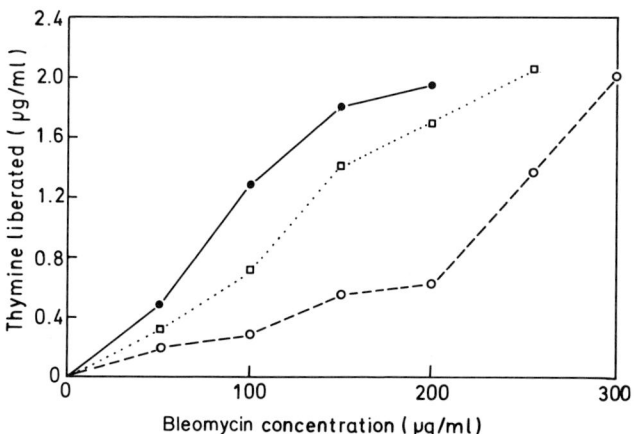

FIG. 15. Accessibility of DNA in different deoxyribonucleoprotein preparations to bleomycin *(91)*. DNA · protein from proliferating cells (●——●), from nonproliferating cells (○---○), and from nonproliferating cells treated with 0.8 M KCl (□······□) was used. The reaction mixture contained 200 µg of bleomycin/ml, 5 mM dithiothreitol and DNA · protein containing 0.5 mg of [^{3}H]thymine-labeled DNA. After incubation for 30 minutes at 37°C, the amount of released thymine was determined.

likely that F1 histone is associated with the DNA in the minor groove. If this model holds true, bleomycin affects DNA in the DNA · protein complex through the minor groove.

Probably as a result of the fragmentation of DNA in cells treated with bleomycin, chromosomal aberrations occur *(94–96)*. The abnormalities observed were mostly gaps, breaks, and fragments with a few dicentric and ring chromosomes.

C. Effect on RNA

As already reported, bleomycin does not bind to RNA *in vitro* (Section II, C). Thus it is not surprising that in intact cells, no binding of the drug to RNA is observed *(80)*. No experiments have been performed to check whether strand-scissions occur in RNA under the influence of bleomycin.

VII. Influence on Nucleic Acid Metabolism *In Vivo*

In intact cell systems or in whole organisms, two different effects of bleomycin are known: first, bleomycin affects the metabolism of endothelial surfaces *(97)*; and second and more important, it influences programmed synthesis. The latter effect is discussed in this section.

A. Influence on DNA Synthesis

Bleomycin inhibits proliferation of both prokaryotic and eukaryotic cells (survey: 77). As shown first by Suzuki *et al.* *(98)* with HeLa cells, this cytostatic effect is due to a selective inhibition of DNA synthesis. In mouse lymphoma cells, up to 10 times the concentration producing an 50% inhibition of cell proliferation reduces drastically only the incorporation of dThd into DNA; Urd and Trp incorporation are not significantly diminished *(4)*. Under these inhibition conditions, the cell volume of the treated cells is 30% higher than that of the untreated controls. This "unbalanced growth" *(99)* of the cells with bleomycin is another hint that RNA and protein synthesis continues, while DNA synthesis and cell division are inhibited. The viability of cells is not changed significantly when treated for 72 hours with twice the Ed_{50} concentration. At higher concentrations, the drug acts cytotoxically; at $3 \times ED_{50}$, the treated cells show a cell viability of 88%.

Terasima and Umezawa *(100)* observed that the effect of bleomycin on HeLa cells depends on the phase of growth cycle. The peak sensitivity is at the transition of late G_1 to the early S stages; G_1, G_2 and especially the mitotic phases are more resistant than log-phase cells *(101)*.

As reported above (Section IV, A), the A · T-specific DNA ligands steffimycin B and Hoechst 33258 potentiate the bleomycin-induced inhibition of DNA synthesis in isolated cell systems (73). The same compounds also synergistically enhance the cytostatic effect of the drug on mouse lymphoma cells (73). This observation strongly suggests that bleomycin directly affects DNA synthesis both *in vitro* and *in vivo*. It is known that DNA is fragmented *in vivo* as well as *in vitro* by the drug (see Sections VI, A and VI, B). The fragmented DNA in the chromatin complex has a lower template capacity for DNA polymerase than the untreated one (91) (Fig. 16A). It is striking that DNA-dependent DNA synthesis in an isolated enzyme system (*E. coli*, pol I) is inhibited more sensitively by bleomycin using as template deoxynucleoprotein from proliferating cells than is synthesis with the same template from nonproliferating cells. It has been suggested that this difference is due to a partial depletion of lysine-rich histones in the DNA · protein complex derived from proliferating cells (91). A similar relationship was noted earlier (102, 103); DNA in the chromatin complex from nongrowing cells is more resistant toward DNase digestion than DNA isolated from proliferating cells, and lysine-rich histones block the accessibility of DNA much more effectively than do arginine-rich histones. It remains to be determined by which mechanism the bleomycin-induced inhibition of DNA synthesis in intact cells is caused, e.g., by a possible noncompetitive inhibition of DNA-dependent DNA polymerase via the aldehyde groups present at the

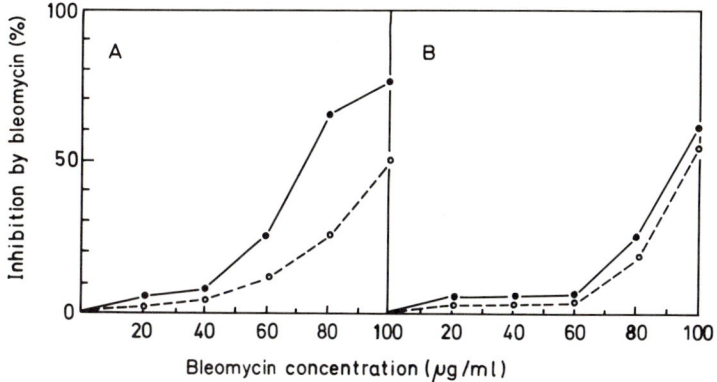

FIG. 16. Influence of bleomycin on template capacity of DNA · protein in the DNA and RNA polymerase systems (91). As template for the polymerases, DNA · protein from proliferating oviduct cells (●——●) and from nonproliferating oviduct cells (○---○) was used. (A) Template capacity of DNA · protein for DNA-dependent DNA polymerase (*E. coli*; pol I). (B) Template capacity of DNA · protein for DNA-dependent RNA polymerase (*E. coli*).

strand scissions, or by a reduction of the binding affinity of the enzyme to the drug-modified DNA.

The studies of the influence of bleomycin on the cell cycle reveal that the cells are reversibly inhibited during the S-phase. Depending on the cell strain, inhibition at the G_1-S *(104)* and S-G_2 boundary *(105)* is observed.

B. DNA Repair

The fragmentation of DNA by bleomycin caused intracellularly is rapidly repaired *(84–86, 96)*. After transfer of drug-treated mouse fibroblasts into a medium free of the antibiotic, single-strand breaks of DNA were rejoined within 3 hours; double-strand breaks of DNA were also rejoined on recovery incubation for 24 hours *(84)*. Also, the chromosomal aberrations caused by bleomycin were rapidly repaired, as shown in Chinese hamster ovary cells *(96)*.

The enzymic processes leading to a repair of the bleomycin-caused strand scission are not sufficiently elucidated. The first report came from Zöllner *et al. (106)*, who demonstrated that after incubation of lymphoma cells with bleomycin in cell culture, a distinct alkaline DNase (pH optimum 9; preference for denatured DNA; endonucleolytic action yielding 3'-OH-terminated polynucleotides) is activated. Evidence was presented showing that this enzyme is involved in a DNA-repair process.

Unscheduled DNA synthesis after modification of the DNA in a chromatin complex by bleomycin has been observed in HeLa cells *(85)* and in rat liver nuclei *(107)*. In rat liver nuclei, a DNA-dependent DNA polymerase activity involved in DNA repair synthesis was observed. This enzyme activity differs from the polymerase activity that increases following partial hepatectomy. In the light of the results of Chang and Bollum *(108)*, DNA polymerase β^2 is the enzyme involved in the repair of the bleomycin-produced DNA damage.

DNA ligase does not seem to be the primary site of bleomycin action in intact cell system. This conclusion must be drawn from the data obtained on HeLa cells, showing that only high concentrations of the drug produce some delay in the joining of short segments of replicating DNA.

C. Influence on Gene Expression

At low concentrations of bleomycin, which affect the cell cytostatically, neither RNA nor protein synthesis is inhibited in either mouse

---

[2] See article by Bollum in Vol. 15 of this series.

lymphoma cells in suspension culture or in oviduct cells in the intact organism *(4)*. The synthesis of mRNA in response to progesterone in estrogen-hormone-stimulated immature quail oviduct is not reduced, nor is the subsequent translation process influenced by the treatment. This report *(4)* underlines the great selectivity of bleomycin in inhibiting only DNA synthesis. The reports *(109, 110)* describing an inhibitory influence of the drug on RNA and on protein synthesis never exclude the possibility that the high concentrations of drug used for these studies are cytotoxic for the cells.

Three attempts have been made to resolve the discrepancy between the finding that in isolated enzyme systems both DNA and RNA synthesis are inhibited (Sections IV, A and IV, B) while in intact cell systems only DNA synthesis is blocked by bleomycin. First, two investigations showed RNA to compete with DNA in the bleomycin reaction *in vitro (4, 62)*. However, the RNA concentrations required for reduction of the cleavage activity of bleomycin *in vitro* seem to be too high to explain the inactivity of the drug on RNA synthesis *in vivo*. Second, Fujimoto *(79)* made the interesting observation that bleomycin accumulates mainly on nuclear membranes. Since various reports (e.g., *111*) show DNA to be connected to the nuclear membrane in the eukaryotic cell, bleomycin could cause DNA strand scissions at the nuclear membrane resulting in an inhibition of both DNA ligase and DNA polymerase. These events could subsequently cause inhibition of complete DNA replication. Third, 30% higher concentrations of bleomycin are needed to reduce the activity of exogenous DNA-dependent RNA polymerase *(E. coli)* to 50% of that obtained with exogenous DNA-dependent DNA polymerase *(E. coli;* pol I) in incubation mixtures using deoxynucleoprotein from proliferating cells as template (Fig. 16). This differential sensitivity of the two polymerases toward bleomycin is no more pronounced if DNA · protein from nonproliferating cells is used as template in the assays. From this finding, it can be assumed that during mitotic cell division as well, the different histones are responsible for the differential inhibition of DNA and RNA synthesis by low concentrations of bleomycin at which only DNA synthesis is inhibited.

VIII. Medical Implications

The antibiotic and cytostatic agent bleomycin is used internationally in medicine as a powerful antitumor agent, especially for squamous cell carcinoma *(112, 113)*. Some of the reasons for the clinical usefulness of this drug are mentioned below.

A. Cell Kinetics

It is well established for Ehrlich ascites carcinoma *(114, 115)*, L cells *(104, 105)* and human malignant melanoma cells *(116)* that bleomycin arrests proliferating, asynchronous cells during the S-phase. It was found, e.g., that 1.5 to 4 times the normal number of cells (human malignant melanoma cells) are in S-phase at the peak times following the first treatment *(116)*. The first consequence of these observations would be that bleomycin becomes a prospective cell-synchronizing agent for *in vivo* work. Additionally, it can be used not only in monotherapy, but also as one component in combination tumor therapy. The most promising substances with which the drug can be applied in cancer polychemotherapy are the DNA-alkylating agents *(117)*, the DNA-intercalating agents *(73)* and the DNA-adducting agents *(73)*. All of these have been shown to stimulate synergistically the antitumor capacity of bleomycin. Antimetabolites acting at the level of the DNA-dependent DNA polymerase, such as 1-β-D-arabinofuranosylcytosine *(118)*, exert only an additive effect when used in combination with bleomycin. However, antimetabolites affecting the *de novo* synthesis of the nucleotides, like methotrexate *(119)*, potentiate the effect *(120)*.

B. Tumor Specificity

The cytostatic or antitumor activity of bleomycin on certain organs or tumors depends upon uptake, inactivation, and activation of the drug. The uptake of bleomycin is tissue- and tumor specific, being highest in urinary bladder, kidney, lung, skin, and carcinoma tissues, and lower in liver, spleen and sarcoma *(51)*. Inactivation by enzymic degradation varies tissue-specifically to a large extent *(51–53)*. The bleomycin-inactivating enzyme activity is high in liver, with lower enzyme concentrations in the testis, spleen, lung, and brain; the enzyme is almost completely absent from skin *(52)*. The efficacy of bleomycin is determined by the ratio between the rate of inactivation and the uptake; this ratio is highest in liver and spleen and lowest in skin and lung. Thus the highest concentration of cytostatically active bleomycin is present in skin and lung, while only low concentrations are present in liver and spleen. The efficacy is not only tissue-specific but also tumor-specific *(51)*; it is much higher in carcinoma than in sarcoma. It also seems likely that bleomycin is activated organ- and tumor-specifically *(80)*. As reported above (Section II, B), the activity *in vitro* is strongly enhanced by thiol-containing compounds. It now appears *(80)* that the level of protein-free thiol compounds in rat ascites hepatoma cells is 1.8-fold lower in bleomycin-resistant cells than in sensitive cells, suggesting a possible enhancement of action by intracellular thiol compounds similar to that found *in vitro*.

From the finding that the efficacy of bleomycin varies markedly among different tissues, a test has been developed to estimate the sensitivity of a certain tumor to bleomycin *(52, 53)*. This sensitivity test is based on determinations of the specific activity of the inactivating enzyme and of the enrichment of the drug in a certain tissue.

Bleomycin differs from most antitumor compounds used in the clinic in lacking immunsuppressive properties *(121)*. The absence of immunsuppression activity, which is a highly desirable characteristic of an antitumor agent, might be due to the low efficacy of the drug in spleen and lymphoid tissues *(51)*.

C. Antiviral Activity

The potential antiviral activity of bleomycin has been checked in two DNA-virus systems, vaccinia virus *(11, 28)* and Rous sarcoma virus *(122, 123)*. The drug inhibits vaccinia virus replication in infected HeLa cells and mice owing to an inhibition of DNA synthesis and core RNA polymerase activity *(11, 28)*. However, this does not seem to be a virus-specific effect because not only is viral-DNA synthesis inhibited, but also that of cellular DNA. Rous sarcoma virus replication as well as virus-induced transformation is inhibited *in vivo* *(122)* as well as *in vitro* *(123)*. In this system, no evidence could be presented showing a more sensitive inhibition of virus replication compared to the inhibition of host cell DNA synthesis *(123)*.

IX. Future Directions

The unique characteristics of bleomycin suggest topics for future studies. Besides the investigations dealing with improved application of it in cancer chemotherapy, the biochemical work will focus especially on the three topics briefly reviewed in the following sections.

A. Tool for Studying Nucleotide Sequences in DNAs

The various methods for studying sequences in DNAs can be subdivided into two groups, according to whether they are based on investigations of properties related to the nucleotide sequences (indirect methods) or on the direct study of the sequence (direct methods) *(124)*. The indirect methods can further be divided into the class of methods determining either the sequence-dependent properties of single-stranded DNA (renaturation kinetics, cyclization of DNA after shearing and exonuclease treatment) or the sequence-dependent properties of double-stranded DNA (buoyant density in CsCl, melting temperature, circular dichroism, silver binding, hydroxylapatite binding). The direct methods include those determining either the sequences themselves (DNA polymerase repair, DNA polymerase ribosubstitution, RNA polymerase copy, direct sequencing) or the frequency of the sequences (depurination, nearest-neighbor analysis, analysis of termini released by DNases).

The bleomycin procedure belongs in the group of direct methods, in particular, in the class of "frequency" methods. It is not a substitute for the depurination procedure, nearest-neighbor analysis, or analysis of termini released by DNases. Under certain conditions, it is specific for removal of thymine from double-stranded DNA or from DNA · RNA hybrids; at low concentrations of the drug, the thymine moieties removed are those adjacent to the guanine (Section II, C). Thus bleomycin shows a higher base specificity than does depurination or depyrimidination, which yield pyrimidine or purine isostichs without discriminating between the two pyrimidine bases on the one hand or the two purine bases on the other (125).

The bleomycin method is relatively specific for double-stranded DNA. In the nearest-neighbor method (126), the frequency of all doublets (irrespective of the nature of the base) in a given DNA is obtained; the method works both with double-stranded and with single-stranded DNA. With the third frequency method, an analysis of termini released by DNases is possible (124). This method uses DNases of different relative base specificity. The sequences recognized by the enzymes are longer than the dinucleotides studied by nearest-neighbor analysis.

The recognition sites for bleomycin involve only two bases. This property is probably due to the low molecular weight ratio (around $M_r = 1500$). The spleen acid DNase ($M_r = 38,000$) recognizes four nucleotides, the pancreatic DNase ($M_r = 31,000$) and the E. coli endonuclease I ($M_r = 24,000$) recognize three nucleotides, and the snail DNase ($M_r = 30,000$) recognizes two nucleotides on both sides of the breaks (124). The restriction enzymes with their higher M_r's recognize up to six nucleotides (127). With bleomycin, a determination of the sequence frequency of dG-dT sets in a given DNA is possible in a way similar to that described by Bernardi with DNases (124). The bleomycin method adds a new feature to the DNase technique.

Bleomycin can also recognize and remove bases from the DNA in a DNA · RNA hybrid and thus makes it possible to detect sequence frequencies in a particular gene as follows: (a) single-stranded total cellular DNA is hybridized with the mRNA of a particular gene; (b) this hybrid is treated with bleomycin; (c) bleomycin should primarily attack bases in the double-stranded DNA · RNA region, the single-stranded, nonhybridized DNA stretches being affected to a much smaller extent. In combination with the electron microscope and the technique of labeling thymine with osmium tetroxide and cyanide ion *(128)*, it should also be possible to localize the gene within the DNA molecule. In addition, the bleomycin-cleavage method seems to have a higher base-specificity than the DNase-cleavage method.

B. Tool for Studying DNA Synthesis

Inhibitors with a high selectivity for DNA synthesis are one tool leading to a better understanding of DNA replication in intact cells. This event involves not only DNA-dependent DNA polymerase(s) but obviously also the action of "site-specific" nucleases on "palindromic sequences," of an RNA-polymerizing enzyme that synthesizes the RNA initiator for DNA synthesis, and of additional protein factors such as "unwinding proteins," which provide the mammalian DNA-dependent DNA polymerases with the capability to recognize the initiation site for DNA synthesis and to unwind or dissociate double-stranded DNA *(129)*. Substrate analogs with a high inhibitory selectivity for eukaryotic DNA-dependent DNA polymerases, like 1-β-D-arabinofuranosylcytosine 5'-triphosphate *(118)* and 9-β-D-arabinofuranosyladenosine 5'-triphosphate *(64)*, are known. Compounds affecting the other procedures mentioned are still lacking. The action of bleomycin on intracellular DNA synthesis is doubtless different from that of the substrate analogs. Thus the elucidation of the intracellular mode of action of bleomycin, a compound that is a highly specific inhibitor of DNA synthesis, hopefully will help to understand a further step in the sequence of the metabolic events leading to DNA replication.

C. DNA Repair Mechanisms

In eukaryotic systems, three DNA-repair mechanisms can be distinguished: first the "short" type of DNA repair of lesions in the DNA caused by ionizing radiation and by some compounds, like ethyl methanesulfonate and propane sultone *(130);* second, the "long" type of DNA repair of damaged nucleotides caused by ultraviolet radiation and compounds like *N*-acetoxy-2-acetylaminofluorene and 2-methoxy-6-chloro-9-[3-(ethyl-2-chloroethyl)aminopropylamino]acridine *(130);* third the repair mechanism specific for apurinic sites, created in DNA by slow spontaneous hydrolysis of purine glycosidic bonds at neutral pH *(32)*. It remains to be determined whether the defects caused by bleomycin in DNA are repaired by one of these repair mechanisms, or whether an additional form of repair is responsible for the reversion of the DNA to its original form that occurs in the cell.

ACKNOWLEDGMENTS

The work reported in this paper was partially undertaken during the tenure of a Yamagiwa–Yoshida Memorial International Cancer Study Grant awarded to W.E.G.M. by the International Union Against Cancer, on a sabbatical leave in the Department of Biochemistry (Professor F. J. Bollum), Albert B. Chandler Medical Center, University of Kentucky, Lexington, Kentucky. Additionally, the research reported was supported by a grant from the Commission for Educational Exchange between the Federal Republic of Germany and the United States of America (Fulbright Commission) given to W.E.G.M.

REFERENCES

1.. J. Meienhofer, H. Maeda, C. B. Glaser, J. Czombos and K. Kuromizu, *Science* **178**, 875 (1972).
2. H. Umezawa, K. Maeda, T. Takeuchi and Y. Okami, *J. Antibiot.* **19**, 200 (1966).
3. M. Ishizuka, H. Takayama, T. Takeuchi and H. Umezawa, *J. Antibiot.* **20**, 15 (1967).
4. W. E. G. Müller, A. Totsuka, I. Nusser, R. K. Zahn and H. Umezawa, *Biochem. Pharmacol.* **24**, 911 (1975).
5. H. Umezawa, *Pure Appl. Chem.* **28**, 665 (1971).
6. H. Umezawa, in "Antibiotics" (J. W. Corcoran and F. E. Hahn, eds.), Vol. 3, p. 21. Springer-Verlag, Berlin and New York, 1975.
7. H. Umezawa, *FP* **33**, 2296 (1974).
8. K. Nagai, H. Suzuki, N. Tanaka and H. Umezawa, *J. Antibiot.* **22**, 569 (1969).
9. K. Nagai, H. Suzuki, N. Tanaka and H. Umezawa, *J. Antibiot.* **22**, 624 (1969).
10. C. W. Haidle, *Mol. Pharmacol.* **7**, 645 (1971).
11. M. Takeshita, S. B. Horwitz and A. P. Grollman, *Virology* **60**, 455 (1974).
12. W. E. G. Müller, Z. Yamazaki, H. J. Breter and R. K. Zahn, *EJB* **31**, 518 (1972).
13. R. Ishida and T. Takahashi, *BBRC* **66**, 1432 (1975).
14. W. E. G. Müller, H. J. Rohde and R. K. Zahn, *Mol. Med.* **1**, 173 (1976).
15. E. P. Geiduschek and A. Daniels, *Anal. Biochem.* **11**, 133 (1965).
16. K. Nagai, H. Yamaki, N. Tanaka and H. Umezawa, *BBA* **179**, 165 (1969).
17. H. Suzuki, K. Nagai, E. Akutsu, H. Yamaki, N. Tanaka and H. Umezawa, *J. Antibiot.* **23**, 473 (1970).
18. W. C. Krueger, L. M. Pschigoda and F. Reusser, *J. Antibiot.* **26**, 424 (1973).
19. M. Chien, A. P. Grollman and S. B. Horwitz, *Am. Chem. Soc. Meet.*, 1975.
20. H. Murakami, H. Mori and S. Taira, *J. Theor. Biol.* **42**, 443 (1973).
21. H. Murakami, H. Mori and S. Taira, *J. Theor. Biol.* **59**, 1 (1976).
22. W. E. G. Müller and R. K. Zahn, *Gann Monogr. Cancer Res.* **19**, 51 (1976).
23. M. Madel, *JMB* **5**, 435 (1962).
24. H. R. Mahler and B. D. Mehrotra, *BBA* **55**, 252 (1962).
25. A. M. Liquori, L. Constantino, V. Crescenzi, V. Elia, E. Giglio, R. Puliti, M. De Santis Savino and V. Vitagliano, *JMB* **24**, 113 (1967).
26. M. Tsuboi, *Bull. Chem. Soc. Jpn.* **37**, 1514 (1964).
27. C. W. Haidle, K. K. Weiss and M. T. Kuo, *Mol. Pharmacol.* **8**, 531 (1972).
28. M. Takeshita, A. P. Grollman and S. B. Horwitz, *Virology* **69**, 453 (1976).
29. L. D. Saslaw and V. S. Waradvekar, in "Methods in Enzymology," Vol. XII, A (L. Grossman and K. Moldave, eds.), p. 108. Academic Press, New York, 1967.

30. D. M. Brown and A. R. Todd, in "The Nucleic Acids" (E. Chargaff and J. N. Davidson, eds.), Vol. 1, p. 409. Academic Press, New York, 1955.
31. A. S. Jones, A. M. Mian and R. T. Walker, *JCS (C)*, 2042 (1968).
32. A. Temperli, H. Türler, P. Rüst, A. Danon and E. Chargaff, *BBA* **91**, 462 (1964).
33. T. Lindahl and A. Andersson, *Bchem* **11**, 3618 (1972).
34. M. T. Kuo and C. W. Haidle, *BBA* **335**, 109 (1973).
35. H. S. Shapiro and E. Chargaff, *BBA* **91**, 262 (1964).
36. M. Geisert, in preparation
37. M. T. Kuo, C. W. Haidle and L. D. Inners, *Biophys. J.* **13**, 1296 (1973).
38. H. Umezawa, H. Asakura, K. Oda, S. Hori and M. Hori, *J. Antibiot.* **26**, 521 (1973).
39. T. Onishi, H. Iwata and Y. Takagi, *BBA* **378**, 439 (1975).
40. T. Takita, Y. Muraoka, T. Yoshioka, A. Fujii, K. Maeda and H. Umezawa, *J. Antibiot.* **25**, 755 (1972).
40a. R. DiCioccio and B. I. Srivastava, *Cancer Res.* **36**, 1664 (1976).
41. H. Netter, "Theoretische Biochemie-Physikalisch-chemische Grundlagen der Lebensvorgänge." Springer-Verlag, Berlin and New York, 1959.
42. C. Zimmer, *Z. Chem.* **11**, 441 (1971).
43. I. Shirakawa, M. Azegami, S. Ishii and H. Umezaea, *J. Antibiot.* **24**, 761 (1971).
44. T. Onishi, H. Iwata and Y. Takagi, *J. Biochem. (Tokyo)* **77**, 745 (1975).
45. R. Shapiro, D. G. Doherty and W. T. Burnett, *Radiat. Res.* **7**, 22 (1957).
46. E. Freese and E. Bautz-Freese, *Radiat. Res., Suppl.* **6**, 97 (1966).
47. H. Asakura, M. Hori and H. Umezawa, *J. Antibiot.* **28**, 537 (1975).
48. J. Bearden and C. W. Haidle, *BBRC* **65**, 371 (1975).
49. F. Reusser, *BBA* **383**, 266 (1975).
50. M. Takeshita, S. B. Horwitz and A. P. Grollman, *Ann. N.Y. Acad. Sci.*, in press.
51. H. Umezawa, T. Takeuchi, S. Hori, T. Sawa, M. Ishizuka, T. Ichikawa and T. Komai, *J. Antibiot.* **25**, 409 (1972).
52. W. E. G. Müller, A. Totsuka, R. K. Zahn and H. Umezawa, *Z. Krebsforsch.* **83**, 151 (1975).
53. W. E. G. Müller and R. K. Zahn, *Prog. Biochem. Pharmacol.* **11**, 119 (1976).
54. H. Umezawa, *Biomedicine* **18**, 459 (1973).
55. H. Umezawa, S. Hori, T. Sawa, T. Yoshioka and T. Takeuchi, *J. Antibiot.* **27**, 419 (1974).
56. P. R. Rony, *JACS* **21**, 6090 (1969).
57. K. Kergomard and M. Renard, *Tetrahedron* **24**, 6643 (1968).
58. N. K. Kochetkov and E. I. Budovskii, "Organic Chemistry of Nucleic Acids," Part B, p. 425. Plenum, New York, 1972.
58a. N. K. Kochetkov and E. I. Budovskii, "Organic Chemistry of Nucleic Acids," Part B, pp. 477–532. Plenum, New York 1972.
59. C. Tamm and E. Chargaff, *JBC* **203**, 689 (1953).
59a. W. E. Cohn and J. X Khym, in "Acides Ribonucléiques et Polyphosphates: Structure, Synthèse et Fonction. Colloques Internationaux du C.N.R.S., Strasbourg 1961," p. 217. C.N.R.S., Paris, 1962.
60. C. W. Haidle and J. Bearden, *BBRC* **65**, 815 (1975).
61. S. Arnott, *Prog. Biophys.* **21**, 265, Pergamon, Oxford, 1970.
62. C. W. Haidle, M. T. Kuo and K. K. Weiss, *Biochem. Pharmacol.* **21**, 3308 (1972).
63. W. E. G. Müller, A. Totsuka, R. K. Zahn and H. Umezawa, in "Bleomycin" (W. Wilmanns, ed.), p. 27. Mack, Illertissen, 1975.
64. W. E. G. Müller, H. J. Rohde, R. Beyer, A. Maidhof, M. Lachmann, H. Taschner and R. K. Zahn, *Cancer Res.* **35**, 2160 (1975).

65. H. Yamaki, H. Suzuki, K. Nagai, N. Tanaka and H. Umezawa, *J. Antibiot.* **24**, 178 (1971).
66. W. E. G. Müller, Z. Yamazaki and R. K. Zahn, *BBRC* **46**, 1667 (1972).
67. Z. Yamazaki, W. E. G. Müller and R. K. Zahn, *BBA* **308**, 412 (1973).
68. W. E. G. Müller, Z. Yamazaki, W. Forster, R. K. Zahn and H. J. Seidel, *Klin. Wochenschr.* **50**, 790 (1972).
69. W. E. G. Müller and R. K. Zahn, *Gann* **67**, 425 (1976).
70. M. Miyaki, J. Kuroda and T. Ono, *Igakunoayumi* **85**, 72 (1973).
71. P. T. Englund, R. B. Kelly and A. Kornberg, *JBC* **244**, 3045 (1969).
71a. W. E. G. Müller, *J. Pharmacol. Therapeut.* **1A**, in press (1977).
72. W. E. G. Müller, W. Hanske, A. Maidhof and R. K. Zahn, *Cancer Res.* **33**, 2330 (1973).
73. W. E. G. Müller, H. J. Rohde, R. Steffen, A. Maidhof and R. K. Zahn, *Cancer Lett.* **1**, 127 (1976).
74. D. E. Comings, *Chromosoma* **52**, 229 (1975).
75. T. Kobayashi, *J. Antibiot.* **24**, 519 (1971).
76. W. E. G. Müller, Z. Yamazaki, J. E. Zöllner and R. K. Zahn, *FEBS Lett.* **31**, 217 (1973).
77. L. J. Eron and B. R. McAuslan, *BBA* **114**, 633 (1966).
78. P. Pietsch and A. Murray, *Biochem. Pharmacol. Suppl.* **2**, 193 (1974).
79. J. Fujimoto, *Cancer Res.* **34**, 2969 (1974).
80. M. Miyaki, T. Ono, S. Hori and H. Umezawa, *Cancer Res.* **35**, 2015 (1975).
81. K. Suzuki, *Japan. J. Genet.* **46**, 277 (1971).
82. T. Onishi, K. Shimada and Y. Takagi, *BBA* **312**, 248 (1973).
83. T. Terasima, M. Yasukawa and H. Umezawa, *Gann* **61**, 513 (1970).
84. M. Saito and T. Andoh, *Cancer Res.* **33**, 1696 (1973).
85. Y. Fujiwara and T. Kondo, *Biochem. Pharmacol.* **22**, 323 (1973).
86. R. Cox, A. H. Daoud and C. C. Irving, *Biochem. Pharmacol.* **23**, 3147 (1974).
87. C. Mittermayer, R. Osieka and H. Madreiter, *Arch. Dermatol. Forsch.* **249**, 401 (1974).
88. T. Ono, M. Miyaki and J. Kuroda, Report at the Annual Meeting of the Japan Cancer Society, Nagoya, November 1972.
89. S. T. Crooke, T. O. Sitz, M. Bannon and H. Busch, *Physiol. Chem. Phys.* **7**, 177 (1975).
90. C. W. Haidle, K. K. Weiss and M. L. Mace, *BBRC* **48**, 1179 (1972).
91. W. E. G. Müller and R. K. Zahn, *Prog. Biochem. Pharm.* **11**, 28 (1976).
92. T. C. Spelsberg and L. S. Hnilica, *BBA* **228**, 202 (1971).
93. K. E. Van Holde, C. G. Sahasrabuddhe and B. R. Shaw, *Nuc. Acids Res.* **1**, 1579 (1974).
94. K. Ohama and T. Kadotani, *Jpn. J. Human Genet.* **14**, 293 (1970).
95. K. D. Paika and A. Krishan, *Cancer Res.* **33**, 961 (1973).
96. W. N. Hittelmane and P. N. Rao, *Cancer Res.* **34**, 3433 (1974).
97. M. Hagedorn, J. Petres and C. Mittermayer, *Arch. Dermatol. Forsch.* **250**, 71 (1974).
98. H. Suzuki, K. Nagai, H. Yamaki, N. Tanaka and H. Umezawa, *J. Antibiot.* **21**, 379 (1968).
99. L. S. Cohen and G. P. Studzinski, *J. Cell. Physiol.* **69**, 331 (1967).
100. T. Terasima and H. Umezawa, *J. Antibiot.* **23**, 300 (1970).
101. F. Mauro, B. Falpo, G. Briganti, R. Elli and G. Zupi, *J. Natl. Cancer Inst.* **52**, 715 (1974).
102. A. E. Mirsky and B. Silverman, *PNAS* **69**, 2115 (1972).
103. T. Pederson, *PNAS* **69**, 2224 (1972).

104. V. Bremerskov, *Eur. J. Cancer* **9**, 25 (1973).
105. M. Watanabe, Y. Takabe, T. Katsumata and T. Terasima, *Cancer Res.* **34**, 878 (1974).
106. E. J. Zöllner, D. Weinblum, J. Overmeier and R. K. Zahn, *Exp. Cell Res.* **99**, 185 (1976).
107. G. P. Sartiano, W. Lynch, S. S. Boggs and G. L. Neil, *PSEBM* **150**, 718 (1975).
108. L. M. S. Chang and F. J. Bollum, *JBC* **247**, 7948 (1972).
109. T. Nakashima, M. Kuwano, K. Matsui, S. Komiyama, I. Hiroto and H. Endo, *Cancer Res.* **34**, 3258 (1974).
110. Y. Daskal, S. T. Crooke, K. Smetana and H. Busch, *Cancer Res.* **35**, 374 (1975).
111. M. G. Ormerod and A. R. Lehmann, *BBA* **228**, 331 (1971).
112. T. Ichikawa, *Proc. Int. Cong. Chemother.*, 5th, A IV-4/35 (1967).
113. R. H. Blum, S. K. Carter and K. Agre, *Cancer* **31**, 903 (1973).
114. M. Nagatsu, R. M. Richart and A. Lambert, *Cancer Res.* **32**, 1966 (1972).
115. J. Shumann and W. Göhde, *Strahlentherapie* **147**, 298 (1974).
116. S. C. Barranco, J. K. Luce, M. M. Romsdahl and R. M. Humphrey, *Cancer Res.* **33**, 882 (1973).
117. H. Sato and H. Ichimura, *Igaku No Ayumi* **69**, 669 (1969).
118. W. E. G. Müller, Z. Yamazaki, H. H. Sögtrop and R. K. Zahn, *Eur. J. Cancer* **8**, 421 (1972).
119. R. Ferone, J. J. Burchall and G. H. Hitchings, *Mol. Pharmacol.* **5**, 49 (1969).
120. R. T. Skeel and J. R. Bertino, *Cancer Res.* **33**, 1028 (1973).
121. H. Yamaki, N. Tanaka and H. Umezawa, *J. Antibiot.* **22**, 315 (1969).
122. M. Takeuchi and T. Yamamoto, *J. Antibiot.* **21**, 631 (1968).
123. A. Totsuka, W. E. G. Müller and R. K. Zahn, *Arch. Virol.* **48**, 169 (1975).
124. G. Bernardi, "Karl-August-Forster-Lectures," Vol. 12. Franz Steiner, Wiesbaden, 1975.
125. H. Türler, *in* "Procedures in Nucleic Acid Research" (G. L. Cantoni and D. R. Davies, eds.), Vol. 2, p. 680. Harper & Row, New York, 1971.
126. J. Josse, A. D. Kaiser and A. Kornberg, *JBC* **236**, 864 (1961).
127. D. Nathans and H. O. Smith, *ARB* **44**, 273 (1975).
128. M. Beer, *in* "Procedures in Nucleic Acid Research" (G. L. Cantoni and D. R. Davies, eds.), Vol. 2, p. 443. Harper & Row, New York, 1971.
129. F. J. Bollum, This Series **15**, 109 (1975).
130. J. D. Regan and R. B. Setlow, *Cancer Res.* **34**, 3318 (1974).

Mammalian Nucleolytic Enzymes

> HALINA SIERAKOWSKA AND
> DAVID SHUGAR
>
> *Institute of Biochemistry and
> Biophysics
> Academy of Sciences
> Warsaw, Poland*

I.	Introduction	60
II.	Endoribonucleases	61
	A. Ribonuclease I (Pancreatic RNase, Alkaline or Neutral RNase), EC 3.1.4.22	61
	B. Nucleolar Ribonuclease	72
	C. Ribonuclease II (Acid Ribonuclease), EC 3.1.4.23	73
	D. Ribonucleases Attacking Double-Stranded RNA	75
	E. "Processing" Ribonucleases	76
	F. Hybrid Nuclease (RNase H), EC 3.1.4.34	78
III.	Exoribonucleases	80
	A. 5'-Exoribonuclease, EC 3.1.4.20	80
	B. Polynucleotide Phosphorylase, EC 2.7.7.8	81
IV.	Endodeoxyribonucleases	82
	A. Deoxyribonuclease I (Pancreatic DNase, Neutral DNase), EC 3.1.4.5	82
	B. Ca, Mg-Dependent Endonuclease	85
	C. Deoxyribonuclease II (Acid DNase), EC 3.1.4.6	87
	D. "Nicking" Enzymes	89
	E. "Nicking-Closing" Enzyme(s) ("Relaxation" Proteins)	91
V.	Exodeoxyribonucleases	93
	A. Deoxyribonuclease III	93
	B. Exodeoxyribonuclease IV, EC 3.1.4.28	94
VI.	Endo-Exodeoxyribonuclease with Preference for Poly(dT)	95
VII.	Single-Stranded-Nucleate Endonuclease (Sugar-Nonspecific Endonuclease; 5'-Endonuclease), EC 3.1.4.21	95
VIII.	Sugar-Nonspecific Exonucleases	97
	A. Phosphodiesterase I (5'-Exonuclease), EC 3.1.4.1, EC 3.1.4.19	97
	B. Phosphodiesterase II (Spleen Exonuclease, Acid 3'-Exonuclease), EC 3.1.4.18	100
IX.	2':3'-Cyclic-nucleotide 3'-Phosphodiesterase, EC 3.1.4.37	102
X.	Localization and Function of Nucleases	102
XI.	Virus-Associated Nucleases	106
	A. Origin of Endodeoxyribonuclease Activity in Polyoma Viruses	107
	B. Human Adenovirus Endonuclease	107
	C. Deoxyribonuclease Activities in Other Viruses	108

D. Viral Ribonucleases 110
E. Viral RNase H 110
XII. Nucleases and Cellular Repair Processes 114
A. Mammalian "Incision" Endonuclease 115
B. Endonuclease Specific for Apurinic Sites 116
C. "Excision" Exonucleases 118
D. Repair Defiency in Xeroderma Pigmentosum 119
References 120

I. Introduction

In the period that has elapsed since we last attempted a reivew of mammalian nucleolytic enzymes (1), the volume of literature is such that comprehensive coverage would require a separate volume. Progress has been truly impressive, largely through the widespread applications of chromatographic and centrifugation techniques for fractionation of cellular components and purification of enzymes, and, particularly in the case of microorganisms, to the ingenious application of basic concepts in molecular genetics to the organized search for enzymes suspected of being involved in such highly specific processes as replication, recombination, repair, RNA maturation, etc. The results of such studies have, in turn, been carried over to investigations on corresponding mammalian systems. Furthermore, the discovery of "reverse transcriptase" in 1970 provided a long-needed stimulus to the investigation of virus-associated and induced enzymes, including nucleases involved in transformation to the tumorous state, a field now expanding so rapidly as to render difficult its updating in a current review.[1]

In what follows, occasional reference is made to enzymes of other vertebrates, microorganisms[2] and plants, when necessary for elucidation of the properties or function of corresponding or related mammalian enzymes. Descriptions of the latter are limited largely to such aspects as specificity, cellular localization, and possible functional role, and supplemented, where necessary, by reference to other specialized reviews on individual enzymes.

For convenience, enzymes are referred to by their most commonly used names, with synonyms in parentheses. The IUB Enzyme Commission number is given where applicable (1a).

The term "exonuclease," originally denoting an activity catalyzing

---

[1] See Green and Gerard in Vol. 14, Gillespie, Saxinger, and Gallo in Vol. 15 of this series.

[2] See Datta and Niyogi in Vol. 17 of this series, concerning bacterial ribonucleases.

stepwise removal, from a free terminus, of mononucleotides, now also includes enzymes that detach oligonucleotides from a free terminus.

The classical terms "endonuclease" and "exonuclease" are now supplemented by "endo-exonuclease," the initial action of which is endonucleolytic. Classification of enzymes in this review is according to this division, which is based on properties no less fundamental than formation of products with 5'- or 3'-phosphate termini, in the latter case either directly or via 2',3'-cyclic phosphates (2). In any event, *neither* of these systems fully reflects the properties of individual enzymes: in the case of DNases, we now have "restriction enzymes," which recognize a particular oligonucleotide sequence, and "repair enzymes," which recognize a defect; and among the RNases are the "processing enzymes," which also recognize a specific sequence and/or some structural feature.

II. Endoribonucleases

A. Ribonuclease I (Pancreatic RNase, Alkaline or Neutral RNase), EC 3.1.4.22[3]

The growing volume of research on bovine pancreatic RNase is periodically summarized in several reviews (2–5). It is, however, difficult to avoid reference to the unusual finding (6) that the physiological form of this enzyme in the pancreas differs from RNase A, in that it exhibits a dual pH optimum, at pH 4.5 and 7.5. The reason for this having been overlooked for more than 30 years is that acid and heat treatment, normally employed in the isolation procedure, as well as proteolytic digestion, lead to an enzyme devoid of activity at pH 4.5, and otherwise indistinguishable from RNase A. Furthermore the two forms of the enzyme display no significant differences in amino acid composition or substrate specificity. It would obviously be of interest to establish whether the foregoing holds also for the pancreatic enzyme of other species. Also of general interest is a comparison of the resistance to proteolytic degradation of bovine RNases A and B, suggesting that the carbohydrate moiety confers significant protection against proteases (7).

The pancreatic RNases from different species exhibit appreciable heterogeneity not only in primary structures (8, 9) and carbohydrate moieties (10), but also in their relative rates of cleavage of RNA and

---

[3] "RNase A" is used here for the major component of bovine pancreatic RNase (RNase I); "RNase B" has been used to indicate a glycosylated form.

uridine and cytidine 2′:3′-cyclic phosphates *(11)*. Their amino-acid sequences have been subjected to systematic analyses *(8, 9)* with a view to elucidation of their phylogeny and the relationship between structure and function.

Pancreatic-type RNases, closely resembling the bovine enzyme, are secreted by the pancreas and salivary glands of all mammals, and are found also in the duodenal contents, blood serum, kidney and urine (Section II, A, 3). They differ in antigenicity, ionic requirements, and to some extent in specificity, from nonsecretory neutral or alkaline RNases found in all mammalian tissues, with which they share such properties as acid-thermostability and endonucleolytic cleavage of RNA via formation of products with terminal pyrimidine nucleoside 2′,3′-cyclic phosphates *(1, 3, 4)*.

The existence of these two types of RNase activities has been known for some time, e.g., in human serum and urine *(4, 12–14)*, and pancreatic-type RNases may be clearly distinguished from other acid-thermostable RNases of the rat by their ability to hydrolyze a synthethic substrate, uridine 3′-(α-naphthyl)phosphate *(15)*. This was subsequently extended to differentiate between the secretory, and other acid-thermostable, RNase activities in man *(16, 17)*. Purified RNases from various human tissues and body fluids formed two distinct groups; (a) those of the pancreas, duodenal contents, and one of two fractions in serum and urine, all with a pH optimum of 8.5, inhibited by Zn^{2+} and Cu^{2+}, active against uridine 3′-(α-naphthyl)phosphate and dsRNA,[4] and approximately 25-fold more active against poly(C) than poly(U); (b) the enzymes of liver and spleen, and the second fraction in serum and urine, optimally active at pH 7, less sensitive to Zn^{2+} and Cu^{2+}, negligibly active against uridine 3′-(α-naphthyl)phosphate and dsRNA, and only severalfold more active against poly(C) than poly(U).

In addition, the two types of activity, exemplified by two acid-thermostable RNases isolated from calf serum *(18)*, differ in sensitivity to the RNase endogenous inhibitor *(3)*. One, similar to RNase A, and sensitive to RNase-A antiserum and the inhibitor, cleaved poly(C) at 10-fold the rate for poly(U); the second, insensitive to RNase-A antiserum and less reactive with the inhibitor, attacked poly(C) at about 2% of the rate for poly(U). A marked preference for poly(C) has also been reported for the enzyme from human pancreas and serum *(17, 19, 20)*, the latter presumably being the fraction of pancreatic origin *(16, 20)*.

Among the nonsecretory acid-thermostable RNases, there is also

---

[4] ds = double-stranded, ss = single-stranded.

heterogeneity with regard to specificity. A cyclizing RNase from beef brain cytosol has been claimed to be active against poly(C), but not against poly(U), although it releases uridine-terminated products from RNA *(21)*; an activity previously isolated from beef brain nuclei was reported to cleave poly(U) at 20% of the rate for poly(C) *(22)*. These differences were regarded as indicating that the two activities from brain are distinct enzymes *(21)*; if so, it is not clear which one corresponds to the RNase from unfractionated beef brain *(23)*, which cleaves RNA to products resembling those generated by pancreatic RNase and whose relative activities against poly(C) and poly(U) were not examined. Two RNases, which attack poly(C) at 7-fold and 0.06-fold the rates for poly(U), have been isolated from bovine aorta *(24)*.

RNases with a marked preference for poly(C) have also been purified from nucleoli of various cells *(25–27)*. Their localization, some differences in specificity relative to RNase A, and their limited cleavage of certain RNA species, point to their possible involvement in RNA maturation (Section II, B).

The increasing use of differential cleavage rates of poly(U) and poly(C) as a criterion for distinguishing between individual enzymes is not without pitfalls. The relative rates of hydrolysis of poly(C) and poly(U) by bovine RNase A depend on the ionic strength of the medium, and are modified by polyamines *(6, 28–31)*. In the presence of spermidine, *Citrobacter* sp. RNase loses its preference for poly(U) relative to poly(C) *(32)*, while spermine stimulates 3-fold the activity of horse submaxillary RNase toward poly(C) and partially inhibits activity toward poly(U) *(31)*. Similar effects of polyamines have been reported with *Escherichia coli* RNases *(33)*.

The mechanism underlying these effects of polyamines is complex. It is probably related to the ability of polyamines to bind to RNA *(31)*,[5] largely at double-stranded regions *(34)*, as well as to poly(U) *(33)*, with concomitant formation of a twin-stranded helical structure *(35)*, a phenomenon not observed with poly(C). A report to the effect that spermine does not bind to poly(U) *(34)* is somewhat surprising in that poly(U) is known to form a twin-stranded helix in the presence of this polyamine *(36)*. Binding to the enzyme is thought to be a factor; spermidine apparently does bind to uridylyl-specific RNase *(37)*, but not to RNase A or horse serum RNase *(31)*.

A higher rate of cleavage of poly(C) relative to poly(U) has also occasionally been taken to indicate a similar preference for hydrolysis of phosphodiester bonds adjacent to cytidine residues in RNA *(20)*.

[5] See Sakai and Cohen in Vol. 17 of this series.

This assumption is not necessarily valid, since calf serum RNases, which hydrolyze poly(C) and poly(U) at markedly differing relative rates, release only slightly different amounts of uridine and cytidine mononucleotides from RNA *(18)*. Human pancreatic RNase, 25-fold more active against poly(C) than poly(U), liberates comparable quantities of cytosine- and uracil-terminated fragments from RNA *(17)*; by contrast, an RNase from human serum, not much different from the foregoing, has been claimed to liberate fragments with 3'-terminated cytidine residues from RNA *(19)*.

The heterogeneity among nonsecretory acid-thermostable RNases is further underlined by an as yet unconfirmed report of an enzyme purified from rabbit reticulocytes *(38)*, optimally active at pH 6.5, and differing from the pancreatic enzyme in that the oligonucleotides liberated from RNA contain guanine, but not adenine, in addition to uracil and cytosine at the 3' termini.

1. DIMERIC FORMS

Bovine pancreatic RNase is a monomeric protein that, under certain conditions, aggregates to form a small percentage of dimers and higher oligomers *(39)*. Human pancreatic and pancreas-derived serum RNases aggregate much more readily, the equilibrium being dependent on the ionic strength of the medium. The aggregated forms are appreciably less active toward single-stranded polynucleotides *(17, 19)*.

Among the nonpancreatic acid-thermostable RNases *(21, 23, 40–44)*, some exhibit a molecular weight ratio (M_r) about twice that of RNase A *(21, 23, 40, 42, 44)*. One of these, bovine semen RNase, consists of two identical subunits linked via two disulfide bridges and noncovalent forces *(45)*. Previously reported differences in the M_r of this enzyme, determined by gel filtration, have been ascribed to pH-dependence of association of dimeric forms to higher aggregates *(46)*.

Dimeric forms of RNases exhibit enhanced activity, relative to the monomers, against dsRNA *(30, 42, 47)*, this increase being 4-fold in the case of dissociable dimers, formed by simple aggregation *(42, 48)*. Covalently cross-linked dimers of RNase A were initially reported to be 400-fold more active against dsRNA than the monomer forms *(30)*, a value subsequently corrected to 8-fold *(48, 49)*. Bovine seminal plasma RNase, a natural dimer *(47)*, also attacks dsRNA more rapidly than does monomeric RNase A *(42)*; this, in conjunction with the results of experiments on dimers formed between RNase A and its inactivated derivative carboxymethylated at histidine residue 119, which was not more active on dsRNA than monomeric RNase A, led to the postulate that hydrolysis of dsRNA required two active sites, one on each subunit of the dimer (50, 51). This interpretation was subsequently rendered invalid by the finding that (a) monomeric bovine seminal RNase cleaves dsRNA as efficiently as the natural dimer (52), and (b) human pancreatic RNase, whether in the monomer or dimer form, cleaves viral dsRNA at least 600-fold more rapidly, relative to ssRNA, than monomeric bovine RNase A, and 60-fold more rapidly than dimeric bovine seminal RNase (17, 42).

At high concentrations (>2.5 µg/ml), the bovine seminal plasma enzyme, as well as aggregated and cross-linked dimers of RNase A, have been reported to attack the RNA strand of a DNA · RNA hybrid (48, 53). The seminal plasma enzyme is also known to exhibit aspermatogenic activity (54). The properties of independently purified acid-thermostable RNases from bull seminal vesicle fluid, vesicle tissues and seminal plasma pointed to their identity (46), further confirmed for the latter two by immunological techniques (55).

One attempt has been made to study the mechanism of cleavage of dsRNA by high concentrations (~1 mg/ml) of RNase A (56). The enzyme apparently catalyzes scission of both strands to release large fragments with pyrimidine nucleotides at the 3' termini. The reaction was also examined at low ionic strength (0.02) on the assumption that the RNA retains its double-stranded structure under these conditions, based on its resistance to RNase T1. However, this is not fully convincing; only a 20-fold excess of RNase T1 was employed relative to the amount required for cleavage of ssRNA, whereas a 1000-fold excess of RNase A, relative to the amount necessary for hydrolysis of ssRNA, was used.

2. INHIBITORS

Highly ordered synthetic polynucleotides like poly(G) and poly(X) have been found to inhibit human pancreatic and pancreas-derived serum RNases, the inhibitory effect being 200-fold higher with poly(C) as substrate than with RNA (19, 20). The inhibitory properties of poly(G) have been successfully applied to enzyme purification by affinity chromatography (19). Less effective inhibition of human serum and other RNase activities by poly(A) has been reported (20); the effect was dependent on the length of the poly(A) chain and was reduced at high ionic strength or by polyamines (57). These observations led to the suggestion that a possible function of the poly(A) tract at the 3' terminus of mRNA[6] is the protection of the molecule against nucleoly-

---

[6] See articles in Vol. 19 of this series (mRNA Symposium).

tic attack *(58)*. At first sight this appears to be an oversimplification. In fact, 56 hours after microinjection of native and poly(A)-free rabbit globin mRNA into *Xenopus* oocytes, the former remained virtually intact, while 85% of the latter was degraded, as measured by ability to direct hemoglobin synthesis and by hybridization to complementary DNA *(59)*. However, native mRNA injected into frog oocytes is degraded at an increased rate in the presence of puromycin, suggesting that its stability in the oocyte is dependent on interaction with ribosomes *(60)*.

The growing interest in isolation of intact RNA species has led to a study of the efficacy of some commonly employed RNase inhibitors, of which Macaloid proved most effective, in limiting both the release of acid-soluble products and decreases in molecular weights of isolated RNA species *(61)*.

3. LOCALIZATION

The immunohistochemical method originally introduced by Marshall *(62)* for intracellular localization of pancreatic RNase has been further developed by Morikawa *(63)*, who found the activity in the apical portion of pancreatic acinar cells, in cells infiltrating the liver sinusoids, in the cytoplasm of intestinal and kidney tubule epithelial cells, in the cytoplasm of mononuclear cells in the intestine, and in red pulp or medullary cord of the spleen and the lymph node. Such studies have been extended to the electron microscope level by use of ultrathin frozen sections of bovine pancreas, and staining with ferritin-rabbit antibovine RNase conjugates *(64)*, with localization of the enzyme in the zymogen granules, the cisternae of the rough endoplasmic reticulum, and the pancreatic secretory ducts, thus confirming and extending previous findings for RNase localization in bovine and rat pancreas *(15, 62, 65)*.

Possible extension of the immunological technique was explored by examining the cross-reactivity of bovine RNase-A antibody with the liver alkaline RNases of man, rabbit, rat, mouse and guinea pig; all these were strongly inhibited. Furthermore, pancreatic sections of all species reacted with fluorescein-labeled antibody to give localization patterns identical with that for bovine pancreas *(66)*. Bearing in mind the differences in specificity between pancreatic and liver RNases in rat and man *(15, 16)*, the foregoing results, if true, indicate either that immunological differences do not parallel those for specificity, or that the liver activities tested consisted predominantly of the pancreas-derived, lysosome-bound RNases *(67)*.

Pancreatic-type RNase has been localized at the optical level by the simultaneous azo-dye coupling technique, using uridine 3'-(α-naphthyl)phosphate as substrate, and was found in the apical cytoplasm of acinar cells and the lumen of excretory ducts of the pancreas and parotid gland. The enzyme was also localized in lysosome-like structures in the epithelial cells of kidney proximal tubules (15). Such localization, and the existence of the enzyme in serum and urine, indicate that it is the extracellular activity adsorbed from the duodenal contents into the serum, and subsequently partially removed by kidney lysosomes prior to its clearance to the urine. Pinocytosis of injected labeled bovine RNase A by mouse kidney pinocytotic vesicles and lysosomes, and its subsequent proteolytic degradation, was earlier reported by Davidson (68, 69). Furthermore, unilateral nephrectomy in rats led to a 3-fold increase in free RNase activity in the contralateral kidney, presumably owing to its adoption of the pinocytotic function of the missing kidney (70). The secretory nature of rat parotid gland RNase has also been established (71).

Cytochemical localization of neutral RNases in rat tissues has been achieved with the aid of the indigogenic method, the substrate being uridine 2'(3')-(4-chloro-3-indolyl)phosphate (72). The enzyme was localized in the cytoplasm of renal proximal and distal tubules, of pancreatic and submaxillary acinar cells, etc. If these findings are indeed valid, they would indicate that this substrate, unlike its α-naphthyl analog, is cleaved not only by extracellular, but also by the remaining, acid-thermostable RNases.

Fractionation procedures have been employed (67) to investigate the subcellular distribution of neutral RNase in rat liver cells, using an assay based on the sensitivity of the enzyme to the endogenous RNase inhibitor from rat liver cytosol. RNase activity was assayed as the difference between activity on RNA in the presence of p-chloromercuribenzoate, which inactivates the inhibitor, and that measured in the presence of excess inhibitor (73). Differential centrifugation indicated that the bulk of activity exists in latent form in the large-granule fraction (67, 73). Density distribution of this activity resembled that of acid hydrolases and was similarly affected by injecting the animals with Triton WR-1339. The decrease in activity of blood plasma and liver provoked by partial pancreatomy and recovery of 0.1% of intravenously injected labeled bovine RNase A in rat liver were interpreted as evidence for the pancreatic origin of rat liver RNase (67). However, in view of differences in sensitivity to inhibitor reported among acid-stable RNases from various cell organelles (74) and calf serum (18), the assay employed may have measured preferentially the extrahepatic activity. An extrahepatic origin of most of the liver neutral RNase activity is inconsistent with other findings (see above, and refs. *15–17, 75*) showing that the pancreatic and liver neutral RNases of mammals are distinct enzymes, and that the pancreas-derived activity does not constitute a significant portion of liver RNase. The lysosomal localization of liver neutral RNase also differs from previous findings *(1)*.

A neutral endoribonuclease, insensitive to EDTA, that hydrolyzes RNA and pyrimidine polyribonucleotides to 3′-phosphate-terminated products, has been localized in rat liver plasma membrane *(76, 77)*. The same activity was found to be effectively inhibited by the natural RNase inhibitor from rat liver cytosol *(78)*, so that it also resembles, in this respect, RNase I. Enrichment of purified plasma membranes with respect to this activity was, however, insignificant in relation to marker enzymes like 5′-nucleotidase and PDase I, suggesting either a multiorganellar localization of the enzyme or partial adsorption of cytoplasmic activity by the membrane. The existence of various inhibitor-sensitive, acid-thermostable, neutral RNases in different cell components of rat hepatocyte has also been proposed *(74)*.

Earlier reports on gross localization of ribonucleolytic activities in normal and malignant tissues, obtained with the film-substrate technique (see refs. *1, 79, 80* for review), have been extended *(81)*. Films containing homopolyribonucleotide substrates *(82)* have been reintroduced with the object of improving the specificity *(83, 84)*. Observed differences in activities against poly(C) and poly(U) were regarded as indicating the presence of distinct enzymes, but this could also be due to differences in local ionic conditions and/or polyamine content (see Section II, A). Of some interest was the high activity noted against poly(A) in the inner medulla of rat kidney *(84)*, but the possibility of artifacts cannot be excluded, e.g., changes in film stainability due to complex formation between substrate and a tissue component *(85)*.

Previous data on variations in RNase levels in different tissues (e.g., ref. *80*) have been supplemented by the finding that nonadherent lymphocytes from mouse peritoneum contain up to 20-fold more RNase than those from other tissues, with an additional 10-fold increase following administration of high doses of cortisone *(86)*. Increases of alkaline RNase activities of thymus and lymph nodes, assayed at alkaline pH, have also been noted following antigen stimulation of rats with sheep erythrocytes *(87)*.

4. Synthetic Substrates

The earlier literature on the specificity of RNase A toward various synthetic substrates has been covered by Richards and Wyckoff *(5)*. A subsequent interesting innovation was the synthesis of uridine 2′,3′-

cyclic phosphorodithioate, which proved to be a substrate (88). This was followed by the preparation of uridine 2′,3′-cyclic phosphorothioate, which is also a substrate and which has been employed in some well-designed experiments to investigate the geometry of the first step in the action of RNase A (89).

Several synthetic chromogenic substrates are now available for RNase assay. These include uridine 3′-(α-naphthyl)phosphate (15, 16, 90), suitable for histochemical work and for colorimetric assays of activity, as well as an improved, PDase-II-resistant version,[6a] 5′-O-benzyluridine 3′-(α-naphthyl)phosphate (91) (see Section VIII, B, 1). Uridine 3′-phenylphosphate is a suitable colorimetric substrate (15), but it should be borne in mind that it is susceptible to PDase II. Uridine 3′-(5-bromo-4-chloro-3-indolyl)phosphate is designed for histochemical applications (92).

5. PATHOLOGICAL ASPECTS

Serum RNase levels have been studied widely under malignant (see refs. 93–95 for literature) and nonmalignant (see refs. 16, 96 for literature) states. Only a few of these have been directed to specific determination of serum RNases from particular sources, e.g., assays favoring certain RNases by suitable choice of incubation medium (14, 16) or the use of more specific substrates (16, 94). A case in point is the plasma of mice bearing transplantable murine tumors, which exhibits elevated activity specifically against poly(U) (97); another is the serum of patients with multiple myeloma, with enhanced activity toward poly(C) (94), and a decrease in activity accompanying successful therapy with alkylating agents, so that the enzyme level might conceivably be a useful diagnostic tool in such chemotherapy (94). By contrast, no change in serum RNase level was found to accompany a decrease in poly(U)-directed activity in white blood cells of leukemic guinea pigs (98).

Reference should be made to Daoust and de Lamirande (81) for a detailed review of the numerous studies on RNase levels associated with neoplasia; it is perhaps worth citing their conclusion that the observed loss of RNase activity in malignant tumor cells, and RNase activation associated with tumor regression, may well be indicative of some role of the enzyme in neoplastic transformation.

Partially purified dimeric bovine seminal, but not pancreatic, RNase was claimed to partially arrest growth of Cocker tumors in mice (99). Inhibition of tumor cell proliferation has also been reported with a cross-linked dimer of RNase A and its inactive alkylated derivative; such a dimer was 50-fold more effective than RNase A itself (100).

[6a] PDase = phosphodiesterase.

High specific uptake of the dimer was observed with three distinct types of tumor cells, but not with fibroblasts. After uptake, the dimeric enzyme was located initially in the nuclear and microsomal fractions, subsequently in the lysosomes and, finally, concomitantly with cell lysis, in the supernatant. The dimer was found to labilize lysosomes *in vitro*, and may exercise a similar effect *in vivo (100)*.

6. Functional Significance

Largely as a result of the study of Barnard *(101)* on the pancreatic RNase level in vertebrate species, ranging from very low in primates to very high in ruminants, some insight has been gained into the physiological function of the enzyme. In species with low pancreatic RNase levels, the enzyme is presumably of minor value to the animal, digesting only a small fraction of dietary RNA, while its aberrant distribution is suggestive of vestigial character. In ruminants and certain herbivores, with microbial fermentation in the stomach or cecum, pancreatic RNase serves to degrade microbial RNA and to salvage its phosphorus and nitrogen, both of which are essential for these vertebrates.

It has also been suggested that variations in the extent of glycosidation of RNases may be explained in terms of selective advantage, viz. that the carbohydrate components prevent rapid adsorption of the enzyme from the gut, thus enabling it to attain the large intestine, where it can digest the RNA of the cecal flora *(10)*; hence the heavily carbohydrated RNases of species with a cecal digestion, like the horse and pig. It is likewise not without significance that the presence of carbohydrates confers protection against proteases *(7)*.

Somewhat surprising is the fact that the extreme differences in pancreatic RNase levels among vertebrates are not reflected in the RNase levels of their sera *(102)*, implying the existence of efficient regulatory mechanism(s) that control the level of serum RNase.

Treatment of chick embryo fibroblasts with poly(I) · poly(C) or with interferon induces a membrane-bound activity that attacks RNA at an alkaline pH, presumably an RNase, and regarded as pointing to the significance of the induced activity for viral interference *(103)*. The induction of membrane-bound ribonucleolytic activity in chick embryo fibroblasts was subsequently confirmed *(104)*, but its possible role in viral interference is rendered questionable by the observation that no such activity was induced under analogous conditions in other types of cells. Furthermore, in neither of the foregoing studies was the induced nucleolytic activity directly identified as a ribonuclease.

7. RNase Endogenous Inhibitor

Earlier literature on the properties and possible physiological significance of the endogenous inhibitor of alkaline acid-thermostable

RNase has been reviewed by Roth (3). Among further attempts to purify the inhibitor from rat liver (105, 106) and other tissues (107–110), the most successful (105) achieved a 3000–4000-fold purification from rat liver cytosol, which was further improved by addition of a final step involving affinity chromatography on CM-cellulose-RNase (111). This latter procedure selectively removed all noninhibitor proteins, but the protein content of the purified inhibitor was too low to permit specific activity determinations. A two-step procedure based on the foregoing technique led to a 1000-fold purification (112). Affinity chromatography has also been applied to obtain a 120-fold purified inhibitor from rat Novikoff hepatoma cytosol (113). A simplified procedure for detection and purification of inhibitors from various sources is based on their similar mobilities on gel electrophoresis (114). Material collected by DEAE-Sephadex chromatography (105) is subjected to analytical or semipreparative polyacrylamide-gel electrophoresis at pH 8.5, with purified inhibitor as a marker. A sample of the gel is stained for protein, and the identified inhibitor is eluted.

Procedures developed for the stabilization and successful storage of purified inhibitor (112, 115) will be welcomed by many researchers in this field.

The M_r's of the inhibitors from rat liver, kidney and Novikoff hepatoma are quite similar, ranging from 42,000 to 50,000, depending on the method of estimation (105, 110, 112, 113).

a. Specificity. The 1000-fold purified inhibitor, tested against several nucleolytic enzymes in various subcellular fractions of rat liver, inhibited only the acid-stable RNase components, but not DNase I, 5'-endonuclease (see Section VII) and PDases I and II (Section VIII) (74). By contrast, a 200-fold purified inhibitor was reported to inhibit RNase I, the so-called RNase III (see Section VII), DNase II, DNase I, and polyadenylase (see Section VII) activities of the large-granule fraction of rat liver, the relative effectiveness being 1.0, 0.6, 0.14, 0.07, and 0.06, respectively (73). With the exception of the results for RNase I, these data require confirmation with highly purified enzyme and inhibitor preparations.

b. Functional Significance. The observed increase in level of RNase endogenous inhibitor in cells engaged in active protein synthesis (106, 116, 117), its decrease in cells with increased catabolism (116, 118–120), and its protective effect on polysome structural integrity (105, 106), point to some essential role of the inhibitor in control of RNA metabolism and protein biosynthesis (3). However, the evidence relating the high ratio of inhibitor to RNase activity with increased protein and/or RNA synthesis, and vice versa, is not unequivocal. Supporting evidence is furnished by a nearly 100-fold higher ratio of inhibitor to RNase level in tumorous and lactating rat mammary glands relative to glands of virgin rats *(117)*; and by an appreciable increase in this ratio in leukemic mouse thymus *(116)*, in thyroids of rats and mice on a low-iodine diet *(121)*, in phytohemagglutinin-stimulated human blood lymphocytes *(122)*, in rapidly proliferating erythropoietic mouse spleens *(123)*; and by a decrease in inhibitor to RNase level in rat liver, thymus and lymph nodes with aging *(116)*, in liver of rats on a protein-free diet *(119)* or suffering from malnutrition *(124)*, in skeletal muscle of mice with muscular dystrophy *(120)*, in liver in hypophysectomized rats *(125)*, in Ehrlich ascites cells following actinomycin D treatment *(126)*, in involuting mouse thymus following X-irradiation *(118)*, etc. Instances of contrary evidence include the decrease in the ratio of free inhibitor to total RNase activity in rat Novikoff hepatoma cytosol *(113)*.

Divergences from the rule have been attributed *(122)* to heterogeneity of cell populations in tissues, with some cells confusing the overall ratio of inhibitor to RNase level. For instance, in the very active spleen, with high total RNase and inhibitor levels, the RNase activity could be due to the relatively fewer phagocytic cells.

Reservations regarding the regulatory role of the inhibitor, based on its restriction only to mammalian species, led to the postulate *(116)* that the inhibitor may actually be more widespread in the animal kingdom, but has escaped detection as a result of variations in its specificity and general lability. It was, in fact, found that chick liver supernatant contains an inhibitor active against chick liver supernatant RNase, but undetectable when tested against bovine pancreatic RNase.

Similarities in the properties of 17β-estradiol receptor protein and RNase inhibitor led to the suggestion that the cytosol hormone receptor protein may itself be the RNase inhibitor in rat uterus *(127)*. While no direct evidence for this was presented, treatment of ovariectomized rats with the hormone did lead to disappearance of inhibitor activity and an increase in level of free RNase activity in the uterus.

B. Nucleolar Ribonuclease

An endoribonuclease activity, isolated from the nucleoli of HeLa *(128)*, Novikoff hepatoma *(25)*, mouse L *(27)* and rat liver *(26)* cells, has been implicated in the sizing of rRNA. The enzyme, purified 200-fold from Novikoff hepatoma ribonucleoprotein particles, attacks poly(C) much more rapidly than poly(U), and is inactive against poly(A), poly(I) and poly(G) *(25)*. It hydrolyzed 5 S RNA slowly to acid-soluble fragments with terminal C and U residues. Its activity, optimal at pH 7, is inhibited by Mg^{2+} and Ca^{2+} *(25)*.

An apparently analogous activity, partially purified (~20-fold) from mouse L cell nuclei, cleaved poly(C) to fragments of average length about 10, with 4% of the products as mononucleotides (27), but was inactive against other synthetic polyribonucleotides. It also cleaved 45 S pre-rRNA, mRNA, hnRNA and, to a limited extent, 28 S rRNA, 18 S rRNA and 45 S pre-rRNA in 80 S ribonucleoprotein particles to yield 3′-phosphate-terminated products. Like the Novikoff hepatoma enzyme, it was inhibited by Mg^{2+}. It was not tested against Ca^{2+}.

The endonuclease activity solubilized from rat liver nucleoli (26) was active against poly(U), poly(A), 16 S RNA, 23 S RNA and MS2 coliphage RNA. It cleaved 45 S pre-rRNA, via fragments of intermediate size, to limit 4 S products, but was inactive against intact reovirus dsRNA. It is possible that this activity (or one of its constituents), and that from HeLa cell nucleoli (128), also a crude extract, may be similar to those purified from Novikoff hepatoma (25) and mouse L cell (27) nucleoli.

Since the limit products of these endonucleases are smaller than rRNA, it has been suggested that their *in vivo* specificity toward pre-rRNA may be regulated by the configuration of the enzyme in nucleolar particles (25) and/or the conformation and protein components of the substrate (27), the latter being regarded as more plausible. Both the intracellular localization of these enzymes and their relation to other acid-thermostable RNases require further study.

An endo-RNase resembling the foregoing nucleolar activities has recently been purified 2400-fold from beef brain nuclei (22). It attacks poly(C) much more rapidly than poly(U) and rRNA, and is inactive on poly(A), poly(I) and poly(G). Exhaustive hydrolysis of poly(C) yielded cytidine 2′, 3′-cyclic phosphate. The intranuclear localization of the enzyme was, however, not established.

C. Ribonuclease II (Acid Ribonuclease), EC 3.1.4.23

An acid-stable, thermolabile endoribonuclease, purified from several mammalian tissues, hydrolyzes RNA, RNA "core," and synthetic polyribonucleotides, the products being oligonucleotides with terminal 2′,3′-cyclic phosphates. It exhibits no base specificity, is optimally active at pH 5–6, and is insensitive to the neutral RNase endogenous inhibitor (see refs. 1, 3, 129, 130 for pertinent literature).

Futai *et al.* (131) claim to have isolated two different acid RNases from rat liver. One, of lysosomal origin, with $M_r = 26,000$, apparently resembles the corresponding splenic enzyme. The other, found in the supernatant, cochromatographed with alkaline RNase, and eluted from a gel column at a position corresponding to $M_r = 13,000$. Some doubts may, however, be entertained with regard to the identity of the latter component.

Another proposed member of this class, partially purified from bovine brain nuclei, attacks rRNA, poly(U), poly(A) and poly(C), releasing 3'-phosphate-terminated oligonucleotides. It has a slightly acidic pH optimum toward all substrates with the exception of poly(C), which is cleaved optimally at pH 8. However, the nonhomogeneity of the enzyme, and the absence of data on the rates of hydrolysis of different substrates, its acid-thermostability, and possible sensitivity to the neutral RNase endogenous inhibitor, raise some questions about its identification and nature. Its intracellular distribution was not examined *(130)*.

LOCALIZATION

The intracellular localization of RNase II initially established as lysosomal *(132)*, has been reinvestigated with the aid of zonal *(133)* and sucrose density gradient *(134)* centrifugations. The results point to the occurrence of the enzyme in lysosome-like particles other than those containing acid phosphatase, in contrast to more recent findings *(135)*, based on differential and isopycnic centrifugation of liver cells of normal and Triton-WR-1339-treated rats, pointing to a distribution pattern similar to that for lysosomal acid phosphatase.

The claim *(131)* for the existence of two distinct acid RNases derives some support from immunoelectrophoretic results showing that RNase II from bovine spleen contains two immunologically distinct activities *(63)*. Fluorescein-labeled antibodies to the unresolved acid RNases showed that, in bovine tissues, these are localized in the cytoplasm of some cells of islets of Langerhans, in leukocytes infiltrating the liver, and in mononuclear cells in the intestinal mucous membrane and in cells of the thymus medulla. The activity was even more pronounced in the cytoplasm of leukocytes and mononuclear cells in the red pulp and the medullary cord of the spleen and lymph nodes. The possibility of localizing heterologous antigens with fluorescent conjugates was investigated by studying the inhibition of acid RNases of man, rabbit, rat, mouse and guinea pig spleen with antibody to bovine spleen RNase. Although only partial inhibition of these activities was noted, spleen sections of man, rat and mouse, but not of guinea pig and rabbit, stained with the fluorescein-labeled antibody, displayed a pattern essentially identical with that for bovine spleen *(66)*. Histochemical localization of acid RNase in the rat has also been examined by the indigogenic method *(72)*.

Modifications in the level of acid RNase activity associated with neoplastic transformation have been reviewed *(81)*. A recent investigation demonstrated elevated acid RNase activity in the thymus and white blood cells of mice genetically predisposed toward leukemia and reticulum cell neoplasms well before the onset of neoplastic transformation. Since the elevation of activity appeared to correlate with a high incidence of spontaneous tumor formation, the assay of acid RNase activity was considered to be a useful diagnostic tool for detection of cancer-prone strains *(136)*.

D. Ribonucleases Attacking Double-Stranded RNA

Considerable effort has been devoted to the search for eukaryotic RNases specific, or highly active, toward dsRNA, stimulated in large part by the demonstration that highly purified *E. coli* RNase III (EC 3.1.4.24), which is specific for dsRNA *(137)*, cleaves HeLa cell nucleolar 45 S RNA and nuclear hnRNA *in vitro* to products with mobilities comparable to those observed *in vivo (138)*. Such cleavage is inhibited by toyacomycin and ethidium bromide, which react with double-stranded sequences. The known RNase III specificity and presumed role in processing of *E. coli* RNA *(139)* and the large RNA transcripts in T3 and T7 phage-infected cells imply that double-stranded regions serve as signals in RNA maturation (see Section X). Furthermore, dsRNA is fairly widespread in mammalian systems, where it constitutes well-defined regions in hnRNA *(140)*. It induces production of interferon *(141)*, and is a highly potent inhibitor of *in vivo* mRNA translation *(142)*. It is also a normal intermediate (as the replicative form) in infection by some RNA viruses *(143)*.

A riboendonuclease activity, purified 50–100-fold from HeLa cell nuclei extracts, with a preference for dsRNA, has been the subject of a preliminary report *(144)*. Its activity against poly(I) · poly(C) was dependent on Mg^{2+} and EDTA, and was inhibited by ethidium bromide. It was also active against poly(C), such cleavage being inhibited by Mg^{2+}, but not by ethidium bromide. It would be of obvious interest to establish whether its localization is indeed nuclear.

Crude supernatant fractions of homogenates of purified Krebs II ascites cells exhibit a marked preference for dsRNA relative to ssRNA, assayed at pH 7.6 with phage f1 and f2 dsRNA and ssRNA, as well as poly(G) · poly(C), as substrates. There was some indication for association of this activity with ribosomes, and it appeared to be most active under conditions of protein synthesis *(142)*. In KB cells, high activity against dsRNA has been found, restricted largely to cytoplasmic membrane or nuclear fractions, which was released from these by detergent treatment; the ratio of this activity against dsRNA, as compared to ssRNA (about 1 : 1) *(143)* suggests the possible existence in these cells of the sought-for RNase specific for dsRNA. A somewhat similar localization of dsRNase activity has been reported in the lung, spleen and some macrophage cells *(145)*. An activity against dsRNA has also been isolated from calf thymus *(146)*.

There have been a number of reports on the existence in mammalian sera of RNases active against dsRNA *(147–149)*, and the existence of such activities has been frequently invoked as an important factor in the diminution of the interferon-inducing activity of intravenously administered dsRNA. In particular, a finding *(148)* that for some years received widespread credence was that of the existence in human and animal sera of an activity specific for dsRNA, with little or no activity against ssRNA. A recent attempt to repeat these results, using the same conditions and the same substrate, poly(I) · poly(C), demonstrated that, in the case of human serum, the presumed dsRNase activity was simply an artifact due to quenching of the radioactivity of the tritium label in the poly(C) component by the high protein content of the serum *(150)*. Presumably the same applies to the results with the animal sera.

An observation of some interest is the fact that, in contrast to all other animal viruses, reovirus particles are transported directly into the lysosomes of L-strain fibroblasts within a few minutes after inoculation. The viral dsRNA liberated after uncoating of the virion is unaffected even after several hours *(151)*, pointing to the absence of dsRNase activity under conditions prevailing in the lysosomes, i.e., pH 4–5 *(152)*.

Acid-thermostable RNases from various mammalian sources exhibit enhanced activity against dsRNA, as compared to ssRNA. These include the enzymes from bovine seminal plasma *(42)*, from the pancreas of man and the whale *(17)*, and natural and synthetic dimers of bovine RNase A *(30, 50)* (see Section II, A, 1). But it should be emphasized that there is at present no conclusive evidence for the existence in mammalian cells of an RNase specifically active against dsRNA, like the *E. coli* RNase III.

E. "Processing" Ribonucleases

The use of radiochemically labeled, homogeneous, cellular species of RNA as substrates for ribonucleases has simultaneously provided important tools for gaining evidence as to the intracellular role of the latter, largely by studies of the cleavage products. This procedure, initially applied to elucidate the role of *E. coli* RNase P, which cleaves a tRNA precursor,[7] and *E. coli* RNase III, which "processes" both an rRNA precursor and a polycistronic mRNA *(153–156)*, is being employed to advantage in mammalian systems.

1. NOMENCLATURE

The enzymes involved in the maturation of RNA, like *E. coli* RNase II *(137)* and RNase P *(153)*, and presumably the mammalian RNase NU *(143)* and RNase P · *Hsa (161)*, are considered to "recognize" a combination of primary and secondary structure information in their substrates *in vivo (137, 163)*. Such specificity requirements render dubious any attempts to classify them within existing schemes on the basis of mode of cleavage, termini released, etc. In accordance with such schemes, RNase P · *Hsa* is a 5'-phosphate-forming endoribonuclease, and RNase NU a 3'-phosphate-forming endoribonuclease. It is clear that an assignment based solely on these criteria conveys little information as to their unique specificity and locates them closer to various functionally unrelated low-M_r enzymes than to one another.

For the moment, a nomenclature patterned in part on that advanced for DNA restriction enzymes has been proposed: RNase P from *E. coli* is RNase P · *Eco*, and the analogous activity from *Homo sapiens* is RNase P · *Hsa (161)*. It appears to us that the proposal to use the designations "P · *Eco*" and "P · *Hsa*" without the prefix RNase *(161)* is an oversimplification, since one would then have to refer to "maturation enzyme" or "RNA maturation enzyme" prior to designating the type of enzyme. The existing nomenclature should therefore be regarded as only provisional.

2. RNASE NU

This thermolabile endonuclease, which cleaves *E. coli* tRNA^{Tyr} precursor at two sites in a region of the molecule not conserved in the tRNA sequence, and active against other RNA species with short *in vivo* lifetimes, has been purified from human KB cells *(143)*. It does not release oligonucleotides from stable RNA species like tRNA, 5 S RNA, rRNA, viral dsRNA or RNA · DNA hybrid. The enzyme probably interacts with some specific structure in unstable RNA species which, from the nature of the cleavage patterns of ϕ80-coded RNA and *E. coli* pre-tRNA^{Tyr} *(157)*, is believed to be a single-stranded region bounded by structured regions. It forms 3'-phosphate terminated oli-

---

[7] See article by J. D. Smith in Vol. 16 of this series.

gonucleotides, and acts optimally at pH 8, with a requirement for a monovalent ion (like NH_4^+) or Ca^{2+}. The enzyme is found in the cytoplasm, loosely associated with ribosomes. Similar activities were detected in monkey kidney cell lines.

3. Enzyme Reducing 4.5 S RNA to 4 S RNA

Another activity, involved in post-transcriptional maturation leading to tRNA, degrades 4.5 S RNA to 4 S RNA by removal of approximately 10–15 nucleotides, but is inactive against other low-M_r nuclear RNAs. It is present in mouse myeloma cell nuclei (158) and probably in homogenates of BHK21/C13 cells (159), and may correspond to an enzyme from E. coli with similar specificity (160).

4. Ribonuclease P · Hsa (tRNA Maturation Enzyme)

This is a thermolabile endo-RNase that cleaves the precursor of E. coli tRNA^{Tyr} at a single site, adjacent to the sequence at the 5' terminus of the mature tRNA, releasing two fragments, one the 5'-phosphate-terminated tRNA^{Tyr} with perhaps several extra residues at the 3' terminus, the other a smaller fragment. It has been isolated and purified from human KB cells (161). Its designation as RNase P · Hsa was based on the fact that its specificity resembles that of RNase P from E. coli, involved in maturation of tRNA (153, 162).

The KB cell enzyme also cleaves other tRNA precursor molecules from E. coli to fragments with electrophoretic mobilities identical to those produced by RNase P. Like the latter enzyme, RNase P · Hsa is inactive toward mature tRNA, and the metabolically stable KB cell 4 S and 5 S RNA species. Unlike the E. coli enzyme, it does not attack M3 RNA, ϕ80 RNA and E. coli pre-4.5 S RNA. It transforms total KB cell pre-tRNA into molecules corresponding electrophoretically to mature tRNA. The enzyme is optimally active at pH 8, requires both Mg^{2+} and a monovalent cation, and is inhibited by mature tRNA. Its occurrence in the cytosol is in accord with its presumed role in tRNA biosynthesis.

F. Hybrid Nuclease (RNase H), EC 3.1.4.34

RNase H attacks the RNA moiety of RNA · DNA hybrids and is normally inactive against ss- and dsRNA and DNA. It has been isolated from calf thymus (164–166), chick embryo (166), rat liver (167), KB cells (168), human leukemia blood cells (169), Ehrlich ascites tumor cells (170), murine myeloma and MOP-21 cells (171), insects (172), oncornaviruses (Section XI), etc.

The cellular enzyme, with M_r's in the range 65,000–90,000, is an endonuclease, optimally active at about pH 8, with liberation of 5'-

phosphate-terminated di- and oligonucleotides; but there are two reports *(165, 168)* of the presence of mononucleotides, and one *(169)* that the leukocyte enzyme yields mono- and oligonucleotides to the exclusion of dinucleotides. With synthetic hybrid homopolymers, the rat liver enzyme is active only when the RNA moiety consists of purine bases *(167)*, the calf thymus enzyme only when adenine residues are present *(165)*, while the KB cell enzyme apparently exhibits no base specificity *(168)*. For the rat liver enzyme, the specificity toward purine bases was found to hold also for a hybrid formed from calf thymus DNA *(167)*.

With the calf-thymus enzyme, as short a deoxyribo strand as a tetranucleotide suffices to form a hybrid with a complementary polyribonucleotide, which is then susceptible to enzymic cleavage *(173)*.

Closer examination has shown the calf-thymus activity to consist of three fractions, separable by column chromatography and in sucrose gradients *(146)*. The fractions, characterized as I, IIa and IIb, differ with respect to ionic requirements (cf. ref. *169*), sensitivity to the -SH reagent N-ethylmaleimide, and molecular weights. All three displayed marked preference for purine residues. Somewhat surprisingly, IIa was active against dsRNA, as well as poly(A). The latter fact may be related to the observation that, on elution with decreasing concentration of aqueous ethanol from a cellulose column, poly(A) elutes like dsRNA at 4°–23°C and between ssRNA and dsRNA at 37°C *(165)*. Attempts to separate these activities from RNase H activity were unsuccessful, so that fraction IIa appeared to resemble RNase III of *E. coli*; however, the RNase H and dsRNase activities of RNase III have now been successfully separated *(174)*. Preliminary experiments suggested that, like *E. coli* RNase III, additionally purified IIa behaves like a processing enzyme, in that it generated characteristic high-M_r fragments from 45 S rRNA precursor *(146)*.

The existence, in calf thymus, of three distinct RNase H activities with different ionic requirements undoubtedly accounts, at least partially, for the conflicting reports in the literature on the properties of eukaryotic RNase H, particularly as regards optimal requirements for Mn^{2+} and/or Mg^{2+}, which vary widely. This subject has been meticulously studied *(173)* for calf thymus RNase H; but, for reasons not clear, without taking into account the reported existence of three fractions in this enzyme *(146)*.

The foregoing, together with the demonstrated activity of fraction IIa of calf thymus on dsRNA, underlines the necessity of adequate controls for establishment of the intrinsic nature of RNase H activity. Such controls should clearly include a hybrid containing labeled

DNA, as well as labeled ss- and dsRNA. It has also been pointed out *(169)* that, whereas the *initial* action of eukaryotic RNase is clearly endonucleolytic (in contrast to viral RNase H, it will attack a closed circular hybrid), it cannot be considered as established that its subsequent action is not also exonucleolytic.

1. LOCALIZATION AND FUNCTION

Little effort has as yet been devoted to the intracellular localization of the enzyme. The bulk of activity in leukemic leukocytes *(169)*, rat liver *(167)* and calf thymus *(164–166)* is found in the cytoplasmic particulate and/or supernatant fraction.

The possible *in vivo* function of cellular RNase H is still somewhat speculative, and not as clear as for the corresponding oncornaviral activities (Section XI, E). It is probably not accidental that at least the initial action of the cellular enzyme is endonucleolytic, whereas that of the viral enzymes is exonucleolytic. Furthermore, with one possible exception *(175)* that requires confirmation, eukaryotic RNase H activites differ from most of the corresponding viral enzymes in that they are not associated with polymerase activities *(146, 169)*, a fact of undoubted significance in relation to function.

The enzyme could play a key role in RNA-primed DNA synthesis. In fact, following *in vitro* DNA synthesis on circular ssDNA initiated by an RNA primer, the latter was effectively removed from the product DNA by the RNase H from KB cells *(176)*. A role for cellular RNase H in the integration of the product of oncornaviral reverse transcriptase into the cellular genome has been proposed *(177)* (Section XI, E).

III. Exoribonucleases

A. 5'-Exoribonuclease, EC 3.1.4.20

An exoribonuclease, partially purified from the nuclei of rat liver and Ehrlich ascites tumor cells *(178, 179)*, is reported to specifically hydrolyze single-stranded polyribonucleotides processively from the 3'-OH terminus to give nucleoside 5'-phosphates. Poly(A) and poly(I) are better substrates than pyrimidine polyribonucleotides, while polyribonucleotides containing about 20% 2'-*O*-methyl residues are attacked at a 7- to 60-fold lower rate *(180)*. Of some interest is the much higher susceptibility of 45 S pre-RNA, relative to 28 S and 18 S mature rRNA *(181)*. The enzyme readily hydrolyzes 5'-phosphate-terminated di- and oligoribonucleotides, is inactive against oligodeoxyribonucleotides and *bis*(*p*-nitrophenyl) phosphate *(182)*, and negligibly active against thymidine 5'-(p-nitrophenyl) phosphate. Its pH optimum is in the range 7.4–9.2, and it requires Mg^{2+}. There is no activity with 3'-phosphate- and 2',3'-cyclic-phosphate-terminated oligonucleotides (178, 179). The enzyme is inhibited by nucleoside 3',5'-, but not 2',5'-, bisphosphates (179), cooperatively inhibited by polydextran sulfate (183), and irreversibly inactivated by low levels of p-bromo- and p-iodoacetamidophenyl nucleotides, and by thymidine 3'-fluorophosphate (184). The enzyme resembles in many respects the earlier reported leukemic cell phosphodiesterase (185), and the apparent differences in specificities could be due to inadequate purification of the latter.

The changes in binding energy and in number of binding sites, associated with processive as compared to random exonucleolytic degradation of polynucleotides, have been evaluated with the aid of a kinetic model derived from steady-state theory. Although the model was applied to study the nuclear exoribonuclease from Ehrlich ascites cells (186), its range of application is much broader.

5'-Exoribonuclease activity also occurs in the nuclei of various mouse tissues, but is apparently more abundant in tumor cells (178, 179). Its occurrence in cell nuclei and the restriction of its activity by 2'-O-methyl residues have been regarded as supporting its role in the processing of rRNA (180, 181), but it should be noted that resistance to nuclease enzymes of 2'-O-methyl residues is by no means uncommon. Isolated Ehrlich ascites cell nuclei do, in fact, contain about 50% of total cell exoribonuclease, and autodegradation of newly synthesized RNA released mainly purine and pyrimidine nucleoside 5'-phosphates, consistent with exoribonuclease activity (187).

B. Polynucleotide Phosphorylase, EC 2.7.7.8

No attempt will be made here to summarize early reports on the existence in some mammalian tissues of polynucleotide phosphorylase capable of polymerizing nucleoside 5'-pyrophosphates, since none of these has been confirmed.

A polynucleotide phosphorylase catalyzing the phosphorolysis of poly(A), poly(C), poly(U) and RNA to nucleoside 5'-pyrophosphates from the 5' termini has been purified severalfold from the inner mitochondrial membrane of rat liver. It is optimally active at pH 8 and requires divalent cations. It exhibits no polymerizing activity, and even its P_i exchange activity is not fully established (188, 189). It has been found in mitochondria of various tissues of mammals, birds and fish. Although localized in the inner mitochondrial membrane of rat liver, its occurrence in rat liver nuclei has not been fully excluded *(190)*.

IV. Endodeoxyribonucleases

A. Deoxyribonuclease I (Pancreatic DNase, Neutral DNase), EC 3.1.4.5

The properties of bovine pancreatic DNase I were last reviewed by Laskowski *(191)*. Since then, the nature of the enzyme and its interaction with metal ions have been widely investigated (see refs. *2, 192–195* for bibliography). For example, it has been established that cations other than Mg^{2+} and Ca^{2+} are activators, although the foregoing combination, or Mn^{2+}, are most effective *(192)*. Trace amounts of Ca^{2+}, or Sr^{2+} or Ba^{2+}, are essential for activity against DNA in the presence of Mg^{2+}, and for binding of the enzyme to DNA *(195)*. Confirming earlier findings, the cleavage sites in DNA vary with the divalent cation(s) present in the incubation medium *(191, 193)*. Cations like Cu^{2+} and Hg^{2+}, which exhibit high affinities for guanine and thymine, respectively, also alter the apparent specificty by binding at sites containing these bases and protecting such sites against hydrolysis *(196)*.

The specificity of bovine DNase I, which, from the data cited above, is relatively complex, has been most extensively investigated by Bernardi *et al.* *(197, 198)* by means of the kinetics of formation of oligonucleotides of defined average length, and the base compositions of the 3'- and 5'-terminal residues, as well as the penultimate residue at the 5' terminus. In general, these base compositions did not vary with the average size of the hydrolysis fragments, testifying to maintenance of specificity during the course of the reaction, but they did differ from those expected on the basis of random cleavage. The results were most readily interpreted in terms of recognition by DNase I of sequences at least three residues in length. Such a specificity, which differs from the single-base specificity of numerous RNases, is, of course, still far below that of restriction nucleases. This sequence specificity of DNase I was further supported in a comparative study of the specificities of five different DNases toward *E. coli* DNA *(199)*, based on analyses of the 5'-terminal dinucleotides and 3'-terminal nucleotides. It has also been proposed that these procedures may be used to compare sequences, in different DNA samples, that are longer than the dinucleotides obtained by nearest-neighbor analyses, so that the method may possess useful analytical applications *(197)*.

It is of interest that poly(2'-chloro-2'-deoxycytidylate) and poly(2'-chloro-2'-deoxyuridylate) *(200)*, and even poly(2'-fluoro-2'-deoxyuridylate) *(201)*, are all resistant to DNase I under conditions where poly(dT) is readily and extensively degraded.

1. INHIBITORS

Apart from its inhibition by endogenous protein inhibitor(s) (see below), DNase-I activity toward duplex DNA is competitively inhibited by bleomycin[8] as a result of decreased affinity for the antibiotic-modified substrate *(202)*. Inhibition is much more pronounced with DNase I than DNase II, and is apparently correlated with the extent of release of thymine from various polydeoxynucleotides by the antibiotic *(203)*. The enzyme has also been reported to be effectively inhibited by hydroxybiphenyls, normally found in some commercial phenols *(204)*; these inhibitors act by hydrogen bonding and intercalation, probably blocking in this way access of the substrate to the tryptophan residues of DNase I.

2. LOCALIZATION

DNase I, previously localized in the mitochondria with about 10% of the activity in the microsomal fraction, was subsequently found to be a latent enzyme, located in the space between the two mitochondrial membranes *(205)*; in nonionic media, the activity is partially adsorbed on the outer surface of the inner membrane. Its localization was reported to correspond to that of 5'-endonuclease *(135)* (Section VII). But it is rather curious that denatured DNA was used as a substrate in these studies, since it is not only less readily attacked than native DNA *(191)*, but also exhibits susceptibility to 5'-endonuclease *(206, 207)*. It had been earlier emphasized that the pH of the incubation medium and a requirement for Mg^{2+} (see also above) are insufficient for ensuring the specificity of DNase-I assay with DNA in crude extracts, and that susceptibility to DNase-I endogenous inhibitor be employed in such cases *(191)*, as has already been done *(208)*. This view is substantiated by the discovery, in rabbit liver mitochondria, of at least two Mg^{2+}-activated enzymes active against DNA *(209)*. The activities are latent and, although insufficiently characterized, the 2000-fold higher specific activity of the extract from lysed mitochondria, relative to that of the cytoplasm, lends credence to their intramitochondrial localization.

Polyacrylamide-gel electrophoresis of homogenates, followed by incubation under controlled conditions and with different types of

[8] See chapter by Müller and Zahn in this volume.

DNA substrates, with densitometry of the reaction products, has been employed to distinguish various DNases in human lymphocytes *(210)* and saliva *(211)*. However, penetration phenomena and artifacts associated with gel electrophoresis are rather numerous, and the overall technique is unlikely to be precise enough for characterization of distinct enzymes in different types of cells. It has, nonetheless, been claimed to detect significant differences in the levels of acid and alkaline DNases in human lymphocytes and in acute lymphocytic leukemic cells *(212)*.

The earlier literature on localization of DNase-like activities in other cell types has been reviewed *(1, 80)*.

3. SYNTHETIC SUBSTRATES

Thymidine 3',5'-bis[di-(p-nitrophenyl)phosphate] has been found suitable as a substrate for the spectrophotometric assay of microgram levels of purified pancreatic DNase I. Also susceptible to the enzyme, but at a 50-fold lower rate, are thymidine 3'-(p-nitrophenyl)phosphate and thymidine 3'-(p-nitrophenyl)phosphate 5'-phosphate. For all three, the release of the p-nitrophenol moiety is from the 3'-phosphate *(213)*. Apart from the interest and novelty of this observation, it is somewhat unexpected to find that the specificity of the enzyme toward these substrates is different from that toward DNA and synthetic polynucleotides, where the monoesterified phosphate is formed at the 5'-position. It is unfortunate that the residual susceptibility of the more rapidly hydrolyzed substrate or its cleavage products to other mammalian nucleases limits its applicability to purified preparations of DNase I. It would nonetheless be of considerable interest to utilize it for kinetic studies and for examining the effects of divalent cations (see above).

4. FUNCTIONAL SIGNIFICANCE

Previous notions regarding the possible functional significance of DNase I *(1, 191)* could not possibly have envisaged the astonishing finding that DNase I is inhibited by actin and that it depolymerizes filamentous F-actin to monomeric G-actin *(208, 214)*, so that it may, perhaps even to a predominant extent, regulate the function of actin (see Section IV, A, 5).

Along more traditional lines, high doses of DNase I were shown to inhibit globin synthesis in rabbit reticulocytes, presumably by interfering with initiation *(215)*, but the mechanism involved has not been elucidated. Various attempts to relate levels of DNase I and DNase II in rat liver following diethylnitrosamine-induced carcinogenesis,[9] as well as spontaneous (human) and experimental (rat) tumors of the nervous system (see refs. *216, 217* for earlier literature), have not provided any clear-cut conclusions regarding the possible relevance of DNases to carcinogenesis.

5. DNase I Endogenous Inhibitor

Mammalian tissues and physiological fluids have long been known to contain a specific protein inhibitor of DNase I *(191, 218, 219)*, present in remarkably high concentration, about 5–10% of the soluble cell protein *(219, 220)*. It has been isolated in a crystalline, homogeneous, state *(218)*. The inhibitor forms high-M_r aggregates, and is active in the monomeric form, combining in a 1 : 1 molar ratio with the enzyme *(218, 219)*. It was subsequently identified as actin, one of the known major structural proteins *(214)*. Phalloidin, which accelerates the polymerization of monomeric G-actin, prevents complexing of the latter with DNase I and abolishes its inhibitory effect *(221)*.

It has now been convincingly shown that DNase I complexes with, and is instantaneously inhibited by, only monomeric G-actin *(208)*. Filamentous F-actin also interacts with DNase I, inhibiting it gradually; DNase I depolymerizes the F-actin with formation of a 1 : 1 complex of DNase I and monomeric G-actin. The depolymerization process is slowed by regulator proteins, such as tropomyosin and tropotonin, and is effectively abolished by heavy meromyosin. In the absence of ATP, the latter complexes with F-actin to prevent binding, and accompanying inhibition, of DNase I. Upon addition of ATP, the complex of F-actin and heavy meromyosin dissociates, the DNase I is immediately inhibited, and the F-actin is depolymerized.

The foregoing results throw entirely new light on the role of DNase I, one of the principal functions of which may well be the regulation of actin filament formation and function, the relation of which to DNA metabolism, if any, remains to be elucidated.

B. Ca,Mg-Dependent Endonuclease

A DNA endonuclease with a requirement for both Ca^{2+} and Mg^{2+}, found strongly associated to rat liver chromatin, and optimally active at pH 7–8, hydrolyzes native, but not denatured, DNA to acid-insoluble products via single-strand scissions, as revealed by alkaline sucrose gradient centrifugation *(222)*. Similar activity has been found

---

[9] See chapter by Lijinsky in Vol. 17 of this series.

in nuclei of other tissues and species *(223, 224)*, with a particularly high level in regenerating liver *(223)*. It is apparently absent from the cytoplasm *(222, 223)*. The enzyme has been purified 370-fold from rat liver *(225)* and 750-fold from bull semen chromatin *(224)*. The corresponding mouse liver enzyme has an M_r of about 40,000; this drops abruptly to about 20,000 on removal of Mg^{2+}, with a concomitant partial loss of Ca^{2+}-dependence *(223)*.

In vitro, the enzyme activates DNA synthesis on templates of isolated liver nuclei *(226)* and chromatin *(227)*. It is inhibited by NAD^+ in the presence of poly(ADP-ribose) synthetase, presumably as a result of ADP ribosylation of the protein molecule *(225)* with resultant blockage of the template-activation system for DNA synthesis in isolated nucleic *(227)*. It is also inhibited by synthetic ribo- and deoxyribopolynucleotides *(225)*. Both the nucleolytic and template activation functions of the enzyme are stimulated severalfold by a nondialyzable, Pronase-sensitive, heat-stable factor from liver and testis nuclei, which has no effect on the DNase-I type of activity *(224)*.

Notwithstanding the differences in response to NAD and to the nuclear stimulation factor between this enzyme and DNase I, the resemblances between the two in ionic requirements and in activation of *in vitro* DNA synthesis on templates of isolated nuclei suggest the need for further comparative studies of the two activities.

The enzyme degrades chromatin in standard preparations of nuclei isolated at low temperatures and stabilized with Mg^{2+} or Ca^{2+} *(228)*. In isolated nuclei incubated in the presence of either cation, there is *in situ* degradation of nucleoproteins to a series of DNA fragments, each of which is a multiple of the molecular weight of the smallest. Analysis of these fragments revealed that the smallest single-stranded class has an M_r about 45,000–63,000, and the smallest double-stranded class 120,000–150,000. The series of double-stranded fragments was found to contain single-stranded fragments, the double-strand breaks occurring as secondary reactions when two single-strand breaks in complementary strands were close to each other *(229)*. Since no such regularity was found among the degradation products of purified DNA, the nuclear-associated activity against nucleoprotein was postulated to be regulated by the positioning of potentially accessible sites controlled by superstructure proteins *(230)*. These sites, which are distributed uniformly along the genome, are also recognized by micrococcal nuclease (EC 3.1.4.7) and DNase I *(231, 232)*; their distribution pattern is not affected by trypsin treatment of the nuclei *(232)*.

Apart from a Ca,Mg-dependent endonuclease, chromatin contains a less tightly bound Mg^{2+}-dependent endonuclease with a pH optimum of 6.2 against duplex DNA *(225)*, but inactive on nuclear DNA *in situ (223)*. It is stimulated severalfold by a prior incubation of isolated nuclei with NAD^+, which appears to affect the enzyme in some indirect manner *(233)*, probably as a result of formation of poly(ADP-ribose). It is inhibited by the nondialyzable, Pronase-sensitive, heat-stable factor from liver and testis nuclei (see above) which stimulates the Ca,Mg-dependent endonuclease *(224)*.

C. Deoxyribonuclease II (Acid DNase), EC 3.1.4.6

The occurrence and properties of this enzyme up to 1971 have been reviewed *(234)* and subsequent material up to 1975 has been succinctly compiled *(2)*.

Early observations on phosphodiesterase activity of purified DNase II against thymidine 3'-(p-nitrophenyl)phosphate and bis(p-nitrophenyl)phosphate were questioned by several authors *(235–237)*, who applied heat treatment to the purification of bovine, sheep and hog spleen DNase II to obtain enzymes inactive against the synthetic substrates. These results could not be reproduced for the hog spleen enzyme *(238)*, suggesting that heat treatment might have preferentially inactivated the phosphodiesterase activity *(234)*. The problem was finally resolved by a mild chromatographic procedure, yielding a 43,000-fold purified preparation of the enzyme from lysosomal membranes of rat liver, with no phosphodiesterase activity against the synthetic substrates *(239)*.

Subsequently, phosphodiesterase-free acid DNases were obtained from human gastric mucosa and cervix uteri *(240)* and from bovine liver *(241)*. In the latter instance, protein denaturing conditions were essential, testifying to the close association of the two activities. It is undoubtedly of significance that the phosphodiesterase activity is susceptible to the protein inhibitor of DNase II *(241)*.

At neutral pH, highly purified DNase II still exhibits activity against DNA, apparently an intrinsic property of the molecule, but differing from the activity at pH 5 in response to ionic strength, inhibitors and activators *(242)*. With the porcine spleen enzyme, sulfate, oxalate and malonate ions, at concentrations one to two orders of magnitude below those that are inhibitory, slightly stimulate activity against DNA *(243)*; with sulfate ions, there is also a change in the mode of cleavage, the number of single-strand breaks increasing markedly relative to the number of double-strand breaks, in agreement with previous studies demonstrating the independence of haplotomic and diplomotic mechanisms, with the former predominating *(244)*. Daunorubicin inhibition of the enzyme, due to intercalation of the antibiotic in the substrate with concomitant binding of the enzyme in an inactive ternary complex, affects the diplotomic mechanism to a greater extent than the haplotomic one (245).

Since the appearance of Bernardi's (234) review, additional activities resembling the hog spleen enzyme have been isolated from a variety of tissues and species (239–241, 246–248). A single-step procedure for obtaining 40-fold and 80-fold purified preparations from bovine and hog spleen, respectively, by antibody affinity chromatography (248) could conceivably be of value for quantitative immunological comparisons of DNase-II-like activities from different sources.

The specificity of hog-spleen DNase II was only recently investigated in some detail (197, 249), by following the kinetics of liberation and the base compositions of the 3'-phosphate and 5'-hydroxy terminal and penultimate nucleotides of the degradation fragments from calf thymus DNA, from three bacterial DNAs with different (G+C)-contents, and from synthetic poly(dA-dT). In general, the specificity of the enzyme, evaluated from the composition of the terminal and penultimate nucleotides of the fragments liberated, did not change during the course of DNA degradation. A linear relation was found to hold between the relative amounts of termini released from different bacterial DNAs and their (G+C)-content, but not for calf thymus DNA, presumably because of the presence of repetitive sequences. With poly(dA-dT), the composition of the 3'-phosphate termini (80% dA and 20% dT) was constant for fragments in the 40 to 15 average size range. These results were subsequently extended to an analysis of the 3'-terminal nucleotides and 5'-terminal dinucleotides in the fragments released from E. coli DNA, and a comparison of the composition of these "trinucleotides" with those calculated on the basis of random attack; these were found to be different for, and characteristic of, different DNases, and to provide a qualitative "specificity spectrum" for each enzyme (199), in general showing that each DNase "recognizes" a different set of short oligonucleotides, with at least four residues in the case of DNase II (250).

In view of the resistance to DNase I of poly(2'-fluoro-2'-deoxyuridylate) (Section IV, A), it is worth noting that this polymer is readily cleaved by DNase II (201).

1. LOCALIZATION

Exclusively lysosomal localization of the enzyme (1, 135, 234) has been questioned by Slor and Lev (251), who found about 10% of the total activity in calf thymus nuclei (252). This nuclear activity apparently did not originate from contamination of the nuclei with lysosomes *(251)*, nor as a result of secondary adsorption of soluble enzyme *(252)*. A 2- to 7-fold increase of nuclear DNase-II activity in the S-phase of synchronized HeLa cells has been interpreted *(253)* as indicative of DNase-II involvement in cellular DNA metabolism. One serious objection to this proposal is that the identity of the nuclear and lysosomal activities has not been unequivocally established *(251)*. Furthermore, purified rat liver nuclei contain less DNase II than other lysosomal hydrolases *(239)*, suggesting that the nuclear activity may be the result of contamination with lysosomes. Relevant to this is the fact that Kupffer cells, isolated from a suspension of rat liver cells by Pronase digestion of parenchymal cells, contain 14-fold more DNase II than the parenchymal cells, and about 7-fold more DNase II than Kupffer cells isolated by other methods *(254)*. It would clearly be desirable to examine the level of other lysosomal enzymes in Kupffer cells isolated by Pronase treatment, since their unusually high DNase-II activity is suggestive of enzyme adsorption, a phenomenon considered, but not conclusively eliminated, in the case of nuclear DNase II *(251)*.

Treatment of rats with hepatocarcinogens, e.g., aflatoxin B_1 or 3'-methyl-N,N-dimethylaminobenzene, led to a marked increase in liver nuclei activity at the time of appearance of regenerative nodules, followed by a decrease at the onset of malignant transformation *(255)*. The relation of these changes in so-called nuclear DNase-II levels to the increase in DNase-II activity in the liver lysosomes and supernatant of aflatoxin-B_1-treated rats *(247, 256)* is not clear.

The localization and functional aspects of DNase II have been discussed elsewhere *(1, 80, 234, 257;* see also Section X).

2. DNase II Endogenous Inhibitor

A natural protein inhibitor of acid DNase II, inactive against pancreatic DNase I, and initially found in mouse liver *(234, 258)*, has been purified to homogeneity from bovine liver by means of affinity chromatography on insolubilized acid DNase II *(241)*. It is a monomeric basic protein, of $M_r = 21,500$, which binds to the enzyme in a 1:1 ratio. It is maximally active at pH 5, and introduces a sigmoid shape in the plot of enzyme activity relative to substrate concentration. Its activity is reversed by RNA and DNA. The inhibitor is also active toward the DNase-II-associated phosphodiesterase activity.

D. "Nicking" Enzymes

These are endo-DNases that introduce one, or a limited number of, single-strand breaks in ss- and/or dsDNA. A number of such enzymes, involved in repair of damaged DNA (and also referred to as incision enzymes), have been isolated in purified forms and reasonably well characterized (Section XII, A, B).

The situation with regard to other such nucleases is, at best, rather confused. There are numerous reports or suggestions on the role of such "nickases" in DNA synthesis and replication (see refs. 257, 259–262 for reviews), but there is as yet no direct evidence to support such a contention, however reasonable it may appear. The difficulties associated with attempts to prove that a nuclease plays a *direct* role in DNA replication have been briefly discussed *(263)*.

Several workers have presented claims for an association of "nickase" activity with DNA polymerase-α[10] from regenerating rat liver cytosol*(264–266)*, this being regarded as evidence for some role in DNA replication. But the two activities can be separated by affinity chromatography on DNA-cellulose *(263)*. The isolated endonuclease activity introduced ss-breaks into supercoiled, duplex linear, and ss-DNAs and was inactive toward RNA. Its cation requirements suggest a close resemblance to the Ca^{2+},Mg^{2+}-dependent endonuclease (Section IV, B), except for inhibition by 4 mM $CaCl_2$. The relationship between the two might be further defined by examination of the behavior of the endonuclease in the presence of EGTA. The fact that prolonged incubation led to increased degradation of the substrate *(263)* raises additional doubts both as to its specific "nickase" properties and its distinctness from other known normal endonucleases.

It may now be considered established that DNA α- and β-polymerases[10] contain no associated endonuclease ("nickase") activities*(261–262)*. In contrast, rat liver mitochondrial DNA polymerase, partially purified, contains significant levels of such activity, and can use duplex circular DNA (in other words, also mitochondrial DNA) as a template, synthesis being initiated at 3'-OH priming points *(259, 260)*. However, further purification is essential to establish whether the presumed activity is an intrinsic property of the mt-polymerase and whether its specificity and mode of action are in accord with its role as a "nickase."

Presumed "nickase" activities have been detected in, and isolated from, highly purified viruses. In the case of polyoma virus, the purified enzyme cleaves the chain at only one site in polyoma DNA I, and at two sites in SV40 DNA. It was subsequently shown that the activities present in purified SV40 and polyoma virions are, in reality, derived by simple adsorption from the fetal calf serum, a constituent of the medium for maintenance of the virus-infected cells *(267, 268)*. The activity from calf serum, partially purified on DEAE-cellulose, exhib-

---

[10] See articles by Gillespie, Saxinger and Gallo (esp. p. 88) and by Bollum in Vol. 15 of this series, regarding DNA polymerases.

ited an absolute requirement for divalent cations at pH 7.4, with Mn^{2+} most effective and Ca^{2+} minimally so, whereas $Ca^{2+} + Mg^{2+}$ gave a marked synergetic effect (267). These properties are quite inadequate for characterization of the activity, which again resembles Ca^{2+},Mg^{2+}-dependent endonuclease, or even pancreatic DNase I.

In general, it is frequently overlooked that very low concentrations of any normal endonuclease will introduce only a limited number of ss-breaks. In fact, the most commonly employed tool for introducing ss-breaks in duplex linear or circular DNA is simply pancreatic DNase I. Establishment of specific single-strand activity would have to be based, at least in part, on adequate concentration of the isolated activity and establishment of the number of breaks introduced in a given substrate with increasing concentrations of the enzyme. Characterization of the fragmentation products would further help to establish the nature of the activity.

E. "Nicking-Closing" Enzyme(s) ("Relaxation" Proteins)

A heat- and Pronase-labile protein, originally found in *E. coli* and referred to as "ω protein" (269), and subsequently isolated from sonicated nuclei of mouse embryo cells (270), possesses the unique property of untwisting (relaxing) positive or negative closed superhelical DNA. Interest in this unique type of activity derives from its apparent combined action as a DNA endonuclease and ligase, which, acting at the growing fork of duplex DNA during replication, would introduce a transient swivel to reduce the length of duplex DNA rotating ahead of the growing fork during replication. In the case of closed circular viral and mitochondrial DNAs, such relaxation activity is essential to account for the existence of closed-circular intermediates and, in fact, Sen and Levine (271) isolated, from SV40-infected monkey cells, a 50 S viral DNA-protein complex capable of relaxing native SV40 DNA to a low superhelix density product.

The foregoing activity has also been referred to as "relaxation protein," "relaxation enzyme," "untwisting protein," "nicking-closing (N-C)" enzyme. It should not be confused with "unwinding" proteins, which act on linear duplex DNA by binding much more strongly to single-stranded than to double-stranded helical DNA (272).

Nicking-closing activities have now been characterized from a variety of sources, including *Drosophila melanogaster* and sea urchin eggs (273), calf thymus (274), cultured HeLa and mouse LA9 cells (275), KB cells (276), and Krebs II ascites cells (277).

There is general agreement on a stringent requirement of 0.2 M NaCl for optimal activity, and no requirement for divalent cations or a dialyzable cofactor. The activities from KB and LA9 cells are strongly inhibited by p-chloromercuribenzoate (275, 276), consistent with their protein character. For the LA9 cell activity, combined sedimentation velocity and gel electrophoresis data suggested that it consists of two 37,000 subunits (275). Zone sedimentation of the KB cell enzyme gave a value of 70,000, whereas gel electrophoresis, which revealed the presence of minor rapidly migrating components, gave a value of about 60,000 for the major component (276). The enzyme from calf thymus exhibited three components with M_r's of 53,000, 51,000, and 32,000 (274).

The major component of highly purified enzyme from LA9 cells, as well as rat liver and calf thymus, comigrate with a subfraction of the lysine-rich histone H1 extracted from homologous sources (278), but no attempt appears to have been made to test purified histone fractions for possible nicking-closing activity. It is perhaps of interest, in this context, that removal of the histones from SV40 "mini-chromosomes" and "nucleosomes" leads to changes in superhelical density of the closed circular SV40 DNA, implying that the free DNA in the DNA-histone complex is relaxed, perhaps as a result of the nicking-closing activity, and that the structure and/or organization of the histone-bound DNA may account for formation of superhelical turns upon DNA purification (277, 279).

1. METHODS OF ASSAY

Progress in this field, originally hampered by the tedious procedure for following relaxation activity by buoyant density centrifugation of the substrate (e.g., SV40 or phage PM2 DNA) in the presence of ethidium bromide in a CsCl gradient, has been impressive since the development of two new assay methods. One of these is based on the fluorescence enhancement of the intercalating agent ethidium bromide when bound to closed circular DNA, the amount of dye bound being dependent on the degree of superhelix density of the DNA (273, 275, 280). The second, technically simpler, employs gel electrophoresis, which gives a striking separation of superhelical DNA relative to the relaxed form, and appears to be capable of clearly resolving DNA samples differing in topological winding number, each species resolved differing from the successive one by a single superhelical turn (281–284).

2. MECHANISM OF ACTION

It is generally assumed that a DNA "relaxing" activity involves a sequence such as cleavage, swiveling and sealing, thus replacing the sequential action of an endonuclease and a ligase in the development of transient swivels in DNA. Attempts to demonstrate a cleaved intermediate have been unsuccessful, while the enzyme is not affected by known mammalian endonuclease I inhibitors, with the exception of denatured DNA; nor does it exhibit conventional ligase activity (275). With the aid of the electrophoresis assay technique, Keller (282) has shown convincingly that the action of the enzyme does not involve a one-hit mechanism, but catalyzes stepwise relaxation of superhelical DNA with formation of a decreasing number of superhelical turns during the course of the reaction. It was estimated that one enzyme molecule from KB cells catalyzes the relaxation of at least ten SV40 molecules. For the mouse L cell enzyme, Vosberg and Vinograd (278) estimate a value of 20 molecules with PM2 DNA-I as substrate.

With the aid of the electrophoretic assay technique, Pulleybank et al. (284) have shown in a very neat, albeit indirect, manner that the action of the enzyme does in fact involve the sequence cleavage, random rotation about a swivel, and resealing. The results are based on the assumption, already referred to above, that the species separated electrophoretically differ from each other by one superhelical turn (281, 282, 284). The enzyme was first allowed to act on closed circular DNA to give limit products, which were separated into a set of species differing in topological winding number, forming a Boltzmann distribution. When any member of this set was isolated, and in turn treated with the nicking-closing enzyme, the original set was regenerated. Finally, when the substrate was treated with DNase I to introduce a single break, and the resulting relaxed DNA molecule was treated with *E. coli* polynucleotide ligase, the closed circular DNA products obtained gave a distribution on gel indistinguishable from that obtained by the action of the N-C enzyme on superhelical DNA.

3. Localization

The cytoplasm of LA9 cells contains only about 2% of the activity found in the nuclei, while the nuclear enzyme accounts for 1–2% of the chromatin protein (278). A sonicated extract of purified mitochondria also exhibits activity, about 0.8% that of the nuclei, adequate for relaxation of mtDNA (275). The activity from *Drosophila* eggs has been isolated from the cytoplasm, but purified nuclei also exhibit activity (273).

V. Exodeoxyribonucleases

A. Deoxyribonuclease III

DNase III is an exonuclease that hydrolyzes DNA in a progressive manner from the 3'-OH terminus to release 5'-mononucleotides and some dinucleotides (285), which resist further hydrolysis. Denatured

DNA and polydeoxynucleotides are 4-fold more susceptible than native DNA, and RNA and polyribonucleotides are not substrates. The enzyme has an M_r of 52,000 and is stabilized by mercaptoethanol. Its pH optimum is 8.5, and Mg^{2+} or Mn^{2+} is required.

The enzyme occurs in nuclei and appears to be a ubiquitous component of mammalian tissues. It probably corresponds to the alkaline DNase activity previously found in mice mammary tumors (286) and Novikoff ascites hepatoma cells (287), and is the most widespread DNase activity at an alkaline pH. Its role is not clear but, at high concentrations, it attacks UV-irradiated DNA and double-stranded polydeoxynucleotides to release small oligonucleotides containing pyrimidine photodimers (288). The low rate of this reaction can hardly be regarded as disqualifying this enzyme from involvement in repair. It is not unlikely that it participates in repair of lesions due to agents other than UV, a possibility worth looking into.

B. Exodeoxyribonuclease IV, EC 3.1.4.28

Exo-DNase-IV designates an exonuclease that degrades double-stranded polydeoxynucleotides from the 5' termini to products consisting of 5'-mononucleotides (80%) and oligonucleotides (20%), principally dinucleotides (288, 289), so that subterminal breaks must occur (288). The enzyme is 10- to 50-fold more active against synthetic double-stranded polydeoxynucleotides than native DNA, and is almost inactive toward single-stranded polydeoxynucleotides and single- and double-stranded polyribonucleotides (289). It can attack DNA not only at free termini, but also at internal breaks (290). Its activity against DNA and poly(dA) · poly(dT) is unaffected by prior UV-irradiation of the polymers, the resulting oligonucleotide products of which are 5 to 8 residues in length containing thymine photodimers. The presence of such lesions increases the proportion of subterminal scissions caused by the enzyme (288).

The mode of cleavage of the enzyme is intermediary between progressive and processive, several adjacent phosphodiester bonds being broken each time the enzyme binds to the substrate. The degree of processiveness increases with increasing temperature; at 37°C, each encounter leads to release of 50–200 residues. This mechanism of action is considered to be applicable to repair functions (290), in which the enzyme fulfills the role of bacterial DNA polymerase I exonuclease activity (see Section XII, C).

DNase IV has an M_r of 42,000, its pH optimum is about 8, and it requires Mg^{2+} or Mn^{2+} (289). It is associated with nuclei, from which it is readily released in aqueous medium (288). It occurs in various tissues of the rabbit, with highest levels in the bone marrow and lymph node.

VI. Endo-Exodeoxyribonuclease with Preference for Poly(dT)

There is as yet no clear-cut mammalian cell counterpart of fungal endo-exonucleases. One potential candidate for this title is an activity purified 950-fold from rat ascites hepatoma cells *(291)*. It hydrolyzes poly(dT) more than 10-fold faster than poly(dA), and has little effect on poly(dC), poly(U), poly(A), or native or denatured DNA. The enzyme appears to attack initially as an endonuclease, and subsequently as an exonuclease, to release 5'-dTMP via oligonucleotides, but its non-homogeneity points to the need for independent confirmation of this finding.

The enzyme requires Mg^{2+}, is optimally active at pH 8.5, and is inhibited by high concentrations of KCl. No obvious function for such an activity is known, notwithstanding the presence of dT tracts in DNA, since its activity against poly(dT) is strongly inhibited when the latter is annealed to a complementary homopolymer. It would be of interest to examine its activity against UV-irradiated poly(dT), which should contain a high proportion of thymine dimers.

One of the nucleases of the mammalian vaccinia virus has been established as an endo-exonuclease *(292)* (Section XI, C,1).

VII. Single-Stranded-Nucleate Endonuclease (Sugar-Nonspecific Endonuclease; 5'-Endonuclease), EC 3.1.4.21

Heat-labile endonucleases, liberating 5'-phosphate-terminated oligonucleotides from single-stranded ribo- and deoxyribopolynucleotides, activated by low concentrations of Mg^{2+} and Mn^{2+}, and inhibited by EDTA and monovalent cations, have been purified from a number of sources *(206, 207, 293–296)*. All of these, with a pH optimum about 7, and variously referred to as pig liver nuclei RNase *(295)*, rat liver nuclei endonuclease *(207, 297)*, rat liver mitochondrial nuclease *(293, 294)*, rat liver 5'-endonuclease *(296)*, polyadenylase *(135, 298)*, sheep kidney endonuclease *(206)*, are probably one and the same enzyme. In all likelihood the so-called lamb brain phosphodiesterase *(299)* is also a member of this class. The enzyme exhibits a progressive mode of action *(206)*.

The enzyme is latent, and is found in mitochondria of rat liver cells and cultured chick embryo fibroblasts *(300, 301)*. In rat liver it was initially localized in the intermembranous space and in the outer membrane *(302)*. The intermembranous space localization was confirmed, but the outer membrane activity was attributed to secondary adsorption *(135, 298)*. The inhibition of the biosynthesis of the enzyme in chick embryo fibroblasts by cycloheximide, and insensitivity of such synthesis to ethidium bromide and chloramphenicol, suggest that the enzyme is coded for by the nuclear genome but translated on cytoplasmic ribosomes *(301)*.

5'-Endonuclease activity is impaired in primary *(303)* and transplantable *(304)* hepatomas. But its level is not significantly affected in rats fed carcinogenic amines, so that loss of such activity does not precede neoplastic transformation *(305)*. Partial hepatectomy is without effect on 5'-endonuclease activity *(306)*.

Two additional activities resembling 5'-endonuclease have been reported. One of these is the so-called "RNase III" from rat liver, optimally active at pH 9, and inhibited by monovalent cations and EDTA. The effect of the latter is reversed by Mg^{2+} and Mn^{2+}. The enzyme is activated by Triton X-100 *(307)*. This enzyme has not been purified, and its specificity and mode of action toward RNA are unknown. Nonetheless, an assay has been proposed *(73, 135)* for this activity, based on its thermolability at 50°C at pH 5; the activity was reported to be sensitive to RNase endogenous inhibitor and to bovine RNase A[3] antisera. Although initially presumed to be diffusely distributed in all subcellular fractions *(308)*, it is now considered to be localized in the intermembranous space of mitochondria *(135)*. In nonionic media, it is partially adsorbed on the outer surface of the inner mitochondrial membrane *(135)*, so that its localization pattern is identical with that for single-strand-specific 5'-endonuclease *(135)*. Since the latter is still quite active at pH 9, and also thermolabile at pH 5 and 50°C *(294)*, reasonable doubts exist as to whether the so-called RNase III and 5'-endonuclease are distinct enzymes. Purification of the former must be awaited to establish this.

The other activity resembling the 5'-endonuclease is an endoribonuclease that attacks synthetic homoribopolymers and poly(A)-rich mRNA,[11] initially in the poly(A) segment, to liberate 5'-phosphate-terminated oligonucleotides. It has been purified over 500-fold from bovine adrenal cortex cytoplasm *(309, 310)*, is active at pH 6.5–9.5, and is stimulated by Mg^{2+} and Mn^{2+} *(310)*. It is inactive against native and denatured DNA, dsRNA, and RNA · DNA hybrids. An apparently similar activity was found to copurify with poly(A)-rich, rapidly labeled RNA from human peripheral lymphocytes *(311)*.

The relationship of the adrenal cortex enzyme to single-strand-specific 5'-endonuclease requires examination *(295)*, since the only major difference between them is the inability of the former to attack denatured DNA.

---

[11] See Vol. 19 in this series, regarding mRNA-poly(A).

Ribosomes of guinea pig adrenals appear to contain a heat- and acid-labile protein that inhibits the action of adrenal and lymphocyte enzymes on poly(A)-rich RNA, but not their action on free poly(A) segments *(309)*. This inhibitor does not affect the action of RNases T1 and T2, nor of *E. coli* RNase II on poly(A)-rich RNA, but it does inhibit cleavage of this substrate by low concentrations of RNase A. More precise data about the specificity of this inhibitor are desirable, e.g., its effect on single-strand-specific 5'-endonuclease *(295)*, since the latter is claimed to be somewhat sensitive to the natural RNase I inhibitor from rat liver supernatant *(73)*.

VIII. Sugar-Nonspecific Exonucleases

A. Phosphodiesterase I (5'-Exonuclease), EC 3.1.4.1, EC 3.1.4.19

Phosphodiesterase I, (PDase I), isolated from numerous tissues *(312–321)*, and resembling venom exonuclease in specificity and other properties, has been discussed in two reviews *(1, 322)*.

The enzymes from liver *(313, 315–320)* and kidney *(312)*, and presumably those from other tissues, exhibit nucleoside pyrophosphohydrolase activity against NAD, nucleoside diphosphate sugars and nucleoside triphosphates, and also attack poly(ADP-ribose). Apparently a single enzyme is responsible for the cleavage of nucleoside 5'-phosphates esterified to oligonucleotides, alkyl and aryl radicals, and pyrophosphates and sugar phosphates, so that it would be more correctly characterized as a nucleotide phosphodiesterase-pyrophosphohydrolase. The liver enzyme is a glycoprotein with an M_r of about 130,000 *(319, 320)*.

1. LOCALIZATION

Both cell fractionation *(323–325)* and histochemical *(326)* studies point to localization of the enzyme in the plasma membrane of rat liver *(323–326)* and, possibly, other *(327)* cells. Release of the enzyme from the membrane requires the use of detergents, organic solvents or proteolytic enzymes *(312, 313, 318–320)*. Earlier reports on nuclear and microsomal localization *(300, 312, 328)* were probably due to contamination by plasma membranes *(317, 323, 324)*. But there is a more recent claim that a portion of the rat liver enzyme is located in the rough endoplasmic reticulum, although the specific and total activities of this fraction were quite low *(320)*. Reported association of some PDase-I activity with lysosomal membranes *(323)* has been ascribed to the partial plasma membrane origin of the latter; in fact, an examination of the fate of the plasma membrane of iodinated mouse fibroblasts actively engaged in phagocytosis showed that 15–30% of the total surface area, along with the same proportion of total PDase activity, were incorporated into the phagolysosomes during phagocytosis (329). More recently, iodination of isolated hepatocytes was utilized to demonstrate that PDase I is an ectoenzyme located on the outer surface of the plasma membrane (330), and that a portion of the activity, isolated from the endoplasmic reticulum, is inaccessible to the extracellular space and consequently is a component of the reticulum, functionally distinct from the plasma membrane activity (331).

Although the liver bile canalicular membranes in rodents are rich in PDase I, very little of this type of activity is found in bile, and even less in serum; the activities from bile and bile canalicular membranes exhibit similar pH-activity profiles, suggestive of their identity (332). The low level of PDase I in the bile, relative to liver homogenate, has been confirmed in the case of the rat, but not for other species, such as sheep and hog, where the enzyme level in the bile exceeded that of the homogenate severalfold (333). The differences in ratio of bile to homogenate activities for several plasma membrane enzymes (PDase I, 5'-nucleotidase, alkaline phosphatase) for various species showed that these enzymes are not released into the bile to equal extents.

Histochemical localization of PDase I has also been attempted at the electron microscope level with the aid of 5-iodo-3-indolyl and 5-nitro-3-indolyl esters of 5-fluorodeoxyuridine 5'-phosphate, both of which yield electron-dense reaction products. The activity in rat liver cells was found in the cytoplasm, largely in association with the Golgi membrane (334), a result at variance with cell fractionation and other histochemical methods (323–326). Possible sources of this discrepancy were not discussed.

2. SYNTHETIC SUBSTRATES

The p-nitrophenyl esters of nucleoside 5'-phosphates (322), which for some years provided convenient and relatively specific substrates for PDase-I assay, are likely to be replaced by the more easily synthesized and less expensive 4-nitrophenyl phenylphosphonate (321), readily hydrolyzed by PDase I and the snake venom enzyme, and fully resistant to PDase II, cAMP-PDase and alkaline phosphomonoesterase. Despite its lower affinity for PDase I, the rate of cleavage of the new substrate by the intestinal enzyme, both at saturation and 1 mM substrate concentrations, exceeds that for thymidine 5'-(p-nitrophenyl) phosphate (321). An analog, 2-naphthyl phenylphosphonate, is suitable for staining of polyacrylamide gels. It would presumably also prove to be suitable as a histochemical substrate (cf. ref. 326).

It is pertinent to note that the specificity of esters of nucleoside 5'-phosphates toward PDase I has been established only for mammalian tissues (see also below) and caution should be exercised in applying them to other material. For example, thymidine 5'-(p-nitrophenyl)phosphate is cleaved by *E. coli* nucleoside phosphoacylhydrolase *(335)* and by nucleotide pyrophosphohydrolase from higher plants *(336)*, both enzymes with no exonucleolytic activity toward oligonucleotides. This susceptibility is due to the acidity of the p-nitrophenyl substituent, which forms a mixed phosphoanhydride type of linkage with the phosphate group, thus eliciting unpredictable responses from some enzymes *(335)*. Attempted use of the new phenylphosphonic esters with materials of plant, fungal or bacterial origin should therefore be preceded by appropriate specificity studies.

The synthesis of thymidine 5'-(α-naphthyl)phosphate for the histochemical localization of PDase I at the light microscope level by the azo-dye coupling technique *(326)* was followed by the preparation of its 5-bromo-4-chloroindolyl analog *(92)* for localization by the indigogenic method. The kinetics of hydrolysis of the latter by snake venom PDase have been examined *(337)*. Subsequently a 5-iodo-3-indolyl derivative was prepared for localization at the electron microscope level; this yielded an electron-dense product, but with formation of microcrystals due to the extensive hydrogen-bonding structure of 5,5'-diiodoindigo *(338)*. Reduced crystal formation was achieved by use of the 5-nitro-3-indolyl ester of 5-fluorodeoxyuridine 5'-phosphate in combination with the 5-iodo-3-indolyl analog *(339)*, but no localization results have been reported. The latter analog has been studied with a view to its possible application in cancer chemotherapy *(339)*.

Ammonium 4-methylumbelliferyl 5'-thymidylate has been synthesized and proposed as a substrate for the fluorometric assay of serum PDase-I activity *(340)*.

3. PATHOLOGICAL ASPECTS

A preliminary study of PDase-I activity in the sera of patients with pathologically high levels of alkaline phosphatase indicated abnormally high activities in patients with necrotic changes of the liver or kidney, and in cases of breast cancer *(341)*. The PDase-I isoenzyme pattern of patients with primary hepatomas, grade 2 and higher, examined by gel electrophoresis with the 5-iodo-3-indolyl esters of nucleoside 5'-phosphate as substrate, exhibited an additional band with high mobility not found in normal individuals *(342, 343)*. Since some noncancerous cirrhosis cases also exhibited this extra band, diagnostic application of this test is feasible only in conjunction with other tests of liver function *(343)*. Rat hepatomas have been reported to exhibit higher PDase-I levels than normal rat liver cells *(344)*.

4. FUNCTIONAL SIGNIFICANCE

The *in vivo* function of PDase I is still largely speculative. Its external plasma membrane localization, and exceptionally high level in actively transporting surfaces (bile canaliculi, intestinal and kidney epithelium brush border) are suggestive of some role in the mediation of transport of metabolites. Nucleoside diphosphate sugars, like UDP-glucose, are rapidly hydrolyzed by the PDase I in liver cell plasma membrane; but they are unable to permeate the membrane even when their cleavage is prevented by inhibition of PDase-I activity *(345)*. Similarly ribo- and deoxyribo- oligo- and polynucleotides exposed to Novikoff ascites hepatoma cells are cleaved by PDase-I type activity at the cell surface; the products, ultimately dephosphorylated by a similarly located phosphatase, are released into the extracellular fluid *(346)*. The absence of PDase-I activity in virus-transformed hamster cells may conceivably point to some role in intracellular adhesion *(347)*.

5. OTHER ACTIVITIES AGAINST ESTERS OF 5'-NUCLEOTIDES

An activity other than PDase I, which cleaves p-nitrophenyl esters of ribo- and deoxyribonucleoside 5'-phosphates, and with a pH optimum of about 8, has been purified 1100-fold from bovine pancreas *(348)*. It is most active against esters of 5'-CMP and 5'-dCMP but, surprisingly, shows relatively little activity against oligonucleotides.

Rat liver lysosomes contain an activity that hydrolyzes the p-nitrophenyl ester of 5'-dTMP optimally at acidic pH *(349)*; this activity could be due to lysosomal acid pyrophosphohydrolase, which is known to hydrolyze the p-nitrophenyl ester of 5'-dTMP *(349)*. Unfortunately, its behavior toward oligonucleotides was not examined.

B. Phosphodiesterase II (Spleen Exonuclease, Acid 3'-Exonuclease), EC 3.1.4.18

This heat-labile exonuclease cleaves preferentially unstructured ribo- and deoxyribo- oligo- and polynucleotides, in a progressive manner from the 5'-OH terminus, to nucleoside 3'-phosphates; 5'-phosphate-terminated chains are resistant. The enzyme readily hydrolyzes thymidine 3'-(p-nitrophenyl)phosphate, widely employed as a standard substrate. Its properties and lysosomal localization were extensively reviewed in 1971 *(350)*, and little significant new information has since appeared. It is, however, widely applied as a tool in sequence studies, e.g., in nearest neighbor analyses.

An activity recently purified from bovine brain white matter *(351)* bears some resemblance to PDase II in ionic requirements, and cleavage of polyribonucleotides to 3'-mononucleotides and thymidine 3'-(*p*-nitrophenyl)phosphate to 3'-dTMP. It differs from PDase II in that it is inactive against single-stranded oligo- and polydeoxynucleotides, pointing to some significant difference in specificity that cannot be accounted for by its reported nonhomogeneity, and clearly requires further investigation.

1. SPECIFICITY ASPECTS

There is some confusion regarding the specificity of PDase II that calls for comment, not only in relation to its physiological role, but also because of its widespread use in sequence studies. At pH 5, poly(C) is fully resistant, whereas poly(A), poly(I) and poly(U) are all hydrolyzed at comparable rates *(350)*. Since poly(C) is a twin-stranded helix at pH 5, this appears to account for its resistance. However, poly(A) is also a twin-stranded helix at pH 5. It consequently appears that cytosine residues themselves confer some resistance *(352)*. It was subsequently found *(200)* that poly(C) is readily hydrolyzed by PDase II at pH 5.5, in fact almost as rapidly as poly(U). Hence, the report *(353)* that C-C, dC-dC, C-U, and higher oligonucleotides with a C at the 5' terminus, are not substrates, is rather curious; these results were obtained with a low enzyme concentration at pH 7, and excessively long incubation periods in the absence of a stabilizer. The enzyme does appear to cleave readily straight-chain alkyl esters of 3'-CMP *(353)*.

It is also of interest that the α-naphthyl ester of 3'-dTMP is hydrolyzed at less than 1% of the rate for the *p*-nitrophenyl ester *(354)*.

The susceptibility to PDase II of esters of nucleoside 3'-phosphates may be reduced by etherification or acylation of the 5'-hydroxyl; e.g., 5'-*O*-methyl, 5'-*O*-ethyl and 5'-*O*-acetyl confer increasing resistance, in the order given. With substituents such as 5'-*O*-tetrahydropyranyl or 5'-*O*-benzyl, resistance to PDase II is total *(91, 354)*, a fact profited from to develop an RNase-I substrate fully resistant to PDase II (see Section II, A, 4).

Poly(2'-chloro-2'-deoxyuridylate) and poly(2'-chloro-2'-deoxycytidylate) are both substrates, but are hydrolyzed at an extremely slow rate *(200)*.

It has been reported that 3',5'-dithymidine phosphorothioate, with the sulfur atom not in the internucleotide linkage, although resistant to PDase I and PDase II, is a competitive inhibitor of both enzymes *(88)*.

By contrast, when the sulfur atom in this compound substitutes for the 5'-O in the internucleotide linkage, hydrolysis by PDase II proceeds readily (355).

2. LOCALIZATION

A preliminary study, by histochemical methods, with the indigogenic substrate thymidine 3'-(5-bromo-4-chloro-3-indolyl)phosphate, demonstrated the presence of the enzyme in the cytoplasmic granules of rat liver and spleen reticulum cells (92). Since the kinetic parameters for hydrolysis of this substrate by PDase II were comparable to those with thymidine 3'-(p-nitrophenyl)phosphate (356), the indolyl derivative appears suitable for histochemistry, a fact that might have been expected to stimulate further research on localization of the enzyme.

IX. 2':3'-Cyclic-nucleotide 3'-Phosphodiesterase, EC 3.1.4.37

Earlier findings on an activity that cleaves the 3' bond of 2',3'-cyclic phosphates of mono- or oligonucleotides to the corresponding 2'-phosphates, and that is found predominantly in the myelin sheath of nervous tissue, have been summarized (357). Solubilization of the enzyme has since been improved to obtain a 3-fold purified extract in good yield, and a spectrophotometric method was described for its assay in RNase-free preparations (358). It is optimally active at pH 6, has an M_r of about 100,000, and is effectively inhibited by p-chloromercuribenzoate (358, 359). It is competitively inhibited by nucleoside 2'-, 3'- or 5'-phosphates, the most potent of which are those of adenosine (358).

Since myelin, which possesses a specific activity exceeding manyfold that of other cell components (360), is a layer of the plasma membrane, it was concluded that the enzyme is associated with plasma membranes (361). It has, in fact, been shown to occur in a masked form in the plasma membrane of human erythrocytes (359) and some other cells (361), and in cultured myelin-free glial cells (362). The enzyme is highly active in calf brain oligodendroglia engaged in production of myelin (363). There is really little evidence to indicate that nucleoside 2',3'-cyclic phosphates are the natural substrates for this enzyme in myelin (357, 358).

X. Localization and Function of Nucleases

Despite the sparsity of data on localization and functional significance of individual mammalian nucleases, we attempt here a recapitulation that includes their possible functional interrelations in the nucleic acid metabolism of the cell and the organism.

The extracellular nucleases secreted by the digestive glands into the intestinal tract, i.e., the pancreatic and salivary gland neutral DNase *(2, 191)* and RNase *(4, 5, 15, 29)*, are undoubtedly involved in the initial stages of degradation of nucleic acids ingested into, or derived from microorganisms inhabiting, the intestinal tract *(101)*.

The intracellular nucleases must be involved in further degradation of foreign nucleic acids, as well as in metabolism of endogenous molecules. It appears logical to group these according to their intracellular localization and to discuss their specificities in relation to each other and to cellular topography.

The plasma membranes contain PDase I (nucleotide pyrophosphohydrolase) *(318, 323)*, 5'-nucleotidase *(364)*, and nonspecific alkaline phosphatase *(364)*. All three activities are located on the membrane with the active sites facing the external medium *(300, 365)* and are considered to function by hydrolysis and dephosphorylation of oligo- and polynucleotides, and nucleotide pyrophosphates, to products that can be transported across the cell membrane. This function is facilitated by their common localization and similar pH and divalent cation requirements. The 5'-phosphate-terminated oligo- and polynucleotides, nucleotide pyrophosphates and nucleotide coenzymes are readily cleaved by PDase I to 5'-nucleotides and sugar phosphates, the 3'-phosphate-terminated oligonucleotides less readily, and only after dephosphorylation by alkaline phosphatase. The resulting 5'-mononucleotides and sugar phosphates are dephosphorylated by 5'-nucleotidase and/or alkaline phosphatase, and the products are transported into the cell; such a function has been established for the 5'-nucleotidase of human lymphocytes by the demonstration that uptake of nucleoside and P_i from AMP in cells containing this enzyme was 4-fold greater than in cells deficient in it *(366)*.

PDase I, 5'-nucleotidase, and alkaline phosphatase are particularly abundant in the brush border region of epithelial cells, i.e., sites where the membrane is well developed. In the duodenum, they undoubtedly participate in absorption of nucleic acid constituents already digested by the secretory pancreas-derived RNase and DNase. However, it is difficult to pinpoint their function at secretory surfaces of the pancreatic and salivary gland serous acini, which are involved in protein transport, and at other cell surfaces, where degradation of oligonucleotides does not appear of primary importance. Perhaps PDase I functions in these instances in another capacity, possibly as a nucleotide pyrophosphohydrolase. Such functional differences may occasionally be reflected by a somewhat altered specificity, as in the case of pancreatic PDase I *(348)*.

Intracellular digestion is known to occur in the lysosomes, the hydrolytic enzymes of which include nucleases *(367)*, e.g., RNase II *(129)*, DNase II *(234)* and PDase II *(350)*, all basic proteins with no SH groups, no requirement for divalent cations, and optimally active at acid pH. They are localized in the intralysosomal fluid, so that they are partitioned off from the rest of the cell by the lysosomal membrane. Since, apart from DNase II, no intracellular inhibitors of these enzymes have been found, it is assumed that this function is fulfilled by the lysosomal membrane, which regulates their access to cellular nucleic acids. Only upon rupture of the membrane, by hypotonicity or detergents, or *in vivo* under pathologic conditions, do the lysosomal enzymes come in direct contact with the cell contents. Under normal conditions, lysosomes are known to internalize foreign bodies and discarded cellular structures *(367)*. The following scheme has been proposed for intralysosomal digestion of polynucleotides *(234)*: ribo- and deoxypolymers are cleaved by acid RNase and DNase, respectively, to 3'-phosphate-terminated oligonucleotides, i.e., optimal substrates for PDase II, which further degrades them to mononucleoside 3'-phosphates. The latter, in turn, are hydrolyzed by lysosomal acid phosphatase or nucleotidase *(368)* to nucleosides and P_i, which are then transferred to the cytoplasm.

The combined secretory, plasma membrane, and lysosomal nucleolytic complexes of mammals correspond both in enzymic content and functional significance to the periplasmic enzyme system of gram-negative microorganisms *(369)*, which furnishes the cell with metabolites by degradation and translocation of extra- and intracellular material, and probably also protects it from biologically active nucleic acids.

RNase I and DNase I, which resemble the pancreatic enzymes in some aspects of their specificity, ionic requirements and susceptibility to endogenous protein inhibitors, occur also in nonsecretory cells, but their role is still largely speculative.

RNase I was considered, until recently, to occur in the mitochondria and cytosol (Section II, A, 3). The presence of an endogenous inhibitor of its activity in the cytosol, and the existence of an inverse correlation between the level of free enzyme and the anabolic activity of the cell, has led to the proposal that the RNase-I–RNase-I-inhibitor system performs a regulatory function in protein metabolism (Section II, A, 7).

DNase I is found, together with 5'-endonuclease in the space between the inner and outer mitochondrial membranes *(135).* Both of these have similar ionic requirements, and may be assumed to produce 5′-phosphate-terminated oligonucleotides from nucleic acids at some unknown stage of their metabolism. However, the finding that DNase I is inhibited by actin and depolymerizes filamentous F-actin to monomeric G-actin *(208, 214)* suggests that one of its major roles may be the regulation of actin filament formation and function. Possibly other endogenous nuclease inhibitors fulfill additional physiological roles (e.g., ref. *127*).

Localized in the inner mitochondrial membrane is polynucleotide phosphorylase. Since, in contrast to the analogous bacterial enzyme, its sole function is the production of nucleoside 5′-pyrophosphates, it is presumed to regulate the intramitochondrial level of NDP's *(190).*

In addition to extra- and intracellular digestive processes, there exist additional key metabolic routes dependent on the participation of an even wider variety of specific nucleases, e.g., the maturation of biologically active RNA, the replication of DNA, the correction of spontaneous or externally induced errors in replication, transcription or translation.

The nuclei appear to contain three enzymes active against DNA at pH 7–8, all requiring divalent cations and producing 5′-phosphate-terminated products. Exo-DNase-IV (Section V, B), which readily releases thymine photodimers from UV-irradiated DNA, is probably involved in repair *(290).* DNase III (Section V, A) also an exonuclease, with a preference for ssDNA, is functionally less well defined *(288).* The third enzyme, the chromatin-bound Ca^{2+},Mg^{2+}-dependent endonuclease, activates *in vitro* DNA synthesis on templates of isolated nuclei, and could be involved in replication of DNA *in vivo (226).* Furthermore, nuclear RNase H, which cleaves endonucleolytically the RNA strand of RNA · DNA hybrids, probably participates in replication of DNA by removal of the RNA fragments that serve as primers for DNA polymerase *(176).*

The mechanisms of maturation of various RNA species, largely dependent on nucleolytic enzymes, are currently well on the way to being elucidated *(370).* The so-called "nucleolar" RNase, which cleaves 45 S RNA to relatively large fragments, and displays only feeble activity against low-M_r RNA, has been proposed as a candidate in the "sizing" of rRNA *(25–27),* but conclusive evidence is lacking. Meanwhile, the finding that *E. coli* RNase III, specific toward dsRNA and undoubtedly involved in the processing of bacterial rRNA and polycistronic phage RNA *(137, 154–156),* can cleave mammalian 45 S precursor RNA to fragments corresponding to the *in vivo* products

(138), has stimulated a search for its mammalian counterpart. Among the various reports on dsRNase activities in mammalian cells, there are at the moment only two that appear sufficiently promising to warrant further investigation: the membrane-bound activity in human KB cells, which cleaves dsRNA as rapidly as ssRNA *(143)*, and the activity isolated by fractionation of calf thymus RNase H *(146)*.

Two RNases involved in post-transcriptional maturation of tRNA, the 4.5 → 4 S RNA sizing activity (Section II, E, 3) *(158)* and RNase P · *Hsa* (Section II, E, 4) *(161)*, have now been detected in various cells. The latter of these is located in the cytosol, where maturation of tRNA is considered to occur. In addition, there is now RNase NU (Section II, E, 2) from human KB and other cells *(143, 157)*, with a preference for unstable RNA and thus a potential candidate for processing of precursors of various RNA species. The identification of these activities will undoubtedly facilitate further efforts directed toward the elucidation of the mechanism of RNA processing.

XI. Virus-Associated Nucleases

The association of neuraminidase activity with influenza virus, first reported in 1942, was long considered an exception to the general rule that animal viruses, like bacteriophages, carry no enzymic activities and are dependent on the machinery of the host cells. Even the discovery of DNA-directed DNA polymerase activity in poxvirus in 1967, and of RNA-directed RNA polymerase in reovirus a year later, aroused little more than passing interest. It was the demonstration of RNA-directed DNA polymerase activity (or "reverse transcriptase") in oncornaviruses in 1970 that provided a powerful stimulus to the search for enzymic activities in purified virions of animal viruses, since this discovery revealed not only the mechanism of transformation to the tumorous state by RNA viruses, but also that these viruses contain, in addition to the genetic information required for such transformation, at least part of the associated enzymic machinery necessary for this purpose.

The subject of virus-associated enzymes in general has been briefly reviewed *(151, 371, 372)*. Establishment of a given activity as part of a viral structure is by no means simple because of possible adsorption not only to the external surface, but even to subviral structures during maturation. Criteria for virion origin of an enzymic activity include: (i) association with highly purified virions; (ii) firm association with some internal viral structure; (iii) enhancement of activity following removal of external viral proteins; (iv) specificity of enzyme activity, particularly in relation to analogous host cell activities; (v) possible function during infection. Even these criteria may not be fully adequate, as shown in the following section. The development of viral and/or cellular mutants may be of key value in this respect (see below).

Our original intention of compiling a table of virion-associated nucleases was abandoned as premature, during the preparation of this review, as a result of the appearance of two papers, which we now summarize.

A. Origin of Endodeoxyribonuclease Acitivity in Polyoma Viruses

Two recent completely independent investigations *(267, 268)* underline the precautions necessary in the interpretation of the origin of virion-associated enzymic activity. A number of laboratories have reported the presence, in highly purified polyoma virions, of a "nicking" activity that converts form I (covalently closed duplex) DNA into a relaxed form; others have found associated with SV40 virions an endonuclease that cleaves at one site in polyoma DNA I, and at two sites in SV40 DNA I (see refs. 267, 268 for literature). Various hypotheses have been advanced for the role of these activities in viral replication.

It has now been shown that the endonuclease activity associated with polyoma virions is derived from the fetal calf serum used in the medium for maintenance of virus-infected cells *(268)*. When a serum-free medium was employed, infectious virions were obtained that were devoid of endonuclease activity. When serum-free virions were treated with commercial DNase I and then banded in a CsCl gradient, the resulting virion preparation exhibited this activity, showing that virion capsid proteins strongly adsorb exogenous DNase-I protein.

In an entirely independent study on SV40 virus, a variety of temperature-sensitive mutants also exhibited associated endonuclease activity; but this activity was not more heat-labile than that of the wild type, raising some question as to its origin. In addition, the SV40-associated activity was found to resemble an activity in the fetal calf serum used in the incubation medium (see Section IV, D). Finally, purification of virus particles by a new procedure, which removed the associated endonuclease activity, yielded particles with normal infectivity *(267)*.

B. Human Adenovirus Endonuclease

Burlingham *et al. (373, 374)* first reported the presence of an endodeoxyribonuclease activity associated with purified adenovirus type 2 and 12 virions. The activity was optimal at pH 7, and exhibited a preference for (G+C)-rich regions in duplex, as compared to denatured, DNA. Under certain conditions, the activity could be found in association with pentons from type 2 virions.

In a follow-up study on representative serotypes of all three subgroups of human adenoviruses, using a highly sensitive ethidium bromide fluorimetric assay for endonuclease activity, Marusyk *et al.* *(375)* could detect no activity with purified virions. By contrast, what was regarded as appreciable activity was found associated with virion-derived pentons, excess pool pentons, and/or dodecons.

Apart from the disagreement between the results of these two groups, several facts suggest that the origin of the endonuclease activity in this instance is suspect. High concentrations of virions led to formation of acid-soluble products from duplex DNA *(373)*. Furthermore, calf serum was an integral constituent of the incubation media for virus-infected cells used by *both* groups. Finally, since purified adenovirus DNA is itself infectious, there appears to be no obvious need for an endonuclease activity in intact virions. In all likelihood, the activity in question was also derived by adsorption from the calf serum in the incubation medium.

C. Deoxyribonuclease Activities in Other Viruses

There are additional reports of virion-associated DNase activities in other viruses. Rous sarcoma virus (RSV) is claimed to possess a DNase that attacks T7 DNA *(375, 376)*, which may account for the small size of the DNA products of the virion polymerase activity. However, Quintrell *et al.* *(377)* could not detect such activity in another strain of RSV, but did find activity against ssDNA. Endodeoxyribonuclease activities have been found in virions of avian myeloblastosis virus and of the Rauscher strain of murine leukemia virus *(372)*, but none of these has been adequately characterized.

Exodeoxyribonuclease activity has been reported in virions of RSV, and its resistance to Pronase treatment was regarded as pointing to its localization inside the virion *(378)*. Attempts to detect exonuclease activities in Rauscher murine leukemia virus were completely negative *(379)*.

Two viruses in which the existence of DNase activities appears to be well established are the oncolytic frog-virus-3 *(380)* and the poxviruses, particularly vaccinia, which forms the subject of the following section.

1. Vaccinia-Associated Deoxyribonucleases

Three apparently distinct DNase activites were originally detected in vaccinia virus-infected cells *(381)*. Two of these were localized in the viral core *(382)*, one characterized as an acid exonuclease (pH 5 optimum), the other as an alkaline (pH 7.8 optimum) endonuclease, both specific against ssDNA on the basis of a filter assay.

The acid DNase was subsequently solubilized from virus cores and purified 200-fold to homogeneity *(292)*. Its M_r is 105,000 and it probably consists of two subunits. Detailed specificity studies showed both endo- and exonuclease activities exclusively toward ssDNA, with a pH optimum of 4.5 *(383)*, so that the enzyme closely resembles the S1 nuclease of *Aspergillus oryzae*. It was furthermore shown that, notwithstanding prompt postinfection inhibition of host cell protein synthesis, the DNase can be labeled with [^{3}H]leucine, indicating that it is virus-induced, and in agreement with earlier findings that an enzyme with similar activity is induced late after vaccinia virus infection *(384)*.

This represents at the moment the only DNase highly purified from a virus. There is as yet nothing known about its functional role. By contrast, the function of the neutral DNase, which has not yet been purified, appears to have been at least partially elucidated.

2. Functional Role of Vaccinia Neutral DNase

The cytotoxic effects of some viruses on their host cells, known for some time, are not contingent on replication, since they are equally expressed by UV-inactivated virions. With vaccinia virus, this is reflected in the arrest of host DNA and protein synthesis, but, contrary to the situation with various phages, there is no effect on preformed cell DNA. Loss of such cytotoxicity of vaccinia virions by heat inactivation suggested denaturation of a viral component, believed to be a nuclease activity *(382)*.

This proposal has derived support from two independent investigations. Vaccinia virus infection of L2 cells, leading to inhibition of host DNA and protein synthesis, was accompanied by the appearance in the nucleus of a neutral DNase similar to that found in the virion core (see above). Further experiments suggested that the inhibition of DNA synthesis was due to hydrolysis of *nascent* ssDNA molecules *(385)*. In accord with this is the observation *(386)* that vaccinia infection of HeLa cells was followed within 90 minutes by the onset of *limited* degradation of some host cell DNA to uniform fragments of M_r about 1 to 2×10^7, a process that went to completion about 7–8 hours after infection. This is probably still the only demonstration of inactivation of a specific host cell function by an enzyme activity originating in the invading particle.

It is pertinent to note, in this context, that another poxvirus, Yaba tumor virus, induces a rapid "switch-on" of host DNA synthesis. It had been earlier hypothesized that this, together with the tumor-inducing capacity of Yaba virus, was linked with the absence of DNase activities in the virions. However, this virus does contain both the acid and neutral DNases, the activities being again localized in the cores (387).

D. Viral Ribonucleases

RNase activities have been found in association with a number of purified viruses. But, with the exception of RNase H of oncornaviruses, the viral origin of these activities is, in most instances, not firmly established.

Degradation of viral RNA in disrupted Rous sarcoma virions has been noted (377), but the origin of the activity responsible was not determined. Subsequently Hung (388) partially purified two such activities, one similar to host cell RNase, the other active against poly(A) and poly(A)·poly(U). The latter was present in infected cells and showed some resemblance to *E. coli* RNase III; no such activity was detected in three strains of avian tumor viruses examined. RNase activities have also been reported associated with purified myxoviruses (389, 390); particularly interesting, and meriting further investigation, was the observation that purified virions of fowl plague and Newcastle disease viruses were inactive against double-stranded poliovirus RNA, but such activity was present in the allantoic fluid of the infected chick embryos from which the virions were isolated (390).

RNase activities against ss- and dsRNA are associated with purified oncogenic frog-virus-3 virions (380, 391). Again, it is the activity against dsRNA that is the more interesting. It was optimally active at pH 7.5–9, exhibited a strict requirement for Mg^{2+}, and was not found in extracts of BHK, HeLa and L-929 cells, but was present in virus-infected BHK cell extracts. Under comparable conditions, neither poliovirus nor reovirus exhibited any RNase activity.

It should be noted that activity against ssRNA may indicate the presence of some PDase, and not just RNase. PDase activity has been reported in purified myxo- and paramyxoviruses (392), but these results are hardly convincing, since the substrate employed, bis(p-nitrophenyl)phosphate, is far from specific for PDases (1, 336, 350).

E. Viral RNase H

Among the various nucleases in animal viruses, most interest has centered around this enzyme, which attacks the RNA component of RNA·DNA hybrids. Like the corresponding cellular enzyme (Section II, F), it appears to be a ubiquitous component of oncornaviruses, and must clearly play some role in the generation of free proviral DNA from the RNA·DNA transcript products of the reverse transcriptase of such viruses. It was first found in purified AMV virions, and copurified with the viral RNA-directed DNA polymerase *(393)*.

Association of RNase H activity with the DNA polymerase of avian leukosis viruses has been confirmed by others *(168, 177, 394–396)*. But, although both activities from AMV virions cochromatographed at all purification steps, and exhibited similar ion requirements for activity *(395)*, they may be differentially inactivated by heat, N-ethylmaleimide, NaF, etc. *(172, 393, 397)*, suggesting that they reside on different molecules, or on different subunits of the same molecule. Phosphocellulose chromatography did, in fact, yield two structurally distinct forms of polymerase, one ($\alpha\beta$) with subunits of M_r 65,000 and 105,000, the other (α) of M_r 65,000 *(395, 396, 398)*. Both forms exhibited RNase H activity. The DNA polymerase of Rous sarcoma virus also consists of two subunits with M_r's of 60,000 and 100,000; in this instance, the polymerase and RNase H activities of the smaller component, α, from a temperature-sensitive mutant, were both 5–7 times more thermolabile than the wild type, indicating that the two activities reside on the same polypeptide chain, and that the subunit α is coded for by viral RNA *(398)*.

The $\alpha\beta$ subunit of AMV was subsequently further dissociated withe aid of Me$_2$SO under conditions that maintained polymerase and nuclease activities. The resulting liberated β subunit exhibited only minimal, if any, activity relative to the liberated α subunit *(399)*, notwithstanding that proteolytic cleavage of the β subunit generates the α subunit *(398, 400)*.

RNase H is also associated with other classes of RNA tumor viruses, including murine and feline viruses *(401)*. In the case of Moloney murine sarcoma virus, the RNase H appears to form an integral part of the viral DNA polymerase *(402)*. A report to the effect that Kirsten murine sarcoma virus lacks detectable RNase H activity *(403)* was shown to be due simply to the lower activity in this virus *(404)*.

Detailed studies on the polymerase and RNase H activities of both Rauscher murine leukemia and Kirsten murine sarcoma viruses *(404)* showed that a fraction of the RNase H activity can be separated from the polymerase by anion-exchange chromatography; the remainder copurifies with the reverse transcriptase. The reverse transcriptase activity from purified viral cores was devoid of RNase H activity, so that the two enzymes not only reside in two different molecules, but probably in different compartments of the virions. Hence the RNase H of some murine viruses is not localized in the core structure, in contrast to AMV, where the nuclease is found in association with the polymerase in the purified core fraction *(393, 404)*.

The RNase H of Moloney murine sarcoma-leukemia virus has been separated into two components, one of which had no detectable polymerase activity *(405)*. The different divalent cation requirements (Mg^{2+} vs Mn^{2+}) of the two fractions, and of avian relative to mammalian viral activities, was believed to account for the earlier failure *(403)* to detect RNase H activity in Kirsten murine sarcoma virus (see above).

1. VIRAL RNASE H SPECIFICITY

In contrast to *E. coli* RNase H *(177)* and mammalian cell RNase H (Section II, F), both of which are endonucleases, viral RNase H is an exonuclease, and will not attack ribonucleotides covalently inserted into closed circular duplex DNA. The AMV enzyme has been most extensively investigated *(168, 177)*, and has been characterized as a purine-specific, processive exonuclease, capable of attack in both the $5' \to 3'$ and $3' \to 5'$ directions, and liberating oligonucleotides containing 2–8 residues with 3'-OH and 5'-phosphate termini. Subsequently the specificity of the $\alpha\beta$ form of AMV (see above) was found to be identical to the foregoing, wheras the isolated α subunit proved to be a progressive (or random) exonuclease *(406)*. It is consequently conceivable that the α subunit of α and $\alpha\beta$ are different molecules; a more reasonable interpretation is that the β subunit of $\alpha\beta$ facilitates binding to the substrate distal to the cleavage site, thus favoring a processive mechanism, an argument supported by the identical molecular weights and common antigenic determinants of the α subunits of α and $\alpha\beta$ DNA-polymerases *(396)*. It would have been of obvious utility to examine the thermolability of both activities relative to polymerase activities.

2. MODE OF ACTION AND FUNCTIONAL ROLE

It may appear obvious that the role of viral RNase H is to generate proviral DNA from the RNA · DNA hybrid transcript of the reverse transcriptase, but the mechanism(s) involved are by no means clear.

In *in vitro* studies, the size of the AMV polymerase reaction products was essentially unchanged when the RNase H activity was selectively inhibited and, conversely, inhibition of polymerase activity did not affect the size of the RNase H products *(397)*. It appears, therefore, that the two activities are not coupled mechanistically, at least *in vitro*. A similar lack of influence of associated RNase H activity prevails with regard to the size, structure, and genetic complexity of the DNA product synthesized *in vitro* by the Rous sarcoma virus polymerase *(407)*.

The results of *in vivo* experiments are, at the moment, also somewhat confusing, perhaps because of the use of two different methods, one based on following the fate of input viral RNA *(408)*, the other of the intracellular DNA transcript *(409, 410)*.

When the fate of labeled input Friend leukemia virus RNA was examined, the genomic RNA appeared to form virus-specific covalently linked RNA–DNA,[12] the presumed precursor of the duplex DNA copy of genome RNA. The hybrids were found exclusively in the cell nucleus in two forms, one in which RNA was covalently linked to viral DNA, the other with viral RNA covalently linked to host cell DNA *(408)*. There are similar results from Rous sarcoma virus-infected chicken embryo fibroblast cells *(411)*, in agreement with earlier autoradiographic studies *(412)*. The elimination of RNA from such substances would then proceed by the combined action of viral and cellular RNase H, as originally proposed *(177)*, since endonucleolytic cellular RNase H would be required to remove viral genome RNA fragments integrated in the cellular chromosome.

Quite different results were obtained when observing the formation of proviral DNA. About 6 hours after infection of mouse bone marrow cells with Moloney murine leukemia virus, free proviral duplex DNA was found exclusively in the cytoplasm; sedimentation under alkaline conditions showed that this DNA, of full genome size (M_r 6×10^6), and capable of transfecting cells with one-hit kinetics *(413)*, contained no alkali-labile linkages, i.e., no RNA, and no ssDNA of whole-genome size was detectable *(410)*. Similar results were obtained on infection of duck embryo fibroblasts with avian sarcoma virus *(409)*. Only subsequently did the viral DNA migrate to the nucleus, where it underwent integration. The cytoplasm as the site of viral DNA synthesis is further indicated by the fact that such synthesis proceded at the same rate in enucleated cells, prepared by prior treatment with cytochalasin B *(409)*. Neither of these two studies provided any information on the fate of the RNA–DNA nor the role of viral RNase H in the transition of such hybrids to the observed duplex viral DNA. However, one cannot exclude the possibility that, in those instances where viral RNase H is associated with the reverse transcriptase, the former may exonucleolytically "excise" the RNA template as the DNA transcript is formed, by analogy with *E. coli* polymerase I, the repair functon of which is accompanied by "excision" of damaged regions in DNA (Section XII). It seems hardly coincidental that viral reverse transcriptase is the only DNA polymerase of mammalian ori-

---

[12] The term "hybrid" has come to mean "hydrogen-bonded" with respect to nucleic acids, hence is better avoided when designating covalently linked RNA and DNA [Ed.].

gin known to possess an associated exonuclease activity, albeit confined to ribo compounds.

3. HERPES VIRUS RNASE H

A purine-specific RNase H activity occurs in association with purified virions of herpes simplex virus types 1 and 2 grown on a variety of cell lines *(414)*. It was not established whether the activity is of cellular origin, and it is pertinent to recall that herpes virions are among the most difficult to free of the outer envelope of cell-derived components. It was not established whether the activity is endo- or exonucleolytic. A possible role for this enzyme is suggested by the known involvement of RNA primers in the replication of certain DNA tumor viruses, and the alkaline lability of herpes simplex virus DNA. Herpes viruses are known to be oncogenic in animals, and there is extensive (but not conclusive) evidence for an intimate association between these viruses and malignancies in man (see ref. *415*). On the other hand, since both single- and double-stranded purified herpes DNA are infectious *(416)*, it is not clear what purpose would be served by a virion-associated RNase H activity.

XII. Nucleases and Cellular Repair Processes

As in the case of microorganisms, mammalian cells are capable of dealing with radiation and other types of damage to cellular DNA *(417–421)*. Until very recently, only enzymes involved in these processes in microorganisms had been isolated, purified, and characterized, and their possible functions examined *in vitro* and in some instances established *in vivo*, largely with the aid of mutants with defective enzyme systems. One such repair system, referred to as "excision repair," involves the sequential activity of at least four enzymes: (i) an endonuclease that "recognizes" the damaged site and introduces a single-strand break ("nick") in its vicinity; (ii) an exonuclease that excises the damaged region; (iii) a DNA polymerase that patches in the resulting gap; (iv) a DNA ligase that seals the final strand interruption.

The past 3–4 years have seen some marked progress in the detection, and occasionally the isolation, of corresponding enzymes of mammalian origin. While the lack of appropriate mutants renders more difficult the establishment of the *in vivo* function of a given enzyme in DNA repair (see below for exception), the evidence accumulated with the aid of some purified enzymes is fully consistent with a repair function. We now briefly review these, and some related, repair enzymes. No attempt is made here to cover the subject of how DNA replication copes with unrepaired lesions in the parental DNA (postreplication repair) (417, 418, 420, 421), in which a number of enzymes may be involved, but regarding which little is as yet known.

A. Mammalian "Incision" Endonuclease

In excision repair, it is the incision endonuclease that recognizes the damaged site and initiates the repair process by introducing a single-strand break ("nick") in the vicinity of the lesion. The presence of such endonucleases in mammalian cells, active against damage caused by UV-irradiation, and referred to as UV-endonucleases, has often been inferred indirectly, by noting that repair of such damage occurs. The major lesion due to UV is the pyrimidine dimer. It is, of course, unlikely that the primary function of repair enzymes in mammalian systems is the liquidation of UV radiation damage (but see Section D below). Photodimers furnish a convenient tool for such studies, because of the technical facility for following photodimer formation and disappearance with the aid of labeled pyrimidines.

An endonuclease purified from rat liver to electrophoretic homogeneity, devoid of exonuclease and phosphatase activities, and inert against denatured DNA, produces single-strand breaks in UV-irradiated or acetylaminofluorene-treated DNA. Subsequent treatment with alkaline phosphatase and the exonuclease-associated activity of DNA polymerase I leads to release of acid-soluble oligonucleotides containing thymine photodimers or acetylaminofluorene-base complexes, respectively (422). Excision of oligomers containing the bound carcinogen is in accord with the notion that excision repair is one of the cellular defense mechanisms against mutation and carcinogenesis (423, 424).

There is evidence for the existence in mammalian cells of more than one type of endonuclease activity against UV damage (421). One such enzyme, purified from human lymphoblasts, with $M_r = 35,000$ and no Mg^{2+} requirement, was found to be "nonspecifically" active against UV-irradiated or untreated DNA, but is not the activity specific for UV-irradiated DNA (425). Another, purified from calf thymus, with $M_r = 30,000$, and independent of Mg^{2+} and Ca^{2+}, is inactive against native and denatured DNA, and produces cleavages with 5'-phosphate termini in UV- and γ-irradiated DNA; with UV-irradiated DNA, its site of action is other than that involving photodimers (426). A partially purified endonuclease from the nuclear acidic protein fraction of hamster plasmocytoma, with activity against UV-irradiated or alkylated DNA (427), is probably a mixture of at least two enzymes (see next section).

Undoubtedly the best characterized incision endonuclease is that specific for apurinic sites, and is the subject of the following section.

B. Endonuclease Specific for Apurinic Sites[13]

Considerable confusion centered for some time around the purported existence of incision endonucleases active against damage due to alkylating agents in DNA (see below). This has now been clarified by the purification from calf thymus of an endonuclease with M_r about 32,000, a pH optimum of 8.5, with a strong dependence on Mg^{2+} or Mn^{2+}. It is inactive against RNA, native and denatured DNA, and ssDNA with apurinic sites; it is also inert toward lesions in native DNA produced by UV or alkylating agents. It will specifically cleave the 3'-phosphate bond of an "apurinic" site in native DNA (428, 429), and is the mammalian counterpart of the *E. coli* enzyme (430) subsequently found in rat liver (431) and in plants (432). Unlike the analogous endonuclease II of *E. coli* and *H. influenzae*, which is associated with DNA 3'-phosphatase–exonuclease (EC 3.1.4.27) activities, the mammalian enzyme does not possess associated exonuclease or phosphatase activities (433).

Such endonuclease activity has been demonstrated in all of a number of cell lines tested, with comparable levels in conditions like progeria (associated with aging and accompanying thermolability of cellular enzymes), and Fanconi's anemia (associated with increased incidence of malignancy) (434).

The *in vivo* role of this enzyme as the incision endonuclease of an excision repair system for repair of apurinic sites is supported by the isolation of a repair-deficient strain of *E. coli* with a defective endonuclease (435). *In vitro* incubation of the enzyme with native DNA containing apurinic sites (most conveniently produced by heating the DNA at neutral pH) leads to formation of single-strand breaks, the number of which correspond to the number of apurinic sites; incubation in the presence of DNA polymerase I and phage T4 ligase, together with the necessary nucleoside 5'-triphosphates, leads to a marked reduction in the number of apurinic sites, measured by the reduction in alkali lability of the DNA (436).

---

[13] Preparations of "apurinic DNase" or "*E. coli* DNase VII" cleave at apyrimidinic sites as well, i.e., at the site of any free glycosyl group (Linn and Lindahl, personal communication).

1. Spontaneous Depurination and Its Repair

It appears entirely reasonable to assume that the "apurinic-site endonuclease" forms part of the cellular machinery designed to cope with errors resulting from spontaneous depurination. From the observed rates of depurination under *in vitro* conditions *(437)*, it has been estimated that up to 10^5 depurinations may occur in a mammalian genome at physiological temperature during the course of one cell generation *(438)*. The widespread occurrence of the enzyme, and its mode of action, provide an interpretation for a puzzling observation on "spontaneous" unscheduled DNA synthesis in G1 HeLa cells, detected by means of autoradiographic techniques *(439)*. It appears logical now to ascribe this to repair of spontaneously formed apurinic sites, i.e., to *in vivo* repair.

2. DNA N-Glycosidases and "Apurinic-Site Endonuclease"

The specificity of the "apurinic-site endonuclease" has led to clarification of the properties of other repair enzymes, and the isolation of new ones. For example, endonuclease II of *E. coli* was for some time claimed to be one of a class of enzymes acting on alkylation damage in DNA *(440)*. It was proposed by Verly *et al. (432)* that this enzyme is, in fact, the apurinic-site endonuclease contaminated with some other activity that removes alkylated base residues prior to incision by the endonculease. This was strikingly confirmed by the demonstration that an *E. coli* endonuclease II preparation contained an N-glycosidase activity that removed 3-methyladenine residues from methylmethanesulfonated DNA *(441)*, leaving an apurinic site then recognized by the endonuclease. The same preparation also contained an N-glycosidase that specifically removed, from ss- and dsDNA, uracil residues resulting from deamination of cytosine residues *(442)*. An additional activity in endonuclease II preparation releases 6-methoxyguanine from DNA treated with the carcinogen N-methyl-N-nitrosourea *(443)*.[14]

Several laboratories have reported the existence of DNases that selectively attack uracil-containing DNA; e.g., Tomita and Takahashi *(444)* claimed that *B. subtilis* contains a DNase that actively degrades phage PBS-1 DNA (containing exclusively uracil in place of thymine), leading to the release of free uridine. However, the solvent system employed did not distinguish between uridine and uracil, and Friedberg *et al. (445)* demonstrated that the enzyme in question was actually a "uracil-DNA N-glycosidase," removing uracil from the DNA.

[14] See chapter by Lijinsky in Vol. 17 of this series.

The same activity has been found in mammalian cells *(441)*, but has not yet been characterized.[15]

It follows from the foregoing that damage involving alkylation of base residues or deamination of cytosine residues in DNA is initially recognized by one of several N-glycosidases; only after this has occurred does the excision repair system commence operation with the incision by the apurinic-site endonuclease. One problem that has apparently not been unequivocally resolved is whether the latter enzyme is also active against an apyrimidinic site. It is not fully excluded that two such enzymes exist, the specificity of each being determined by the presence in the complementary strand of a purine or pyrimidine residue (see footnote 13 on p. 116).

C. "Excision" Exonucleases

In prokaryotes, excision of the damaged region is performed by the exonuclease activity associated with the DNA polymerase, which refills the gap, or by exonuclease VII of *E. coli (445a)*. In eukaryotes, the known DNA polymerases, with one recently reported exception *(445b)*, do not possess exonuclease activities, but excision of induced lesions does occur. Mammalian cells must consequently contain free exonucleases capable of fulfilling this function. To date, three potential candidates for the exonuclease function of an excision repair system have been described, viz. DNase IV (Section V, B) and, to a lesser extent, DNase III (Section V, A), and an exonuclease purified from human placenta *(446)*. The latter enzyme, which attacks at both the 3' or 5' termini of single-stranded regions in duplex DNA with the release of 5'-phosphate-terminated oligonucleotides 4–7 residues in length, excises pyrimidine dimers from UV-irradiated DNA previously incised with endonuclease from *M. luteus* specific for UV-irradiated DNA.

It is further worth noting that an appreciable portion of the damage resulting from exposure to ionizing radiation includes single-strand breaks in DNA. In these conditions, the repair process is obviously initiated by an excision exonuclease. Such repair is known to be relatively efficient, but the exonucleases involved have not as yet been adequately identified.

Exonuclease Activity and Fidelity of DNA Replication

Both bacteria and bacteriophages possess exonuclease activity, usually associated with their purified DNA polymerases. Although

[15] Bleomycin removes thymine from DNA; see chapter by Müller and Zahn in this volume.

$5' \to 3'$-exonuclease activity is exhibited by only some prokaryotic polymerases (e.g., *E. coli* polymerase I), $3' \to 5'$-exonuclease has been found in every prokaryotic polymerase tested for this activity *(417)*. In the case of polymerase I and T4 DNA polymerase, the $3' \to 5'$-exonuclease activity fulfills a "proofreading" function during replication by selectively removing mismatched nucleotides from the 3' termini of primer chains *(262, 417, 447)*. By contrast, mammalian polymerases α and β extend primers terminated in mismatched nucleotides *in vitro* so that the latter are internally located in the DNA chains *(262, 448)*. The same holds for the DNA polymerase of avian myeloblastosis virus, and RNA tumor virus *(449)*. Since mutation frequencies in eukaryotes are no higher than in prokaryotes, additional enzymes must be involved *in vivo* to prevent incorporation of mismatched nucleotides during or following DNA replication. The exonucleases referred to in the previous paragraph are at least potential candidates for this function *in vivo*.

D. Repair Deficiency in Xeroderma Pigmentosum

Reference was made above to the difficulties of establishing directly the *in vivo* function of a suspected repair enzyme in mammalian systems, because of the lack of suitable mutants defective in repair. One exception is provided by the fibroblast cells of patients suffering from xeroderma pigmentosum (XP), a disease inherited as an autosomal mutation, the clinical symptoms being hypersensitivity to sunlight, with a high incidence of multiple cutaneous carcinomas; in a more severe type, the De Sanctis–Cacchioni syndrome, neurological disorders accompany tumor formation *(450)*.

Fibroblasts from XP patients, after UV-irradiation, are defective in excision repair, in that they are unable to excise photodimers. This is due to the absence of the incision UV-endonuclease *(451, 452)*, which recognizes photodimer damage. While both clinical forms of the disease, with a few exceptions *(453)*, reflect this defect in excision repair, the two syndromes are probably due to mutations in different genetic complementation groups coding for initiation repair, since binuclear cells that harbor a nucleus from each variant exhibit normal levels of repair of UV damage *(454)*. XP fibroblasts are therefore the mammalian equivalent of *E. coli uvr*$^-$ mutants. Five complementation groups of XP (A–E) are now known, with differing excision repair abilities, and ability to complement each other to remove the repair defect *(455, 456)*.

Phage T4 UV-endonuclease, specific for photodimers in DNA, can be transported into XP cells with the aid of Sendai virus, thus leading to virtually full restoration of repair ability, so that the repair defect in such cells is due to the absence of the incision endonuclease *(457)*. Since all five complementation types of XP cells gave the same result, it follows that the incision UV-endonuclease must be coded for by at least 5 genes.

It has been proposed *(417)* that assay for the excision repair defect may be of clinical value, by testing whether UV-irradiated fibroblasts from the skin of a suspected XP patient fail to show incorporation of 5-bromo[^{3}H]uracil into small patches of DNA. Possible applications along such lines have been reported *(457–460)*.

Acknowledgments

We should like to acknowledge the support of the Polish Academy of Sciences (Project 09.7.1) during the preparation of this review.

References

1. D. Shugar and H. Sierakowska, this series **7**, 369 (1967).
1a. IUPAC-IUB, "Enzyme Nomenclature, Recommendations, Elsevier, Amsterdam; Am. Elsevier, New York, 1972; Suppl. *BBA* **429**, 1 (1976).
2. D. Kowalski and M. Laskowski, Sr., *in* "Handbook of Biochemistry and Molecular Biology," 3rd ed. (G. D. Fasman, ed.), Vol. II: Nucleic Acids, pp. 491–535.
3. J. S. Roth, *in* 'Methods in Cancer Research' (H. Busch, ed.), Vol. III, p. 153. Academic Press, New York, 1967.
4. E. A. Barnard, *ARB* **38**, 677 (1969).
5. F. M. Richards and H. W. Wyckoff, *in* "The Enzymes" (P. D. Boyer, ed.), Vol. IV, p. 647. Academic Press, New York, 1971.
6. J. Bartholeyns, S. Moore and W. H. Stein, *Int. J. Pept. Protein Res.* **6**, 407 (1974).
7. A. J. Birkeland and T. G. Christensen, *J. Carbohydr.* **2**, 83 (1975).
8. J. J. Beintema, A. J. Scheffer, H. van Dijk, G. W. Welling and H. Zwiers, *Nature NB* **241**, 76 (1973).
9. G. W. Welling, G. Leijenaar-Van den Berg, B. Van Dijk, A. Van den Berg, G. Groen, W. Gaastra, M. Emmens and J. J. Beintema, *BioSystems* **6**, 239 (1975).
10. J. J. Beintema, W. Gaastra, A. J. Scheffer and G. W. Welling, *EJB* **63**, 441 (1976).
11. G. J. Ronda, W. Gaastra and J. J. Beintema, *BBA* **429**, 853 (1976).
12. R. Delaney, *Bchem* **2**, 438 (1963).
13. J. Naskalski, *Przegl. Lek.* **29**, 394 (1972).
14. J. Sznajd, *Folia Med. Cracov.* **14**, 297 (1972).
15. M. Zan-Kowalczewska, H. Sierakowska, A. Bardon and D. Shugar, *BBA* **341**, 138 (1974).
16. A. Bardon, H. Sierakowska and D. Shugar, *Clin. Chim. Acta* **67**, 231 (1976).
17. A. Bardon, H. Sierakowska and D. Shugar, *BBA* **438**, 461 (1976).
18. R. G. von Tigerstrom and J. M. Marchak, *BBA* **418**, 184 (1976).
19. M. Schmukler, P. B. Jewett and C. C. Levy, *JBC* **250**, 2206 (1975).
20. K. K. Reddi, *BBRC* **67**, 110 (1975).
21. H. Okazaki, M. E. Ittel, C. Niedergang and P. Mandel, *BBA* **391**, 84 (1975).
22. C. Niedergang, H. Okazaki, M. E. Ittel, D. Munoz, F. Petek and P. Mandel, *BBA* **358**, 91 (1974).

23. M. Elson and D. G. Glitz, *Bchem* **14**, 1471 (1975).
24. R. A. Lewis and W. Gamble, *BJ* **115**, 95 (1969).
25. A. W. Prestayko, B. C. Lewis and H. Busch, *BBA* **319**, 323 (1973).
26. A. Boctor, A. Grossman and W. Szer, *EJB* **44**, 391 (1974).
27. I. Winicov and R. P. Perry, *Bchem* **13**, 2908 (1974).
28. S. B. Zimmerman and G. Sandeen, *Anal. Biochem.* **10**, 444 (1965).
29. W. D. Ball, *BBA* **341**, 305 (1974).
30. J. Bartholeyns and S. Moore, *Science* **186**, 444 (1974).
31. K. Igarashi, H. Kumagai, Y. Watanabe, N. Toyoda and S. Hirose, *BBRC* **67**, 1070 (1975).
32. C. C. Levy, W. E. Mitch and M. Schmukler, *JBC* **248**, 5712 (1973).
33. K. Igarashi, Y. Watanabe and S. Hirose, *BBRC* **67**, 407 (1975).
34. T. Ikemura, *BBA* **195**, 389 (1969).
35. R. Kedracki and W. Szer, *Acta Biochim. Pol.* **14**, 163 (1967).
36. W. Szer, *JMB* **16**, 585 (1966).
37. C. C. Levy, P. A. Hieter and S. M. Le Gendre, *JBC* **249**, 6762 (1974).
38. W. Farkas and P. A. Marks, *JBC* **243**, 6464 (1968).
39. A. M. Crestfield, W. H. Stein and S. Moore, *ABB, Suppl.* **1**, 217 (1962).
40. S. W. Melbye and I. M. Freedberg, *BBA* **384**, 466 (1975).
41. R. Kuciel and W. Ostrowski, *BBA* **402**, 253 (1975).
42. M. Libonati and A. Floridi, *EJB* **8**, 81 (1969).
43. D. F. Goldspink and R. J. Pennington, *BJ* **118**, 9 (1970).
44. E. Z. Rabin and V. Weinberger, *Biochem. Med.* **14**, 1 (1975).
45. G. D'Alessio, M. C. Malorni and A. Parente, *Bchem* **14**, 1116 (1975).
46. R. Stanek, J. Matousek and J. Dostal, *Int. J. Biochem.* **6**, 43 (1975).
47. G. D'Alessio, A. Parente, C. Guida and E. Leone, *FEBS Lett.* **27**, 285 (1972).
48. M. Libonati, S. Sorrentino, R. Galli, R. la Montagna and A. di Donato, *BBA* **407**, 292 (1975).
49. D. Wang, G. Wilson and S. Moore, *Bchem* **15**, 660 (1976).
50. M. Libonati, *BBA* **228**, 440 (1971).
51. G. D'Alessio, J. Doskocil and M. Libonati, *Biochem. J.* **141**, 317 (1974).
52. M. Libonati, M. C. Malorni, A. Parente and G. D'Alessio, *BBA* **402**, 83 (1975).
53. T. Taniguchi and M. Libonati, *BBRC* **58**, 280 (1974).
54. E. Leone, L. Greco, R. K. Rastogi, L. Iela, *J. Reprod. Fert.* **34**, 197 (1973).
55. R. de Prisco and L. del Giudice, *Ital. J. Biochem.* **23**, 222 (1974).
56. V. G. Edy, M. Szekely, T. Loviny and Ch. Dreyer, *EJB* **61**, 563 (1976).
57. C. C. Levy, M. Schmukler, J. J. Frank, T. P. Karpetsky, P. B. Jewett, P. A. Hieter, S. M. LeGendre and R. G. Dorr, *Nature* **256**, 340 (1975).
58. C. C. Levy, *Life Sci.* **17**, 311 (1975).
59. G. Marbaix, G. Huez, A. Burny, Y. Cleuter, E. Hubert, M. Leclercq, H. Chantrenne, H. Soreq, U. Nudel and U. Z. Littauer, *PNAS* **72**, 3065 (1975).
60. C. C. Allende, J. E. Allende and R. A. Firtel, *Cell* **2**, 189 (1974).
61. G. H. Jones, *BBRC* **69**, 469 (1976).
62. J. M. Marshall, *Exp. Cell Res.* **6**, 240 (1954).
63. S. Morikawa, *J. Histochem. Cytochem.* **15**, 662 (1967).
64. R. G. Painter, K. T. Tokuyashu and S. J. Singer, *PNAS* **70**, 1649 (1973).
65. L. J. Greene, C. H. Hirs and G. E. Palade, *JBC* **238**, 2054 (1963).
66. S. Morikawa, M. Yamamura, T. Harada and Y. Hamashima, *J. Histochem. Cytochem.* **16**, 410 (1968).
67. J. Bartholeyns, Ch. Peeters-Joris and P. Baudhuin, *EJB* **60**, 385 (1975).
68. S. J. Davidson, W. L. Hughes and A. Barnwell, *Exp. Cell Res.* **67**, 171 (1971).
69. S. J. Davidson, *J. Cell Biol.* **59**, 213 (1973).

70. P. Rosso, J. Diggs and M. Winick, *PNAS* **70**, 169 (1973).
71. M. R. Robinovitch, L. M. Sreebny and E. A. Smuckler, *JBC* **243**, 3441 (1968).
72. P. L. Wolf, J. P. Horwitz, J. Freisler, J. Vazquez and E. von der Muehll, *Experientia* **24**, 1290 (1968).
73. J. Bartholeyns, C. Peeters-Joris, H. Reychler and P. Baudhuin, *EJB* **57**, 205 (1975).
74. C. Gagnon, G. Lalonde and G. de Lamirande, *BBA* **353**, 323 (1974).
75. J. Gordon, *ABB* **112**, 429 (1965).
76. A. Yannarell and N. N. Aronson, *BBA* **311**, 191 (1973).
77. N. N. Aronson, Jr. and A. Yannarell, *BBA* **413**, 135 (1975).
78. D. Gavard, C. Gagnon and G. de Lamirande, *BBA* **374**, 207 (1974).
79. R. Daoust and H. Amano, *J. Histochem. Cytochem.* **8**, 131 (1960).
80. R. Daoust, *Int. Rev. Cytol.* **18**, 191 (1965).
81. R. Daoust and G. de Lamirande, *Sub-Cell. Biochem.* **4**, 185 (1975).
82. H. Sierakowska and D. Shugar, *Acta Biochim. Pol.* **8**, 427 (1961).
83. R. Daoust and R. Morais, *J. Histochem. Cytochem.* **20**, 350 (1972).
84. R. Daoust, *J. Histochem. Cytochem.* **23**, 51 (1975).
85. R. Daoust and J. G. Durocher, *J. Histochem. Cytochem.* **17**, 350 (1969).
86. D. Maor, N. Vardinon, E. Eylan and P. Alexander, *Experientia* **31**, 634 (1975).
87. D. Maor and I. P. Witz, *Immunology* **20**, 259 (1971).
88. F. Eckstein, *JACS* **92**, 4718 (1970).
89. D. A. Usher, E. S. Erenrich and F. Eckstein, *PNAS* **69**, 115 (1972).
90. R. Kole and H. Sierakowska, *Acta Biochim. Pol.* **18**, 187 (1971).
91. R. Kole, H. Sierakowska and D. Shugar, *BBRC* **44**, 1482 (1971).
92. P. L. Wolf, J. P. Horwitz, J. V. Freisler, J. Vazquez and E. Von der Muehll, *BBA* **159**, 212 (1968).
93. J. Aleksandrowicz, J. Naskalski, J. Sznajd and J. Urbanczyk, *Acta Med. Pol.* **7**, 299 (1966).
94. K. Fink, W. S. Adams and W. A. Skoog, *Am. J. Med.* **50**, 450 (1971).
95. P. B. Chretien, W. Matthews, Jr., and P. L. Twomey, *Cancer* **31**, 175 (1973).
96. J. H. Connolly, R. M. Herriott and S. Gupta, *Brit. J. Exp. Pathol.* **43**, 402 (1962).
97. W. P. Drake, L. P. Kopyta, C. C. Levy and M. R. Mardiney, Jr., *Cancer Res.* **35**, 322 (1975).
98. W. P. Drake, D. R. Pokorney, J. D. Ruckdeschel, C. C. Levy and M. R. Mardiney, *J. Natl. Cancer Inst.* **54**, 1475 (1975).
99. J. Matousek, *Experientia* **29**, 858 (1973).
100. J. Bartholeyns and P. Baudhuin, *PNAS* **73**, 573 (1976).
101. E. A. Barnard, *Nature* **221**, 340 (1969).
102. M. C. Lechner and M. C. Doque-Magalhaes, *Experientia* **29**, 1479 (1973).
103. P. I. Marcus, T. M. Terry and S. Levine, *PNAS* **72**, 182 (1975).
104. H. Maenner and G. Brandner, *Nature* **260**, 637 (1976).
105. A. A. M. Gribnau, J. G. G. Schoenmakers and H. Bloemendal, *ABB* **130**, 48 (1969).
106. T. Moriyama, T. Umeda, S. Nakashima, H. Oura and K. Tsukada, *J. Biochem.* **66**, 151 (1969).
107. H. von Priess and W. Zillig, *Hoppe-Seyler's Z. Physiol. Chem.* **348**, 817 (1967).
108. Y. Takahashi, K. Mase and Y. Suzuki, *J. Neurochem.* **17**, 1433 (1970).
109. B. J. Ortwerth and R. J. Byrnes, *Exp. Eye Res.* **12**, 120 (1971).
110. E. S. Bishay and D. M. Nicholls, *ABB* **158**, 185 (1973).
111. A. A. M. Gribnau, J. G. G. Schoenmakers, M. van Kraaikamp and H. Bloemendal, *BBRC* **38**, 1064 (1970).
112. C. Gagnon and G. de Lamirande, *BBRC* **51**, 580 (1973).
113. D. Gauvreau, C. Gagnon and G. de Lamirande, *Cancer Res.* **34**, 2500 (1974).

114. W. G. M. Van Den Brock, M. A. G. Koopmans and M. Bleemendal, *Mol. Biol. Rep.* **1**, 295 (1974).
115. A. A. M. Gribnau, J. G. G. Schoenmakers, M. van Kraaikamp, M. Hilak and H. Bloemendal, *BBA* **224**, 55 (1970).
116. N. Kraft and K. Shortman, *Aust. J. Biol. Sci.* **23**, 175 (1970).
117. D. K. Liu, G. H. Williams and P. J. Fritz, *Biochem J.* **148**, 67 (1975).
118. N. Kraft, K. Shortman and D. Jamieson, *Radiat. Res.* **39**, 655 (1969).
119. C. Quirin-Stricker, M. Gross and P. Mandel, *BBA* **159**, 75 (1968).
120. B. W. Little and W. L. Meyer, *Science* **170**, 747 (1970).
121. P. V. N. Murthy and J. M. McKenzie, *Endocrinology* **94**, 74 (1974).
122. N. Kraft and K. Shortman, *BBA* **217**, 164 (1970).
123. L. G. Pools, *BBA* **432**, 245 (1976).
124. P. Rosso and M. Winick, *J. Nutr.* **105**, 1104 (1975).
125. E. N. Brewer, L. B. Foster and B. H. Sells, *JBC* **244**, 1389 (1969).
126. R. G. von Tigerstrom, *Can J. Biochem.* **50**, 244 (1972).
127. M. Zan-Kowalczewska and J. S. Roth, *BBRC* **65**, 833 (1975).
128. M. E. Mirault and K. Scherrer, *FEBS Lett.* **20**, 233 (1972).
129. A. Bernardi and G. Bernardi, *BBA* **129**, 23 (1966).
130. M. E. Ittel, C. Niedergang, D. Munoz, F. Petek, H. Okazaki and P. Mandel, *J. Neurochem.* **25**, 171 (1975).
131. M. Futai, S. Miyata and D. Mizuno, *JBC* **244**, 4951 (1969).
132. C. de Duve, B. C. Pressman, R. Gianetto, R. Wattiaux and F. Appelmans, *BJ* **60**, 604 (1955).
133. Y. E. Rahman, J. F. Howe, S. L. Nance and J. F. Thomson, *BBA* **146**, 484 (1967).
134. Y. E. Rahman and E. A. Cerny, *BBA* **178**, 61 (1969).
135. P. Baudhuin, C. Peeters-Joris and J. Bartholeyns, *EJB* **57**, 213 (1975).
136. W. P. Drake, D. R. Pokorney, S. Chipman, C. C. Levy and M. R. Mardiney, Jr., *J. Exp. Med.* **141**, 918 (1975).
137. H. D. Robertson, R. E. Webster and N. D. Zinder, *JBC* **243**, 82 (1968).
138. S. Gotoh, N. Nikolaev, E. Battaner, C. H. Birge, and D. Schlessinger, *BBRC* **59**, 972 (1974).
139. H. D. Robertson and J. J. Dunn, *JBC* **250**, 3050 (1975).
140. W. Jelinek and J. E. Darnell, *PNAS* **69**, 2537 (1972).
141. E. De Clercq, *Top. Curr. Chem.* **52**, 173 (1974).
142. H. D. Robertson and M. B. Matthews, *PNAS* **70**, 225 (1973).
143. A. L. M. Bothwell and S. Altman, *JBC* **250**, 1451 (1975).
144. C. H. Birge and D. Schlessinger, *FP* **33**, 1275 (1974).
145. R. Stern and J. Wilczek, *FP* **32**, 620 (1973).
146. W. Busen and P. Hausen, *EJB* **52**, 179 (1975).
147. J. J. Norlund, S. M. Wolff and H. B. Levy, *PSEBM* **133**, 439 (1970).
148. R. Stern, *BBRC* **41**, 608–613 (1970).
149. P. F. Torrence, J. A. Waters, C. E. Buckley and B. Witkop, *BBRC* **52**, 890 (1973).
150. A. M. Korner, *BBRC* **68**, 699 (1976).
151. S. Dales, *Bacteriol. Rev.* **37**, 103 (1973).
152. C. deDuve, Th. de Barsy, B. Poole, A. Truet, P. Tulkens and F. van Hoof, *Biochem. Pharmacol.* **23**, 2495 (1974).
153. H. D. Robertson, S. Altman and J. D. Smith, *JBC* **247**, 5243 (1972).
154. J. J. Dunn and F. W. Studier, *PNAS* **70**, 1559 (1973).
155. J. J. Dunn and F. W. Studier, *PNAS* **70**, 3296 (1973).
156. N. Nikolaev, L. Silengo and D. Schlessinger, *PNAS* **70**, 3361 (1973).
157. A. L. M. Bothwell and S. Altman, *JBC* **250**, 1460 (1975).

158. W. F. Marzluff, Jr., E. C. Murphy and R. C. C. Huang, *Bchem* **13**, 3689 (1974).
159. E. J. Smillie and R. H. Burdon, *BBA* **213**, 248 (1970).
160. G. S. Chen and M. A. Q. Siddiqui, *PNAS* **70**, 2610 (1973).
161. R. A. Koski, A. L. M. Bothwell and S. Altman, in press.
162. P. Schedl, P. Primakoff and J. Roberts, *Brookhaven Symp. Biol.* **26**, 53 (1974).
163. A. L. M. Bothwell, B. C. Shark and S. Altman, *PNAS* **73**, 1912 (1976).
164. P. Hausen and H. Stein, *EJB* **14**, 278 (1970).
165. R. C. Haberkern and G. L. Cantoni, *Bchem* **12**, 2389 (1973).
166. J. G. Stavrianopoulos and E. Chargaff, *PNAS* **70**, 1959 (1973).
167. W. Roewekamp and C. E. Sekeris, *EJB* **43**, 405 (1974).
168. W. Keller and R. Crouch, *PNAS* **69**, 3360 (1972).
169. M. G. Sarngadharan, J. P. Leis and R. C. Gallo, *JBC* **250**, 365 (1975).
170. S. Natori, K. Takeuchi and D. Mizuno, *J. Biochem. (Tokyo)* **74**, 1177 (1974).
171. G. O'Cuinn, F. J. Persico and A. A. Gottlieb, *BBA* **324**, 78 (1973).
172. D. Doenecke, V. J. Marmaras and C. E. Sekeris, *FEBS Lett.* **22**, 261 (1972).
173. J. G. Stavrianopoulos, A. Gambino-Giuffrida and E. Chargaff, *PNAS* **73**, 1087 (1976).
174. H. D. Robertson and T. Hunter, *JBC* **250**, 418 (1975).
175. B. J. Weimann, J. Schmidt, N. Kluge, W. Ostertag and D. E. Wolfrum, *EJB* **59**, 581 (1975).
176. W. Keller, *PNAS* **69**, 1560 (1972).
177. J. P. Leis, I. Berkower and J. Hurwitz, *PNAS* **70**, 466 (1973).
178. H. M. Lazarus and M. B. Sporn, *PNAS* **57**, 1386 (1967).
179. M. B. Sporn, H. M. Lazarus, J. M. Smith and W. R. Henderson, *Bchem* **8**, 1698 (1969).
180. S. E. Stuart and F. M. Rottman, *BBRC* **55**, 1001 (1973).
181. R. P. Perry and D. E. Kelly, *JMB* **70**, 265 (1972).
182. A. D. Skridonenko, *Biokhimiya* **40**, 13 (1975).
183. H. M. Lazarus and M. B. Sporn, *Bchem* **10**, 505 (1971).
184. M. B. Sporn, D. M. Berkowitz, R. P. Glinski, A. B. Ash and C. L. Stevens, *Science* **164**, 1408 (1969).
185. E. P. Anderson and L. A. Heppel, *BBA* **43**, 79 (1960).
186. H. M. Lazarus, M. B. Sporn and D. F. Bradley, *PNAS* **60**, 1503 (1968).
187. E. A. Speers and R. G. von Tigerstrom, *Can J. Biochem.* **53**, 79 (1975).
188. Y. P. See and P. S. Fitt, *BJ* **130**, 343 (1972).
189. T. Godefroy-Colburn and M. Grunberg-Manago, in 'The Enzymes' (P. D. Boyer, ed.), 3rd ed., Vol. VII, p. 533. Academic Press, New York, 1972.
190. Y. P. See and P. S. Fitt, *BJ* **130**, 355 (1972).
191. M. Laskowski, Sr., in 'The Enzymes' (P. D. Boyer, ed.), 3rd ed., Vol. IV, p. 289. Academic Press, New York, 1971.
192. E. Junowicz and J. H. Spencer, *BBA* **312**, 72 (1973).
193. E. Junowicz and J. H. Spencer, *BBA* **312**, 85 (1973).
194. A. Douvas and P. A. Price, *BBA* **395**, 201 (1975).
195. P. A. Price, *JBC* **250**, 1981 (1975).
196. P. Clark and G. L. Eichhorn, *Bchem* **13**, 5098 (1974).
197. G. Bernardi, S. D. Ehrlich and J. B. Thiery, *Nature NB* **246**, 36 (1973).
198. S. D. Ehrlich, U. Bertazzoni and G. Bernardi, *EJB* **40**, 143 (1973).
199. A. Bernardi, G. Gaillard and G. Bernardi, *EJB* **52**, 451 (1975).
200. J. Hobbs, H. Sternbach, M. Sprinzl and F. Eckstein, *Bchem* **11**, 4336 (1972).
201. B. Janik, M. P. Kotick, T. H. Kreiser, L. F. Reverman, R. G. Sommer and D. P. Wilson, *BBRC* **46**, 1153 (1972).

202. W. E. G. Muller, Z.-I. Yamazaki, J. E. Zollner and R. K. Zahn, *FEBS Lett.* **31**, 217 (1973).
203. W. E. G. Muller, Z-I. Yamazaki, H. J. Breter and R. K. Zahn, *EJB* **31**, 518 (1972).
204. J. M. Gottesfeld, N. H. Adams, A. M. El-Badry, V. Moses and M. Calvin, *BBA* **228**, 365 (1971).
205. P. Baudhuin, E. Hertoghe-Lefevre and Ch. de Duve, *BBRC* **35**, 548 (1969).
206. K. Kasai and M. Grunberg-Managa, *EJB* **1**, 152 (1967).
207. G. A. Cordis, P. J. Goldblatt and M. P. Deutscher, *Bchem* **14**, 2596 (1975).
208. S. E. Hitchcock, L. Carlsson, and U. Lindberg, *Cell* **7**, 531 (1976).
209. M. Durphy, P. N. Manley and E. C. Friedberg, *J. Cell Biol.* **62**, 695 (1974).
210. E. J. Zollner, H. Storger, H. J. Breter and R. K. Zahn, *Z. Naturforsch.* C **30**, 781 (1975).
211. E. J. Zollner, D. H. Kelpsch, R. K. Zahn and R. Knepper, *Enzyme* **19**, 60 (1975).
212. E. J. Zollner, J-D. Beck, E. M. Lemmel, and R. K. Zahn, *Cancer Lett.* **1**, 119 (1975).
213. T-h. Liao, *JBC* **250**, 3721 (1975).
214. E. Lazarides and U. Lindberg, *PNAS* **71**, 4742 (1974).
215. L. Felicetti and C. Urbani, *BBA* **414**, 146 (1975).
216. K. Tempel and R. Hollatz, *Zbl. Vet. Med.* A **22**, 619 (1975).
217. K. Tempel, D. Stavrou and W. Weidenbach, *J. Neurol. Sci.* **26**, 335 (1975).
218. U. Lindberg, *JBC* **241**, 1246 (1966).
219. M. U. Lindberg and L. Skoog, *EJB* **13**, 326 (1970).
220. U. Lindberg and S. Eriksson, *EJB* **18**, 474 (1971).
221. A. Schafer, J. X. de Vries, H. Faulstich and T. Wieland, *FEBS Lett.* **57**, 51 (1975).
222. R. Ishida, H. Akiyoshi and T. Takahashi, *BBRC* **56**, 703 (1974).
223. D. R. Hewish and L. A. Burgoyne, *BBRC* **52**, 475 (1973).
224. Y. Tanigawa, K. Yoshihara and S. S. Koide, *BBRC* **59**, 935 (1974).
225. K. Yoshihara, Y. Tanigawa and S. S. Koide, *BBRC* **59**, 658 (1974).
226. L. A. Burgoyne, M. A. Wagar and M. R. Atkinson, *BBRC* **39**, 254 (1970).
227. L. Burzio and S. S. Koide, *BBRC* **53**, 572 (1973).
228. A. J. Marshall and L. A. Burgoyne, *NARes.* **3**, 1101 (1976).
229. L. A. Burgoyne, D. R. Hewish and J. Mobbs, *BJ* **143**, 67 (1974).
230. D. R. Hewish and L. A. Burgoyne, *BBRC* **52**, 504 (1973).
231. M. Noll, *Nature* **251**, 249 (1974).
232. L. A. Burgoyne and J. Mobbs, *NARes* **2**, 1551 (1975).
233. E. Ohtsuka, Y. Tanigawa and S. S. Koide, *Experientia* **31**, 175 (1975).
234. G. Bernardi, in "The Enzymes" (P. D. Boyer, ed.), 3rd ed., Vol. IV, p. 271. Academic Press, New York, 1971.
235. M. K. Swenson and M. E. Hodes, *JBC* **244**, 1803 (1969).
236. H. Slor, *BBRC* **38**, 1084 (1970).
237. H. Slor and M. E. Hodes, *ABB* **139**, 172 (1970).
238. P. J. Sicard, A. Obrenovitch and G. Aubel-Sadron, *FEBS Lett.* **12**, 41 (1970).
239. J. T. Dulaney and O. Touster, *JBC* **247**, 1424 (1972).
240. M. Yamanaka, Y. Tsubota, M. Anai, K. Ishimatsu, M. Okumura, S. Katsuki and Y. Takagi, *JBC* **249**, 3884 (1974).
241. P. Lesca, *JBC* **251**, 116 (1976).
242. H. Slor and T. Lev, *JBC* **247**, 2926 (1972).
243. R. G. Oshima and P. A. Price, *JBC* **249**, 4435 (1974).
244. P. J. Sicard, A. Obrenovitch and G. Aubel-Sadron, *BBA* **268**, 468 (1972).
245. V. Barthelemy-Clavey, G. Serros and G. Aubel-Sadron, *Mol. Pharmacol.* **11**, 640 (1975).
246. M. J. Pitout, P. G. Kempff and J. C. Schabort, *Int. J. Biochem.* **5**, 241 (1974).

247. M. J. Pitout, P. G. Kempff and J. C. Schabort, *Int. J. Biochem.* **5**, 263 (1974).
248. K. W. Ryder and M. E. Hodes, *J. Chromatogr.* **80**, 128 (1973).
249. J. P. Thiery, S. D. Ehrlich, A. Devillers-Thiery and G. Bernardi, *EJB* **38**, 434 (1973).
250. A. Devillers-Thiery, S. D. Ehrlich and G. Bernardi, *EJB* **38**, 416 (1973).
251. H. Slor and T. Lev, *BJ* **123**, 993 (1971).
252. H. Slor, *BJ* **136**, 83 (1973).
253. H. Slor, H. Bustan and T. Lev, *BBRC* **52**, 556 (1973).
254. T. Berg and D. Boman, *BBA* **321**, 585 (1973).
255. M. J. Pitout, J. J. Vander Watt, P. G. Kempff and J. C. Schabort, *Chem.-Biol. Interact.* **6**, 227 (1973).
256. A. A. Pokrovskii, L. V. Kravchenko, N. A. Tutel'an, *Biokhimija* **36**, 690 (1971).
257. I. R. Lehman, *ARB* **36**, 645 (1967).
258. P. Lesca and C. Paoletti, *PNAS* **64**, 913 (1969).
259. M. Fry and A. Weissbach, *Bchem* **12**, 3602 (1973).
260. J. B. C. Tibbetts and J. Vinograd, *JBC* **248**, 3367, 3380 (1973).
261. A. Weissbach, *Cell* **5**, 101 (1975).
262. H. J. Edenberg and J. A. Huberman, *Annu. Rev. Genet.* **9**, 245 (1975).
263. M. Mechali and A.-M. De Recondo, *EJB* **58**, 461 (1975).
264. E. Baril, O. Brown and J. Laszlo, *BBRC* **43**, 754 (1971).
265. M. Crerar and R. E. Pearlman, *JBC* **249**, 3123 (1974).
266. E. C. Wang, D. Henner and J. J. Furth, *BBRC* **65**, 1177 (1975).
267. D. C. Dooley, M. T. Ryzlak and H. L. Ozer, *J. Virol.* **17**, 352 (1976).
268. J. McMillen, M. S. Center and R. A. Consigli, *J. Virol.* **17**, 127 (1976).
269. J. C. Wang, *JMB* **55**, 523 (1971).
270. J. J. Champoux and R. Dulbecco, *PNAS* **69**, 143 (1972).
271. A. Sen and A. J. Levine, *Nature* **249**, 343 (1974).
272. B. Alberts, L. Frey and H. Delius, *JMB* **68**, 139 (1972).
273. W. A. Baase and J. C. Wang, *Bchem* **13**, 4299 (1974).
274. D. E. Pulleyblank and A. R. Morgan, *Bchem* **14**, 5205 (1975).
275. H.-P. Vosberg, L. I. Grossman and J. Vinograd, *EJB* **55**, 79 (1975).
276. W. Keller, *PNAS* **72**, 2550 (1975).
277. J. E. Germond, B. Hirt, P. Oudet, M. Gros-Bellard and P. Chambon, *PNAS* **72**, 1843 (1975).
278. H.-P. Vosberg and J. Vinograd, *BBRC* **68**, 456 (1976).
279. J. D. Griffith, *Science* **187**, 1202 (1975).
280. A. R. Morgan and D. E. Pulleyblank, *BBRC* **61**, 346 (1974).
281. R. E. Depew and J. C. Wang, *PNAS* **72**, 4275 (1975).
282. W. Keller, *PNAS* **72**, 4876 (1975).
283. W. Keller and I. Wendell, *CSHSQB* **39**, 199 (1975).
284. D. E. Pulleyblank, M. Shure, D. Tang, J. Vinograd and H.-P. Vosberg, *PNAS* **72**, 4280 (1975).
285. T. Lindahl, J. A. Gally and G. M. Edelman, *JBC* **244**, 5014 (1969).
286. J. G. Georgatsos, *BBA* **129**, 204 (1966).
287. P. M. K. Ip and S. G. Sung, *Can. J. Biochem.* **46**, 1121 (1968).
288. T. Lindahl, *EJB* **18**, 407 (1971).
289. T. Lindahl, J. A. Gally and G. M. Edelman, *PNAS* **62**, 597 (1969).
290. T. Lindahl, *EJB* **18**, 415 (1971).
291. T. Tsuruo, *ABB* **171**, 533 (1975).
292. H. Rosemond-Hornbeak and B. Moss, *JBC* **249**, 3292 (1974).
293. P. J. Curtis, M. G. Burdon and R. M. S. Smellie, *BJ* **98**, 813 (1966).
294. P. J. Curtis and R. M. S. Smellie, *BJ* **98**, 818 (1966).

295. L. A. Heppel, *in* "Procedures in Nucleic Acid Research" (G. L. Cantoni and D. R. Davies, eds.), p. 31. Harper, New York, 1966.
296. R. Morais, M. Blackstein and G. de Lamirande, *ABB* **121**, 711 (1967).
297. A. D. Skridonenko and G. A. Gorchakova, *Izv. Akad. Nauk USSR* 936 (1975).
298. P. Baudhuin, J. Bartholeyns and C. Peeters-Joris, *Arch. Int. Physiol. Biochim.* **78**, 985 (1970).
299. J. W. Healy, D. Stollar, M. I. Simon and L. Levine, *ABB* **103**, 461 (1963).
300. D. de Lamirande, R. Morais and M. Blackstein, *ABB* **118**, 347 (1967).
301. L. Leblond-Larouche, C. Dupuis and R. Morais, *EJB* **65**, 423 (1976).
302. R. Morais, *BBA* **189**, 38 (1969).
303. C. Dupuis and R. Morais, *Rev. Can Biol.* **32**, 177 (1973).
304. G. de Lamirande, *Cancer Res.* **27**, 1722 (1967).
305. R. Pilon, C. Dupuis and R. Morais, *Chem. Biol. Interact.* **8**, 371 (1974).
306. R. Morais and G. de Lamirande, *BBA* **209**, 145 (1970).
307. Y. E. Rahman, *BBA* **119**, 470 (1966).
308. Y. E. Rahman, *BBA* **146**, 477 (1967).
309. L. A. Perkins, I. B. Abrass, H. I. Miller and M. G. Rosenfeld, *JBC* **249**, 6999 (1974).
310. H. I. Miller, L. A. Perkins and H. G. Rosenfeld, *Bchem* **14**, 1964 (1975).
311. M. G. Rosenfeld, I. B. Abrass and L. A. Perkins, *BBRC* **49**, 230 (1972).
312. W. E. Razzell, *JBC* **236**, 3031 (1961).
313. M. Futai and D. Mizuno, *JBC* **242**, 5301 (1967).
314. J. G. Georgatsos and S. Ierokleous, *ABB* **121**, 739 (1967).
315. M. Futai, D. Mizuno and T. Sugimura, *JBC* **243**, 6325 (1968).
316. W. E. Razzell, *Can J. Biochem.* **46**, 1 (1968).
317. O. Touster, N. N. Anderson, J. T. Dulaney and H. Hendrickson, *J. Cell Biol.* **47**, 604 (1970).
318. K. Decker and E. Bischoff, *FEBS Lett.* **21**, 95 (1972).
319. W. H. Evans, D. O. Hood and J. W. Gurd, *BJ* **135**, 819 (1973).
320. E. Bischoff, T. A. Tran-Thi and K. F. A. Decker, *EJB* **51**, 353 (1975).
321. S. J. Kelly, D. E. Dardinger and L. G. Butler, *Bchem* **14**, 4983 (1975).
322. W. E. Razzell, *in* "Methods in Enzymology" (S. P. Colowick and N. O. Kaplan, eds.), Vol. VI, p. 236. Academic Press, New York, 1963.
323. M. Erecinska, H. Sierakowska and D. Shugar, *EJB* **11**, 465 (1969).
324. D. Thines-Sempoux, A. Amar-Costesec, H. Beaufay and J. Berthet, *J. Cell Biol.* **43**, 189 (1969).
325. A. I. Lansing, M. L. Belkhode, W. E. Lynch and I. Lieberman, *JBC* **242**, 1772 (1967).
326. H. Sierakowska, H. Szemplinska and D. Shugar, *Acta Biochim. Pol.* **10**, 399 (1963).
327. H. B. Bossmann, A. Hapogian and E. H. Eylar, *ABB* **128**, 51 (1968).
328. G. de Lamirande, S. Boileau and R. Morais, *Can. J. Biochem.* **44**, 273 (1966).
329. A. L. Hubbard and Z. A. Cohn, *J. Cell Biol.* **59**, 152a (1973).
330. W. H. Evans, *Nature* **250**, 391 (1974).
331. E. Bischoff, J. Wilkening, T.-A. Tran-Thi and K. Decker, *EJB* **62**, 279 (1976).
332. W. H. Evans, T. Kremmer and J. G. Culvenor, *BJ* **154**, 589 (1976).
333. G. Holdsworth and R. Coleman, *BBA* **389**, 47 (1975).
334. K. C. Tsou, J. Hendricks, P. D. Gupta, K. W. Lo, *Histochem. J.* **6**, 327 (1974).
335. P. F. Spahr and R. F. Gesteland, *EJB* **12**, 270 (1970).
336. R. Kole, H. Sierakowska and D. Shugar, *BBA* **438**, 540 (1976).
337. J. P. Horwitz, C. V. Easwaran, P. L. Wolf and L. S. Kowalczyk, *BBA* **185**, 143 (1969).
338. D. Rabiger, M. Chang, S. Matsukawa and K. C. Tsou, *J. Heterocycl. Chem.* **7**, 307 (1970).

339. K. C. Tsou, S. Aoyagi and E. E. Miller, *J. Med. Chem.* **13**, 765 (1970).
340. K. W. Lo, W. Ferrar, W. Fineman and K. C. Tsou, *Anal. Biochem.* **47**, 609 (1972).
341. I. Hymie, M. Meuffels and W. J. Poznanski, *Clin. Chem.* **21**, 1383 (1975).
342. K. C. Tsou, S. Ledis and M. G. McCoy, *Cancer Res.* **33**, 2215 (1973).
343. K. C. Tsou, M. G. McCoy, H. T. Enterline, R. Hiberman and H. Wahner, *J. Natl. Cancer Inst.* **51**, 2005 (1973).
344. K. C. Tsou, H. P. Morris, K. W. Lo and J. J. Muscato, *Cancer Res.* **34**, 1295 (1974).
345. E. Bischoff, J. Wilkening and K. Decker, *Hoppe-Seyler's Z. Physiol. Chem.* **354**, 1112 (1973).
346. S. T. Crooke, S. Okada and H. Busch, *Cancer Res.* **32**, 1745 (1972).
347. B. Sela and L. Sachs, *FEBS Lett.* **30**, 100 (1973).
348. T. Terao and T. Ukita, *J. Biochem.* **58**, 153 (1965).
349. R. Brightwell and A. L. Tappel, *ABB* **124**, 325, 333 (1968).
350. A. Bernardi and G. Bernardi, *in* "The Enzymes" (P. D. Boyer, ed.), 3rd ed., Vol. IV, p. 329. Academic Press, New York, 1971.
351. A. Guha, *Brain Res.* **83**, 65 (1975).
352. F. Sanger and G. G. Brownlee, *in* "Methods in Enzymology," Vol. XII, Part A, (L. Grossman and K. Moldave, eds.), p. 361. Academic Press, New York, 1967.
353. A. Holy, *Coll. Czech. Chem. Commun.* **39**, 310 (1974).
354. H. Sierakowska and D. Shugar, *Acta Biochim. Pol.* **18**, 143 (1971).
355. A. F. Cook, *JACS* **92**, 190 (1970).
356. J. P. Hortwitz, V. Easwaran and P. L. Wolf, *BBA* **276**, 206 (1972).
357. G. I. Drummond and M. Yamamoto, *in* "The Enzymes" (P. D. Boyer, ed.), 3rd ed., Vol. IV, p. 355. Academic Press, New York, 1971.
358. T. E. Hugli, M. Bustin and S. Moore, *Brain Res.* **58**, 191 (1973).
359. T. Sudo, M. Kikuno and T. Kurihara, *BBA* **255**, 640 (1972).
360. H. C. Agrawal, N. L. Banik, A. H. Bone, A. N. Davison, R. F. Mitchell and M. Spohn, *BJ* **120**, 635 (1970).
361. T. Kurikara, J. L. Nussbaum and P. Mandel, *Life Sci.* **10**, 421 (1971).
362. J. P. Zanetta, P. Benda, G. Gombos and I. G. Morgan, *J. Neurochem.* **19**, 881 (1972).
363. S. E. Poduslo and W. T. Norton, *J. Neurochem.* **19**, 727 (1972).
364. J. W. de Pierre and M. L. Karnovsky, *J. Cell Biol.* **56**, 275 (1973).
365. J. W. de Pierre and M. L. Karnovsky, *JBC* **249**, 7111 (1974).
366. H. Fleit, M. Conklyn, R. D. Stebbins and R. Silber, *JBC* **250**, 8889 (1975).
367. C. de Duve, *in* "Lysosome in Biology and Pathology" (J. T. Dingle and H. B. Fell, eds.), Vol. 1, p. 3. North-Holland Publ., Amsterdam, 1969.
368. C. Arsenis and O. Touster, *JBC* **243**, 5702 (1968).
369. L. A. Heppel, *Science* **156**, 1451 (1967).
370. J. J. Dunn (ed.), *Brookhaven Symp. Biol.* **26** (1974).
371. B. R. McAuslan, *Life Sci.* **14**, 2085 (1974).
372. H. M. Temin and D. Baltimore, *Adv. Virus Res.* **17**, 129 (1972).
373. B. T. Burlingham, W. Doerfler, U. Pettersson and L. Philipson, *JMB* **60**, 45 (1971).
374. B. T. Burlingham and W. Doerfler, *Virology* **48**, 1 (1972).
375. R. G. Marusyk, A. R. Morgan and G. Wadell, *J. Virol.* **16**, 456 (1975).
376. S. Mizutani, D. Boettiger and H. M. Temin, *Nature* **228**, 424 (1970).
377. N. Quintrell, L. Fanshier, B. Evans, W. Levinson and J. M. Bishop, *J. Virol.* **8**, 17 (1971).
378. S. Mizutani, H. M. Temin, M. Kodama and R. D. Wells, *Nature NB* **230**, 232 (1971).
379. J. Hurwitz and J. P. Leis, *J. Virol.* **9**, 116 (1972).
380. H. S. Kang and B. R. McAuslan, *J. Virol.* **10**, 202 (1972).
381. A. M. Aubertin and B. R. McAuslan, *J. Virol.* **9**, 554 (1972).

382. B. G. T. Pogo and S. Dales, *PNAS* **63**, 820 (1969).
383. H. Rosemond-Hornbeak, E. Paoletti and B. Moss, *JBC* **249**, 3287 (1974).
384. B. R. McAuslan and J. R. Kates, *Virology* **33**, 709 (1967).
385. B. G. T. Pogo and S. Dales, *PNAS* **70**, 1726 (1973).
386. J. R. Parkhurst, A. R. Peterson and C. Heidelberger, *PNAS* **70**, 3200 (1973).
387. J. Schwartz and S. Dales, *Virology* **45**, 797 (1971).
388. P. P. Hung, *Virology* **51**, 287 (1973).
389. M. Rosenbergova, E. Matisova and S. Pristasova, *Acta Virol.* **15**, 515 (1971).
390. K. J. Wiegers and R. Drzeniek, *Z. Naturforsch. C* **28**, 346 (1973).
391. P. Palese and G. Koch, *PNAS* **69**, 698 (1972).
392. S. Pristasova and M. Rosenbergova, *Acta Virol.* **18**, 293 (1974).
393. K. Moelling, D. P. Bolognesi, H. Bauer, W. Busen, H. W. Plassmann and P. Hausen, *Nature NB* **234**, 240 (1971).
394. D. Baltimore and D. Smoler, *JBC* **247**, 7282 (1972).
395. K. F. Watson, K. Moelling and H. Bauer, *BBRC* **51**, 232 (1973).
396. D. P. Grandgenett, G. F. Gerard and M. Green, *PNAS* **70**, 230 (1973).
397. L. C. Brewer and R. D. Wells, *J. Virol.* **14**, 1494 (1974).
398. I. M. Verma, *J. Virol.* **15**, 121 (1975).
399. D. P. Grandgenett, *J. Virol.* **17**, 950 (1976).
400. K. Moelling, *Virology* **62**, 46 (1974).
401. B. J. Weimann, J. Schmidt and D. I. Wolfrum, *FEBS Lett.* **43**, 37 (1974).
402. D. P. Grandgenett, G. F. Gerard and M. Green, *J. Virol.* **10**, 1136 (1972).
403. L. Wang and P. H. Duesberg, *J. Virol.* **12**, 1512 (1973).
404. A. M. Wu, M. G. Sarngadharan and R. C. Gallo, *PNAS* **71**, 1871 (1974).
405. G. F. Gerard and D. P. Grandgenett, *J. Virol.* **15**, 785 (1975).
406. D. P. Grandgenett and M. Green, *JBC* **249**, 5148 (1974).
407. M. S. Collett and A. J. Faras, *J. Virol.* **17**, 291 (1976).
408. M. M. Sveda, B. N. Fields and R. Soeiro, *J. Virol.* **18**, 85 (1976).
409. H. E. Varmus, R. V. Guntaka, W. J. W. Fan, S. Heasly, and J. M. Bishop, *PNAS* **71**, 3874 (1974).
410. A. M. Gianni, D. Smotkin and R. A. Weinberg, *PNAS* **72**, 447 (1975).
411. J. Leis, A. Schincariol, R. Ishizaki and J. Hurwitz, *J. Virol.* **15**, 484 (1975).
412. S. Dales and H. Hanafusa, *Virology* **50**, 440 (1972).
413. D. Smotkin, A. M. Gianni, S. Rozenblatt and R. A. Weinberg, *PNAS* **72**, 4910 (1975).
414. C. Kleinicke, H. Fischer and R. M. Flugel, *BBRC* **60**, 1491 (1974).
415. D. Shugar, *FEBS Lett.* **40**, Suppl., S 48 (1974).
416. P. Sheldrick, M. Laithier, D. Lando and M. L. Ryhiner, *PNAS* **70**, 3621 (1973).
417. A. Kornberg, "DNA Synthesis." Freeman, San Francisco, 1974.
418. A. R. Lehmann, *Life Sci.* **15**, 2005 (1974).
419. P. A. Cerutti, *Life Sci.* **15**, 1567 (1974).
420. B. S. Strauss, *Life Sci.* **15**, 1685 (1974).
421. L. Grossman, A. Braun, R. Feldberg and I. Mahler, *ARB* **44**, 19 (1975).
422. J. L. Van Lancker and T. Tomura, *BBA* **353**, 99 (1974).
423. R. W. Hart and R. B. Setlow, *in* "Molecular Mechanism for Repair of DNA" (P. C. Hanawalt and R. B. Setlow, eds.), Part 5b, p. 719. Plenum, New York, 1975.
424. P. D. Lawley and W. Warren, *Chem.-Biol. Interact.* **12**, 211 (1976).
425. T. P. Brent, *BBA* **407**, 191 (1975).
426. S. Bacchetti and R. Benne, *BBA* **390**, 285 (1975).
427. A. Caputo, G. Zupi and A. Cianciulli, *NARes* **2**, 905 (1975).
428. S. Ljungquist and T. Lindahl, *JBC* **249**, 1530 (1974).

429. S. Ljungquist, A. Andersson and T. Lindahl, *JBC* **249**, 1536 (1974).
430. W. G. Verly and Y. Paquette, *Can. J. Biochem.* **50**, 217 (1972).
431. W. G. Verly and Y. Paquette, *Can. J. Biochem.* **51**, 1003 (1973).
432. W. G. Verly, Y. Paquette and L. Thibodeau, *Nature NB* **244**, 67 (1973).
433. S. Ljungquist, B. Nyberg and T. Lindahl, *FEBS Lett.* **57**, 169 (1975).
434. G. W. Teebor and N. J. Duker, *Nature* **258**, 544 (1975).
435. D. M. Yajko and B. Weiss, *PNAS* **72**, 688 (1975).
436. W. G. Verly, F. Gossard and P. Crine, *PNAS* **71**, 2273 (1974).
437. S. Greer and S. Zamenhof, *JMB* **4**, 123 (1962).
438. T. Lindahl, in "Molecular and Cellular Repair Processes" (R. F. Beers, Jr., R. M. Herriot and R. C. Tilghman, eds.), p. 3. Johns Hopkins University, Baltimore, Maryland, 1972.
439. B. Djordjevic, R. G. Evans, A. G. Perez and M. K. Weill, *Nature* **224**, 803 (1969).
440. S. M. Hadi, D. Kirtikar and D. A. Goldthwait, *Bchem* **12**, 2747 (1973).
441. T. Lindahl, *Nature* **259**, 64 (1976).
442. T. Lindahl, *PNAS* **71**, 3649 (1974).
443. D. M. Kirtikar and D. A. Goldthwait, *PNAS* **71**, 2022 (1974).
444. F. Tomita and I. Takahashi, *J. Virol.* **15**, 1073 (1975).
445. E. C. Friedberg, A. K. Ganesan and K. Minton, *J. Virol.* **16**, 315 (1975).
445a. J. W. Chase and C. C. Richardson, *JBC* **249**, 4553 (1974).
445b. J. J. Byrnes, K. M. Downey, V. L. Black and A. G. So, *Bchem* **15**, 2817 (1976).
446. J. Doniger and L. Grossman, *JBC* **251**, 4579 (1976).
447. T. Kornberg and A. Kornberg, in "The Enzymes" (P. D. Boyer, ed.), Vol. X, p. 119. Academic Press, New York, 1974.
448. L. M. S. Chang, *JBC* **248**, 6983 (1973).
449. N. Battula and L. A. Loeb, *JBC* **251**, 982 (1976).
450. J. E. Cleaver and D. Bootsma, *Annu. Rev. Genet.* **9**, 19 (1975).
451. R. B. Setlow, J. D. Regan, J. German and W. L. Carrier, *PNAS* **64**, 1035 (1969).
452. J. E. Cleaver, *PNAS* **63**, 428 (1969).
453. M. C. Paterson, P. H. M. Lohman and M. L. Sluyter, *Mutat. Res.* **19**, 245 (1973).
454. M. C. Paterson, P. H. M. Lohman, A. Westerveld and M. L. Sluyter, *Nature* **248**, 50 (1974).
455. K. H. Kraemer, H. G. Coon, R. A. Petinga, S. F. Barrett, A. E. Rahe and J. H. Robbins, *PNAS* **72**, 59 (1975).
456. K. H. Kraemer, E. A. De Weerd-Kastelein, J. H. Robbins, W. Keijzer, S. F. Barrett, R. A. Petinga and D. Bootsma, *Mutat. Res.* **33**, 327 (1975).
457. K. Tanaka, M. Sekiguchi and Y. Okada, *PNAS* **72**, 4071 (1975).
458. C. A. Ramsay, *Nature* (Conf. Rep.) **252**, 742 (1974).
459. J. E. Cleaver, *Adv. Radiat. Biol.* **4**, 1 (1974).
460. M. Stefanini, L. Dalpra, G. Zei, R. Giorgi, A. Falaschi and F. Nuzzo, *Mutat. Res.* **34**, 313 (1976).

Transfer RNA in RNA Tumor Viruses[1]

LARRY C. WATERS AND
BETH C. MULLIN[2]

Carcinogenesis Program
Biology Division
Oak Ridge National Laboratory
and the
University of Tennessee—
Oak Ridge Graduate School
of Biomedical Sciences
Oak Ridge, Tennessee

I. Introduction	131
II. Nucleic Acid Components of RNA Tumor Viruses	134
A. 35 S and 70 S RNA	134
B. 4 S to 28 S RNA	136
III. Transfer RNA in the Free 4 S RNA of the Virus	138
IV. Transfer RNA Associated with the 70 S RNA of the Virus	141
V. Function of Transfer RNA in RNA Tumor Viruses	146
VI. Summary	155
References	155
Addendum	160

I. Introduction

Even before extensive physical and chemical analysis of RNA tumor virus RNAs had begun, there was evidence suggesting that these viruses contained tRNA. From initial studies, it was concluded that most, if not all, of the viral tRNA is of host-cell origin, but that inclusion of the cellular tRNA within the virion is a selective, rather than a random, process. Recent studies have defined a function for particular viral tRNAs as primers,[3] or initiators, of viral DNA synthesis. In this review, we discuss what is currently known about the tRNA content and specific tRNA composition within RNA tumor viruses and

---

[1] The studies conducted in this laboratory and reported in this review were supported by the Virus Cancer Program of the National Cancer Institute and by the Energy Research and Development Administration under contract with the Union Carbide Corporation. B. C. M. was supported by National Institutes of Health Postdoctoral Training Grant No. CA05296.

[2] Present address: Biology Department, Wilmington College, Wilmington, Ohio.

[3] In this review, "primer" refers to the RNA molecule to which the initial deoxynucleotide of a nascent DNA chain is covalently attached, and "template" refers to the RNA from which the DNA is transcribed.

present views relating to probable and possible functional roles for the viral tRNAs.

For more than 10 years, the notion that tRNA has a regulatory function in cell growth and differentiation has been a major impetus for research in many laboratories, including our own. Early theories of how tRNA might function to regulate cellular protein synthesis both quantitatively and qualitatively include the "modulation" (1) and the "adapter modification" hypotheses (2, 3). In both cases, the availability of a tRNA to translate efficiently each codon within an mRNA is the critical issue. Anderson et al. clearly demonstrated the feasibility of these theories using systems reconstructed in vitro (4, 5). To elaborate upon the studies that support the possibility that cell growth and differentiation might be controlled at the translational level by tRNA would be a mammoth undertaking and is not the intent of this review. Suffice it to state that the results consist of demonstrating quantitative and/or qualitative differences in tRNAs in systems ranging from phage-infected bacteria, to specialized cells producing specific proteins, to normal versus malignant mammalian cells and tissues. The problem has become not whether the observed tRNA changes are real or artifactual, but rather whether the changes in tRNAs are responsible for the biochemical or physiological alteration being investigated, or are a result of the alteration.

Transfer RNA has been implicated in the transcriptional regulation of the operons coding for the biosynthesis of the enzymes involved in amino-acid biosynthesis in bacteria (6, 7). Mutants that contain tRNA^{His} lacking two uridine-to-pseudouridine modifications are non-repressible (8). This is strong evidence for tRNA involvement in repression of the histidine biosynthetic pathway. Furthermore, it implicates the involvement of elements of the tRNA structure that do not appear to be required for its translational function. Specific binding of leucyl-tRNA to the first enzyme in the leucine–isoleucine–valine biosynthetic pathway has been shown (9) and might represent the active repressor form of the tRNA. Results suggesting that a similar mechanism of repression may exist in mammalian cells were recently presented (10). Transfer RNA has also been implicated in the regulation of branched-chain amino-acid transport in Escherichia coli (11). Jacobson reported that a specific tRNA can interact with tryptophan pyrrolase in extracts of a vermilion mutant of Drosophila melanogaster and render the enzyme inactive (12). Although this observation has been challenged (13), it represents an intriguing mechanism by which tRNA could regulate enzyme activity.

Although the evidence for tRNA involvement in the regulatory roles indicated above is not conclusive, other functions not directly related to protein synthesis are well documented. These include involvement in aminoacyl phosphatidylglycerol synthesis *(14, 15)*, terminal addition of amino acids to proteins *(16)*, and microbial cell-wall biosynthesis *(17)*.

The tRNA functions already indicated would seem to depend on the intracellular state of the tRNA. The T bacteriophage provide an example of how cellular tRNA pools might be altered both quantitatively and qualitatively by external sources. The DNA of these phages contains genes for certain tRNAs that are structurally different and therefore distinguishable from those in the host *E. coli* cells *(18)*. Bacteriophage lacking these genes are viable *(19)*, so that a functional role for these tRNAs is not obvious. Wilson *(20)* considers that the bacteriophage-specified tRNAs most probably function to ensure optimum rates of viral protein synthesis by supplementing those tRNAs that might be at rate-limiting levels in certain hosts. Certainly the presence of tRNA genes in T phages has enabled us to learn much about tRNA synthesis and processing *(18)*. Likewise, the ability to transduce tRNA genes with $\phi 80$ bacteriophage has increased our knowledge of tRNA structure and function *(18)*.

There is yet another way in which tRNA is associated with viruses. Several plant viruses contain tRNA-like structures as an integral part of the viral RNA *(21)*. Plant virus RNAs specifically acylatable with valine *(22)*, tyrosine *(23)* or histidine *(24)* have been demonstrated. Mengovirus RNA can be acylated with histidine *(25)*. Prochiantz *et al.* *(26)* presented evidence suggesting that the RNA bacteriophages Qβ, MS2 and R17 also contain tRNA-like structures within their genomes. Demonstration that amino-acylated turnip yellow mosaic virus and tobacco mosaic virus RNAs interact *in vitro* with plant elongation-factor-I suggests that the tRNA-like structures in the plant virus RNAs might function in their translation into protein *(27)*. Since the involvement of bacterial elongation-factor-Tu in Qβ replication has been implicated, it might be that the tRNA-like structures in these RNA virus genomes are required for viral RNA replication *(27, 28)*. Recent information regarding structure and reactivity of these tRNA-like structures has been summarized by Haenni *et al.* *(28a)*.

Another functional role for tRNA other than in protein synthesis is that of primer for DNA synthesis. This recently discovered function is thus far confined to the RNA-dependent DNA synthesis in RNA tumor viruses. In succeeding sections (III–V) we discuss what is currently known about tRNA in RNA tumor viruses with emphasis on primer function.

II. Nucleic Acid Components of RNA Tumor Viruses

A. 35 S and 70 S RNA

Not all the viruses mentioned in this review have been shown to be tumorigenic, but, because they contain the basic features of RNA tumor viruses and, as discussed in a recent review *(29)*, are potentially tumorigenic, they are referred to, for convenience, as RNA tumor viruses. Two distinguishing features of these viruses are the presence of a high-M_r RNA ($>5 \times 10^6$), or 70 S RNA[4] *(30–33)*, and an enzyme capable of the reverse transcription of RNA into DNA *(34, 35)*. Conditions that disrupt hydrogen bonds dissociate the 70 S RNA into its basic subunit structure, which has a sedimentation value of about 35 S *(36–38)*. Discrete low-M_r RNA components (4–9 S) are also released from the 70 S RNA complex by this treatment (see Sections III and IV). These RNA components are referred to as 70 S-associated.[4]

Initial estimates of the M_r of the 70 S and 35 S RNA components based on velocity sedimentation analysis were 10 to 12×10^6 and 3×10^6, respectively, suggesting a 70 S RNA complex composed of 3–4 35 S RNA molecules. Recent analyses of Moloney murine leukemia virus (Mo-MuLV) and Rous sarcoma virus (RSV) 70 S and 35 S RNAs by sedimentation and gel electrophoresis techniques *(39)*, and of Mo-MuLV RNAs by sedimentation equilibrium analysis *(40)* showed M_r's nearer 7×10^6 for the 70 S RNA complex. Both analyses suggested a two-subunit structure, which is also favored on the basis of M_r's determined by electron microscopy *(41–43)*. The study of RD-114 viral RNA is particularly convincing in this regard *(43)*.

The 70 S RNA contains the viral genome but the data conflict as to whether the 35 S viral RNA subunits are identical or different. Estimates of genetic complexity based on kinetics of hybridization of RNA to complementary DNA suggested that the genomes of RSV *(44)* and Mo-MuLV *(45)* are haploid. However, Baluda *et al.* *(46)*, using the same method, obtained results indicating polyploidy of the avian myeloblastosis virus (AMV) RNA genome. Conclusions from the analysis of molar yields of purified oligonucleotides obtained by T1 ribonuclease digestion of viral RNAs are very consistent, and all indicate polyploidy *(47–49)*. A recent and exciting development is the use of oligonucleotide mapping techniques to map the viral RNA genes *(50, 51)*. Additional supporting evidence for identical RNA subunits has been presented *(43, 52, 53)*. The simplest interpretation of the

---

[4] For convenience, we refer to the largest RNA complexes (50–70 S) isolated from any of the tumor viruses discussed as "70 S RNA."

present evidence is that the 70 S RNA complex contains two identical subunits of M_r about 3×10^6. As discussed by Baltimore (54), this is probably sufficient to code for the known viral specific proteins.

The 35 S RNA subunits of many tumor viruses contain polyadenylate sequences (55–57), located at the 3' end of the molecules (50, 52, 58). From inhibition studies using cordycepin, Wu et al. (59) proposed that polyadenylylation is a posttranscriptional modification of 35 S RNA that is required for virus production.[5] The recent demonstration of a polyadenylylating enzyme copurifying with type-A virus particles and utilizing an endogenous 35 S RNA primer is suggestive evidence that polyadenylylation of viral RNA may occur in the cytoplasm (60). The presence of a "cap" structure, $m^7GpppGm-Cp$ at the 5' end of B77 avian sarcoma virus (ASV) 35 S RNA has been reported (61).[6] This feature together with poly (A) at the 3' end gives viral 35 S RNA the appearance of an extra-large eukaryotic mRNA. The absence of a negative viral RNA strand in virus-producing cells (62), the presence of 35 S and related smaller RNAs on polyribosomes (63), and the fact that viral 35 S RNA can be effectively translated in vitro (64), all suggest that viral RNA, in a form similar or identical to that found in the virion, is the message for virus-specific proteins. Levin et al. (65) have interpreted their observations on 70 S RNA-deficient MuLV particles as indicating that the viral RNA utilized as mRNA and the RNA included in the virion are derived from separate intracellular pools. Whether there are subtle but important structural and functional differences between the 35 S RNA translated intracellularly and the RNA encapsulated within the virion remains to be determined.

Since the 70 S RNA complex has not been demonstrated in cells producing virus, and since its dissociation into subunits does not involve covalent bond breakage, it would be interesting to know when, how, and why the complex is formed. The RNA in 3-minute-old RSV appears to be predominantly 35 S, and it can be converted to 70 S by in vitro incubation of the immature virus (66). The evidence suggests that conversion involves association of 35 S RNA with RNAs of the 4–12 S variety. RNA with sedimentation properties intermediate between 35 and 70 S has been observed in other young or immature virions (67, 68).

An alternative explanation for these observations was given in a

---

[5] See articles by Edmonds et al. and others in Vol. 19 of this series; also by Brawerman in Vol. 17.

[6] See articles by Furuichi et al., Rottman et al., Busch et al., and Moss et al. in Vol. 19 of this series.

recent study of B77 ASV RNA *(69)*, which showed that with extraction conditions designed to minimize dissociation of hydrogen bonds, predominantly 70 S RNA was isolated from immature (up to 5 minutes) virions. Under more stringent conditions (0.1 M NaCl instead of 0.5 M), 70 S RNA was readily obtained from mature (3–5 hour) virions, yet a significant amount of 35 S RNA was obtained from immature virions. These results suggest that a relatively unstable form of the 70 S RNA complex exists even in immature virions, and that it is stabilized as the virus matures. Stabilization of the 70 S RNA complex was demonstrated by the *in vitro* incubation of intact virions, but could not be demonstrated using disrupted virions. Again involvement of low-M_r RNA components as linkers was suggested. Stoltzfus and Snyder *(69)* propose that the reason 70 S, but not 35 S, RNA has consistently been demonstrated in immature murine viruses is that the murine virus 70 S complex is inherently more stable than that from avian viruses. This is consistent with our own observation that the AMV 70 S complex is more completely dissociated at 60°C in 0.1 M NaCl than is the complex from AKR-MuLV (60–75% versus 30–40%). Although these studies *(66, 69)* disagree as to when the 35 S RNA subunits become linked, they both show that viral RNA is undergoing physical change within the maturing virion.

B. 4 S to 28 S RNA

In addition to the 70 S RNA complex, RNA extracted from RNA tumor viruses consistently contains 4–28 S components.[7] Discrete 4 S *(37, 70, 71)*, 5 S *(72–74)*, 7–9 S *(75–78)* and 18 S and 28 S *(70, 76)* species have been identified. In all probability, these RNA components are of cellular origin *(37, 71, 76, 79, 80)*. However, a cellular origin of a viral component does not rule out a functional role for that component in the life cycle of the virion (see Section V).

Estimates of the amounts of the various 4–28 S RNA components associated with the virion have varied considerably and this has raised doubts as to their biological significance. Two main factors have contributed to these variable findings. First, most experiments to characterize viral RNA components of necessity utilize radioisotopic labeling techniques. For this to be quantitative, all components must be allowed to reach an equal degree of labeling prior to sampling. As an example, from confluent mouse cell cultures the virus-contained 35 S RNA becomes labeled with a half-life of 2–4 hours, whereas the viral free 4 S RNA and the 4 S RNA in the cells become labeled with a

[7] These RNA components are referred to as "free" RNAs to distinguish them from "70 S-associated" RNAs.

half-life of more than 70 hours (J. N. Ihle, W.-K. Yang and L. C. Waters, unpublished data). This factor has previously been considered (76) as a possible explanation for the reported absence of 4 S RNA in Rauscher-MuLV (81). A second factor is that the RNA components of lower M_r are subject to contamination with degradation products derived from the 70 S RNA complex. This is especially true in cases in which the virus is derived from a cell culture and has had prolonged contact with the culture medium. For these reasons, accurate quantitative estimates of viral RNA components should be made using plasma-derived viruses or short-term harvests from cell cultures. Also, the cells should have an equal degree of labeling in all components if viral components are to be quantitated on the basis of labeling.

With the exception of certain components of the viral 4 S RNA (see Section V) no function has been defined for the 4–28 S RNA components of RNA tumor viruses in their life cycle. The 5 S RNA in RSV is identical to that found in uninfected chick cells (72). It is found both free and associated with the 70 S RNA in avian and murine viruses (72–74) and will reassociate with avian viral 35 S *in vitro* (82). Thermal melting curves show that, in RSV, 5 S RNA dissociation and the conversion of 70 S to 35 S RNA occur concomitantly (72), suggesting a possible linker function for 5 S RNA in the 70 S RNA complex.

The 7–9 S viral RNA components (75–78) are also found in cells (80, 83, 84). Within viruses they are found both free and in association with the 70 S RNA complex (73, 74, 78), and they too have been proposed as linkers (74). We are unaware of any data to support such a function.

The presence of 18 and 28 S RNA in virus preparations does not appear to be a result of external contamination (70, 76). In our own experience in isolating spectrophotometrically measurable quantities of RNA from as many as 10 different viruses, including some from plasma, we always observe 18 and 28 S RNA. Quantitative estimates of these RNAs suggest that there could be on the average no more than one ribosome per virion (72, 76). Ribosomelike structures have been observed in AMV (85).

In addition to the several RNA components there are small amounts of DNA (86–88) contained in these viruses. The DNA found in RSV is homologous to avian cell DNA and not to viral RNA and there is evidence that the viral DNA is not involved in the life cycle of the virus (89). Bromodeoxyuridine incorporation into DNA can be very convincing when an effect is shown, but the absence of an effect should not be, and probably was not considered to be, conclusive. The DNA in the virion is a very efficient template for reverse transcriptase in viral lysates *(89, 90)*. The DNA has recently been localized in the core component of AMV *(91)*. It would be premature at this time to exclude a functional role for this DNA in virus–cell interactions.

III. Transfer RNA in the Free 4 S RNA of the Virus

After the 70 S RNA complex, the next most abundant RNA component has a sedimentation value of 4 S and is termed viral "free" 4 S RNA. Murine viruses with intact 35 S RNA molecules, indicative of minimal degradation, contain about 60% as much free 4 S RNA as 70 S RNA on the basis of absorbance at 260 nm. A similar distribution occurs in the AMV from plasma as supplied by Life Sciences Research Laboratories. Generally, there is a much larger proportion of 4 S RNA in preparations in which the 35 S RNA is extensively degraded.

An early concern about the viral free 4 S RNA was whether it is indeed an integral component or merely represents cellular material adhering to the outside of the virion. Experiments to exclude the latter possibility included ribonuclease or phosphodiesterase treatment of the virus, mixing radioactively labeled cellular components with unlabeled virus as a measure of adventitious binding, and varying the techniques for virus purification *(70, 71, 92, 93)*. From such studies, it was concluded that most if not all the free 4 S RNA is contained within the virion. In fact, it has been reported that most of the viral 4 S RNA is in the viral core *(94)*. In contrast, other studies indicate that nearly all the 4 S RNA is outside the core *(77, 95)*. This discrepancy could be due to differences in the core preparations *(94)*, but it is clear that the free 4 S RNA can be separated from a subviral component containing the 70 S RNA *(77, 95)*.

Beaudreau *et al.* *(96)* were first to report that viral RNA can be amino-acylated. Since then, a number of studies have shown that tRNA is the major component of viral free 4 S RNA *(92, 93, 97–101)*. It was proposed rather early that the tRNA is cellular in origin and is included in the virus during budding from the cell *(70)*. Evidence for this includes methyl labeling to the same extent and at the same rate as cellular tRNA *(37, 71)*, similar nucleotide composition *(37, 71, 101)*, and the capacity to hybridize to cellular DNA and to be competed for by cellular tRNA *(79)*.

Although the viral free 4 S RNA is derived from the cell, its tRNA distribution does not reflect that of total cellular tRNA *(92, 93, 97–100, 102)*. A tabulation of the differences we have found is shown in Table I. In the free 4 S RNA from AMV, tRNA changes relative to the amounts

TABLE I

COMPARISON OF THE tRNA COMPOSITION OF VIRAL FREE 4 S RNAs WITH THAT OF THE HOST CELL[a]

Source of RNA	Total radio-activity[b] (cpm)	Percentage of total radioactivity identified as:															
		Trp	Lys	His	Arg	Asp	Thr	Ser or Ser and Glu	Pro	Gly	Ala	Val	Met	Ile	Leu	Tyr	Phe
Chicken myeloblast 4 S RNA	33,690	1.4	11.8	4.0	18.9	4.1	13.0	1.4	5.7	2.4	5.0	10.2	4.5	2.9	9.2	2.0	3.3
AMV-free 4 S RNA	64,520	32.3	13.4	5.9	7.5	2.8	3.2	0.1	12.3	0.7	6.2	0.1	10.6	1.7	1.4	0.3	1.4
AKR mouse fibroblast 4 S RNA	73,120	2.3	11.8	3.3	18.2	3.6	17.4	2.2	3.0	2.4	2.5	14.6	1.8[c]	3.3	9.4	1.7	2.5
AKR-MuLV free 4 S RNA	43,210	1.1	22.0	11.1	24.3	2.5	5.1	1.2	13.3	8.2	2.2	3.0	0.5[c]	3.1	1.4	0.5	0.5
Human tumor cell 4 S RNA	86,120	1.7	16.2	7.6	17.4	4.0	8.9[d]	0.5	5.9	3.0	0.3	7.1	12.3	4.0	6.7	1.8	2.8
RD-114 virus free 4 S RNA	113,350	1.3	23.7	22.2	10.0	3.0	1.7[d]	0.5	24.3	3.2	0.2	2.5	2.9	1.6	1.5	0.4	0.9

[a] Determined by the method described in Section IV and illustrated in Fig. 2. Portions of these data were taken from Waters (100) and Waters et al. (99) with permission from Academic Press and the American Society of Microbiology.
[b] Sum of radioactivities eluted from the amino-acid analyzer in peak positions corresponding to the indicated amino acids.
[c] Nonsaturating concentrations of methionine in the aminoacylation reaction.
[d] Serine only.

present in the cell include increases in the tRNAs for Trp, Pro, and Met, and decreases in those for Arg, Ser, Thr, Glu, Gly, Val, Leu, Tyr, and Phe *(99)*. These results are in excellent agreement with those of Travnicek *(92)* as well as of Gallagher and Gallo *(98)*, who assayed for the tRNAs individually. One study shows an increase of tRNA^{Val} in free 4 S RNA from AMV over that in avian cells *(102)*, but this result is clearly inconsistent with findings in four other laboratories using AMV *(92, 93, 98, 99)*. From our data *(99)*, the number of tRNAs in AMV free 4 S RNA per 35 S RNA molecule can range from 3 to 7 for tRNA^{Lys} and tRNA^{Trp} down to 0.02 for tRNA^{Glu} and tRNA^{Val}. These values are in good agreement with those of Sawyer and Dahlberg, who analyzed RSV free 4 S RNA by a completely different technique *(73)*.

Additional evidence for the selectivity with which cellular tRNAs are incorporated into virions is seen for the AKR-murine and RD-114 viruses (Table I). There is much less difference in the tRNA composition of the three host cells than in the free 4 S RNA from the three viruses. Faras finds that reticuloendotheliosis virus grown in chick embryo cells contains very little tRNA^{Trp} whereas in RSV grown in the same cells, tRNA^{Trp} comprises about 30% of the total 4 S RNA (A. J. Faras, personal communication; compare data for AMV in Table I). These observations support the interpretation that the viral genome, in some manner, controls which tRNAs are included in the virion *(97)*. However, it is still possible that the tRNA in the viral free 4 S RNA might be a reflection of the compartmentalization of tRNA within the virus-producing cell. Analysis of the tRNA in 70 S RNA-deficient virus particles *(65)* and in cellular membrane components should help to resolve this question.

The amino-acid acceptances discussed above tell nothing about the relative content of the various isoacceptor tRNAs in the virus, nor would they detect any virus-specific tRNAs. This aspect has been investigated in considerable detail in the avian viruses. For many tRNAs, there is very little quantitative difference in isoacceptor patterns of viral free tRNA and cellular tRNA (J. W. Carnegie and Waters, unpublished data; *98, 103*). An exception is tRNA^{Met}, certain isoacceptors of which are preferentially included in the viral free tRNA *(97, 98, 103–106)*. Such a complete study of tRNA isoacceptors in other viruses has not been reported; however, for AKR-MuLV, there does appear to be some selectivity as to which proline tRNA isoacceptors are included in the virus (Waters and Mullin, unpublished data). The available evidence indicates that viral tRNAs are qualitatively the same as cellular tRNAs, and that probably none are unique to the virus.

IV. Transfer RNA Associated with the 70 S RNA of the Virus

As it became increasingly apparent that the viral tRNAs are cellular in origin and that there are no apparent special translational requirements for establishing infection, the impetus for viral tRNA studies diminished. Interest was again stimulated by a series of observations (for details see Section V) indicating that there is tRNA associated with the viral 70 S RNA complex and that certain tRNA species can function to prime viral DNA synthesis *in vitro*. Another function for tRNA had been discovered.

Our own involvement and interest in the tRNAs in RNA tumor viruses began in 1969 when John Carnegie brought AMV RNA samples to us to be analyzed by RPC column chromatography. In 1973, in collaboration with W.-K. Yang, we did a systematic chromatographic analysis of the free and 70 S-associated 4 S RNAs in the AKR-MuLV system. The RPC-5 chromatographic results (Fig. 1A) clearly indicate a difference in the cellular and viral free 4 S RNAs that is consistent with amino-acid-acceptor analyses *(92, 93, 97–100, 102)* and two-dimensional gel electrophoretic analyses *(107)*. There is a further selection of the 4 S RNA associated with the 70 S RNA complex (Fig. 1, B, C, and D), a characteristic indicated first by chromatographic analyses of oligonucleotides *(78, 108)* and later verified by two-dimensional gel electrophoretic analyses *(73)* and aminoacylation studies *(99, 100)*. It is difficult to obtain large quantities of radiolabeled 70 S-associated 4 S RNA from murine viruses produced by currently available cell culture systems. The amount of labeled 4 S RNA available in the fraction released from 70 S RNA between 60° and 80°C is small, and excludes all but the simplest subsequent analyses (Fig. 1 D). From analyses of labeled viral RNA, we could show differences between the various subfractions of viral 4 S RNA but could not identify them. Additional methods to study viral 4 S RNAs, without the need for highly labeled viruses, were needed.

It had been suggested that most of the 70 S-associated 4 S RNA components were of cellular origin and to some extent had base complementarity with viral 35 S RNA. It is fairly easy to radiolabel cellular 4 S RNA in cell culture to a high specific activity ($\sim 6 \times 10^5$ cpm/μg). With such material, we were able to show that unique 4 S RNA molecules from the cell can hybridize *in vitro* with viral 35 S RNAs *(109)*, and that AMV 35 S RNA can hybridize with certain species of 4 S RNA of which tRNA^{Trp} was the predominant component *(110)*. Similar experiments suggest that the tRNAs that hybridize with AKR-MuLV 35 S RNA do not include tRNA^{Trp}.

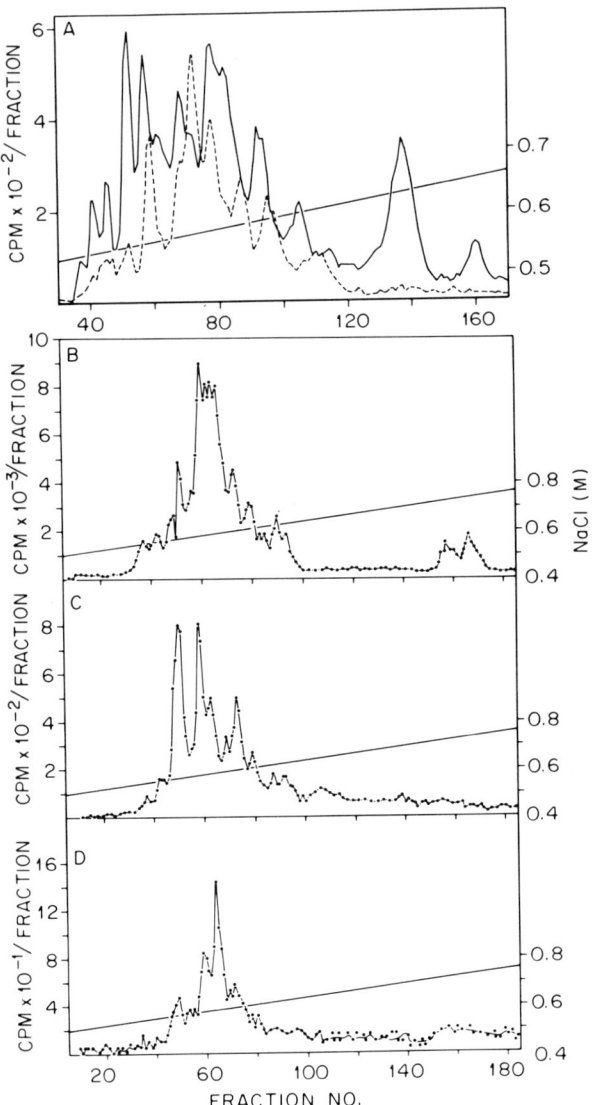

FIG. 1. RPC-5 column chromatographic analysis of radioactively labeled AKR mouse fibroblast and AKR/MuLV 4 S RNAs. (A) [^{3}H]Uridine-labeled AKR-MuLV free 4 S RNA (----) cochromatographed with [^{14}C]uridine-labeled AKR fibroblast 4 S RNA (——); (B) ^{32}P-labeled AKR-MuLV free 4 S RNA; (C) ^{32}P-labeled AKR-MuLV 60°C 70 S-associated 4 S RNA; (D) ^{32}P-labeled AKR-MuLV 60°–80°C 70 S-associated 4 S RNA. (B)–(D) represent the total amount of labeled material isolated from nine 4-hour virus harvests from ~9600 cm^2 (8 roller bottles) of nearly confluent AKR-MuLV-producing mouse cells.

In vitro hybridization studies can be very useful to demonstrate potentially important tRNA-35 S RNA interactions, but identifying the tRNAs can be difficult. A more direct and biologically significant approach would be to show which tRNA-35 S RNA interactions actually occur within the virion. Toward this goal, we have developed a procedure that enables us to qualitatively and semiquantitatively identify the tRNA species in small amounts of 4 S RNA. The rationale for these experiments is that the tRNAs most tightly bound to the viral 35 S RNA are likely to be those most functionally important to the virus. The technique has been described in detail (99) and is presented schematically in Fig. 2. Free 4 S RNA and 70 S RNA fractions are obtained from the virus. The 70 S-associated 4 S RNA is obtained from 70 S RNA by a stepwise heat treatment releasing sequentially those fractions that are less or more strongly bound to the 35 or 70 S RNA. The 4 S RNA fractions are then aminoacylated with a mixture of ^{3}H-labeled amino acids, each at the same concentration and the same specific

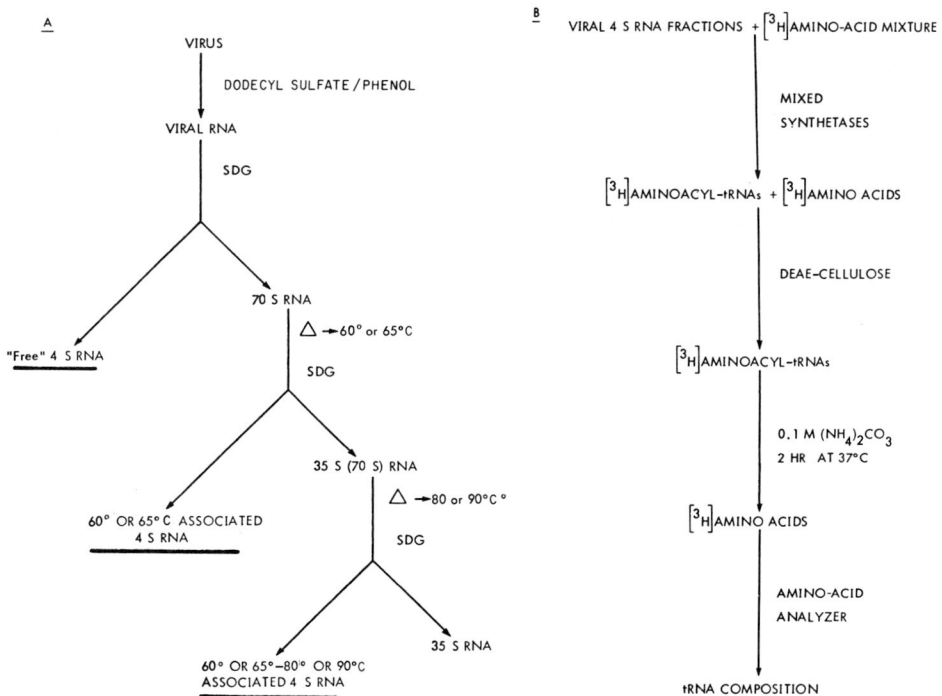

FIG. 2. Method used to identify the tRNAs in various viral 4 S RNA fractions. (A) Isolation of the 4 S RNA fractions. (B) Identification of the tRNAs in the 4 S RNA fractions. SDG = sucrose density gradient fractionation.

activity, using mixed aminoacyl-tRNA synthetase preparations. The [^{3}H]aminoacyl-tRNA is freed of unbound labeled amino acids by DEAE-cellulose chromatography, deacylated and the composition of the tRNA in the 4 S RNA fraction determined by analysis of the released amino acids. A similar approach has been used to study tRNA patterns in *Drosophila melanogaster (111)* and to identify the amino acid bound to tobacco mosaic virus RNA *(24)*.

Figures 3 and 4 illustrate the results of such experiments with AMV and AKR-MuLV. Tryptophan tRNA was essentially the only tRNA remaining in association with AMV 35 S RNA at temperatures above 60°C. In AKR-MuLV, 50–60% of the tRNA that dissociates from 35 S RNA at temperatures above 60°C is tRNA^{Pro}. Additional insight into the specificity of tRNA association with viral 35 S RNA can be gained by analysis of tRNAs from other viruses. A tabulation of our results from a total of nine viruses is given in Table II *(112)*. All the murine and feline leukemia viruses (FeLV) examined and the simian sarcoma virus (SSV) have tRNA^{Pro} most tightly associated with their

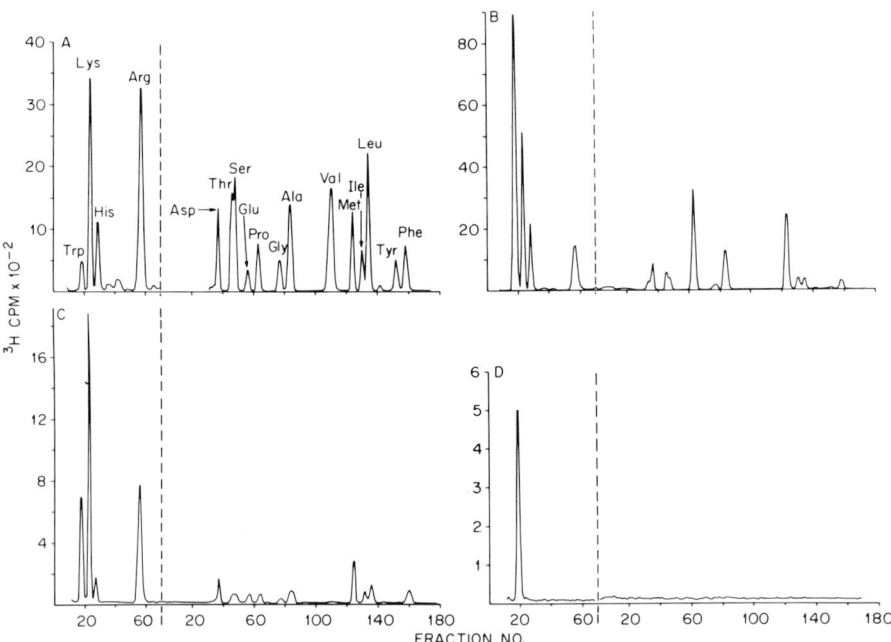

FIG. 3. Analysis of the amino acids obtained from *in vitro* aminoacylated cellular and viral 4 S RNAs. (A) Chicken embryo cell 4 S RNA; (B) AMV free 4 S RNA; (C) AMV 60°C 70 S-associated 4 S RNA; (D) AMV 60°–80°C 70 S-associated 4 S RNA. Reproduced, with permission, from Waters *et al.* *(99)*.

TABLE II

AMINO ACID tRNAs IN THE 60°–90° FRACTION OF THE 70 S-ASSOCIATED 4 S RNA FROM VARIOUS VIRUSES

Virus[a]	Total radioactivity[b] (cpm)	Trp	Lys	His	Arg	Asp	Ser or Ser and Thr	Glu	Pro	Gly	Ala	Val	Met	Ile	Leu	Tyr	Phe
AMV	1210	98.6[c]	1.4	ND[d]	ND	ND	ND	ND	ND	ND	ND	ND	ND	ND	ND	ND	ND
R-MuLV	2360	ND	6.0	1.7	3.3	1.6	1.5[e]	2.3	81.6	ND	ND	ND	0.8	1.2	ND	ND	ND
AKR-MuLV	890	3.8	6.5	1.3	10.8	ND	6.4	6.8	54.5	0.9	3.9	ND	ND	2.0	3.0	ND	ND
Mo-MuLV	4470	0.7	8.0	1.3	8.0	1.3	2.9[e]	1.2	66.1	1.1	0.5	ND	1.9	4.8	0.9	ND	1.1
F-MuLV	3610	0.7	3.4	1.0	5.8	1.2	1.8[e]	1.0	75.2	0.6	1.2	1.0	0.8	5.5	0.6	ND	ND
FeLV (R)	7710	0.7	17.9	3.2	7.4	1.3	1.4[e]	1.3	58.9	0.8	0.5	0.9	0.8	2.1	2.8	ND	ND
FeLV (T)	3000	1.4	28.7	3.7	6.5	1.8	2.4[e]	2.9	43.0	2.3	ND	ND	1.9	2.8	2.6	ND	ND
SSV	3170	1.6	14.5	4.3	7.9	ND	2.0[e]	4.7	59.0	1.5	ND	ND	1.0	2.3	1.1	ND	ND
RD-114	3100	ND	21.2	3.2	6.8	18.2	3.2	2.9	3.8	36.9	ND	ND	1.4	1.4	0.9	ND	ND

[a] AMV = avian myeloblastosis virus; MuLV = murine leukemia virus; R = Rauscher; AKR = AKR mouse; Mo = Moloney; F = Friend; FeLV = feline leukemia virus; (R) = Rickard strain; (T) = Theilen strain; SSV = simian sarcoma virus; RD-114 = feline xenotropic virus.
[b] Sum of radioactivities eluted from the amino-acid analyzer in peak positions corresponding to the indicated amino acids.
[c] Italicized figures indicate major components.
[d] ND = None detected.
[e] Serine only.

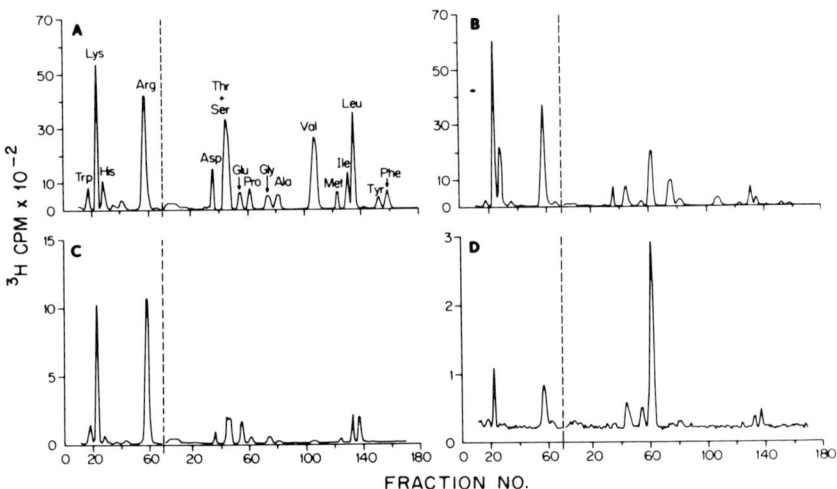

FIG. 4. Analysis of the amino acids obtained from *in vitro* aminoacylated cellular and viral 4 S RNAs. (A) AKR mouse fibroblasts 4 S RNA; (B) AKR-MuLV free 4 S RNA; (C) AKR-MuLV 60°C 70 S-associated 4 S RNA; (D) AKR-MuLV 60°–80°C 70 S-associated 4 S RNA. Reproduced, with permission, from Waters *(100)*.

genome. Neither tRNA^{Trp} nor tRNA^{Pro} are tightly associated with the 35 S RNA of the feline xenotropic virus, RD-114. Three tRNAs (those specific for Lys, Asp, and Gly) are consistently present in the 4 S RNA fraction that dissociates from RD-114 70 S RNA above 65°C. Below 65°C, tRNA^{Lys} and tRNA^{Asp} are the major components, and tRNA^{Gly} represents less than 3% of the total tRNA; therefore, at temperatures above 65°C, the relative enrichment of tRNA^{Gly} is much the greatest.

Although this method does not define the functions of the tRNAs in the various viral 4 S RNA subfractions, it does identify them and should facilitate further functional studies. We must emphasize that the technique detects only biologically active tRNAs and that nonacylatable tRNAs and/or non-tRNA 4 S RNA molecules of potential importance will not be detected. In principle, the sensitivity of the method is limited only by the specific activity of the labeled amino-acid mixture. Presently we can obtain significant results with 4 S RNA samples derived from approximately 10 μg of 70 S RNA.

V. Function of Transfer RNA in RNA Tumor Viruses

The discovery of an RNA-dependent DNA polymerase in RNA tumor viruses *(34, 35)* opened the way for a host of experiments on the transcription of viral RNA into DNA. The first indication that the low-M_r RNAs of RNA tumor viruses might play an important role in the transcription of 70 S RNA came from demonstrations that the template activity of thermally inactivated 70 S RNA can be restored by the addition of oligodeoxynucleotides to the assay mixture *(113)*. It was known that 4 S RNA structurally *(108)* and functionally *(114)* related to tRNA was released upon heat dissociation of the 70 S RNA complex, but not until the experiments of Canaani and Duesberg *(115)* was it known that template activity was dependent upon the presence, on the 35 S RNA, of 70 S-associated 4 S RNA molecules. This study defined viral 35 S RNA and certain 70 S-associated 4 S RNA molecules as the basic functional template-primer required for *in vitro* viral DNA synthesis.

A function for the 4 S RNA molecules in viral DNA synthesis was suggested by the fact that both a template and a primer molecule with a free 3' hydroxyl were required for reverse transcriptase activity *(116, 117)*. Analysis of DNA transcribed from 70 S RNA *(118, 119)* and from synthetic polynucleotides *(120)* by reverse transcriptase revealed the presence of RNA and DNA in covalent linkage.[8] With 70 S RNA as the template-primer, the RNA–DNA covalent bond was shown to be a phosphodiester linkage and, of the several possible combinations, only rA-dA *(121–124)*, rU-dC and rC-dC *(121)* have been observed. The combinations observed may depend on the method used to disrupt the virus *(124)* but rA-dA is consistently found.

The experiments of Canaani and Duesberg *(115)* showed that only about 20% of the total 70 S-associated 4 S RNA could have a primer function. Primer molecules from RSV were separated from nonprimer RNA by virtue of their increased size due to the covalent attachment of nascent DNA, and this RNA had a slightly higher (G + C)-content than other 70 S-associated 4 S RNA molecules *(125)*. The simplicity of the oligonucleotide chromatograms from an RNase T1 digestion of the primer RNA fraction suggested that the primer is a single species of 4 S RNA *(126)*. Analysis of the 4 S RNAs of RSV by two-dimensional gel electrophoresis revealed an RNA species, "Spot 1," with an RNase T1 map identical to the primer RNA. Spot 1 RNA was present at a level of 1–2 copies per virion in the 70 S-associated 4 S RNA and 6–8 copies in the free 4 S RNA fraction. It contained several modified nucleosides characteristic of tRNA *(73)*. The structural characteristics of primer molecules tagged with labeled deoxynucleotides, of 4 S RNA selectively dissociated from 70 S RNA between 63° and 80°C, and of Spot 1 RNA from total 70 S-associated and free 4 S RNA *(127, 128)* were

---

[8] The term "hybrid" has come to mean "hydrogen-bonded" and is better avoided when covalently linked RNA and DNA is meant [Ed.].

determined, and in each case, the fractions appeared to contain a single species of tRNA. The primary sequence of this tRNA is identical with that of chick cell tRNA^{Trp} purified from uninfected cells (107, 129–131). Further, the RNA can be selectively aminoacylated with tryptophan (129). Cell tRNA^{Trp} hybridized to RSV 35 S RNA (131) or AMV 35 S RNA (110) restores the template activity of thermally inactivated 35 S RNA. There is a report that the primer RNA, the major 4 S RNA component dissociated from the 70 S RNA in AMV, is distinct from that observed in RSV on the basis of oligonucleotide chromatographic analysis (132). This observation is difficult to reconcile with our own, which showed tRNA^{Trp} to be one of the major tRNAs associated with AMV 70 S RNA and the last to be dissociated from it by heat (99; see Fig. 3).

As one might predict, it is the 3' end of tRNA^{Trp} that forms a stable hybrid with 35 S RNA (133), and it is the first 17 nucleotides, excepting possibly the terminal adenosine, of the tRNA that are complementary to the viral RNA (134). Since native primer and reconstituted primer are physically and functionally dissociated from 35 S RNA with identical melting curves (82, 127; Mullin and Waters, unpublished observations), it is likely that the *in vivo* association is very similar to that defined by these *in vitro* studies (133, 134). This observation is of fundamental importance in that it eliminates the requirement that a primer be totally hybridized with the template, a point to be remembered as primer studies are extended to other systems. Furthermore, the involvement of a limited number of nucleotides at the 3' end eliminates a requirement for extensive structural changes in tRNA for template-primer formation to occur. The extent to which other portions of the primer tRNA molecule might be involved in the transcription process is not known. However, quarter molecules containing the 3' end of tRNA^{Trp} can hybridize with 35 S RNA and prime DNA synthesis *in vitro* (J. M. Bishop, personal communication). This would indicate that the 5' three-quarters of the primer tRNA is not essential for polymerase to successfully interact with the template-primer complex *in vitro*.

Considering the 5'-terminal to 3'-terminal direction of transcription, it is logical to assume that the primer tRNA is located at or near the 3' end of the 35 S RNA molecule. However the region of the 35 S RNA complementary to tRNA^{Trp} has been located near the 5' end of the molecule (135, 136) and d-A-A-T-G-A-A-G-C, the initial DNA sequence transcribed from RSV (137) or AMV (132) RNA, is known to be transcribed at the 5' end (135).

Early observations indicated that DNA synthesized in an *in vitro* assay using detergent-disrupted virions, although small, when hybridized with saturating amounts of DNA, could protect 75% of the viral genome from ribonuclease digestion *(138)*. Furthermore, at least 85% of the DNA sequences represented no more than 5% of the genome *(139)*. A majority of these multiple-copy DNA sequences are transcribed from the 5' terminus of the viral RNA *(140)*. Because all the DNA transcripts appeared to be small, it was suggested that there must be multiple initiation sites along the 70 S RNA *(126)*. Demonstrations that all, or nearly all, DNA transcripts are initiated on primer tRNA^{Trp} *(127)* and that there is only one tRNA^{Trp} binding site per 35 S RNA molecule *(73, 82, 99, 135, 136)* are clearly inconsistent with multiple initiation sites.

The use of all four deoxynucleoside triphosphates at high concentrations ($\geq$ 60 μM) in *in vitro* DNA synthesis reactions has made it possible to obtain a greater proportion of long DNA transcripts, which contain between 300 and 5000 nucleotides *(141, 142)*. In a detailed analysis of DNA transcripts made using both avian and murine endogenous and reconstituted systems, Haseltine *et al.* observed that even at high deoxynucleoside triphosphate concentrations, 25% or more of the product is found as a single DNA transcript about 100 nucleotides long in the avian system and about 135 nucleotides long in the murine system *(143)*. These were referred to as "structural stop" transcripts. In assays where individual deoxynucleoside triphosphates were present in limiting concentrations, "sequence stop" transcripts were formed that differed in size depending upon which deoxynucleotide was limiting and which template was used. They suggest that the short "structural stop" sequences represent DNA chains initiated on the primer tRNA and extending to the 5' terminus of the 35 S RNA molecule.

Localization of primer tRNA^{Trp} approximately 110 bases from the 5' terminus of the RSV 35 S RNA molecule has been reported *(144)*. Taylor and co-workers *(144)* further showed that most of the DNA transcripts greater than 2500 nucleotides in length begin with a similar sequence of at least 160 nucleotides. Using Mo-MuLV, Haseltine *et al.* *(143)* found molar equivalents of two different oligopyrimidines, unique to initial transcripts less than 90 nucleotides long, in transcripts of discrete sizes up to 1000 nucleotides in length. These observations suggest a model for *in vitro* viral DNA synthesis in which synthesis begins at a single and unique primer site and proceeds by linear extension of these initial transcripts. In its simplest form, this model would predict that DNA transcripts approximately 10,000 nucleotides in length would be required for complete protection of a hybrid formed with 35 S RNA from nuclease digestion. This prediction is not substantiated by experimental evidence *(141, 142)*, as discussed below.

Collet and Faras noted that their large DNA transcripts represented more of the RSV genome than would be predicted by their size *(141)*, and Rothenberg and Baltimore reported that at a DNA/RNA ratio of one, DNA transcripts averaging 1000 nucleotides in size protected more than 70% of the viral genome *(142)*. Clearly the situation is more complex than the above model would predict. A suggested mechanism *(143)*, in which DNA synthesis initiates at a single site, but at some point beyond the first 160 nucleotides the growing DNA chain skips to different sites at which chain elongation proceeds, could explain this paradox. Evidence has been presented *(144)* that suggests that RNase H, at least *in vitro*, begins to remove the RNA at the 5' terminus after DNA synthesis reaches that point. This could provide a mechanism for releasing, from the 5' end of the genome, nascent DNA transcripts that would reassociate at different sites on the genome and provide new secondary "initiation" sites.

An alternative and likely explanation for the extensive genetic complexity of short DNA transcripts is that, in the studies reported *(141, 142)*, artificial initiation sites are present in the template as a result of nuclease action (A. J. Faras, personal communication). Initiation at these sites could generate short transcripts, a composite of which would be complementary to most of the genome. DNA transcripts, up to 2000 nucleotides in length, produced in a reconstructed system containing the 35 S RNA · tRNA^{Trp} complex and purified reverse transcriptase, protect the genome in direct proportion to their length, i.e., to a maximum of 20% protection (A. J. Faras, personal communication). Haseltine *et al.* *(143)* suggested that the DNA transcripts comprising the background material in their gels might represent transcripts initiated at multiple sites on the genome. They did not report on the genetic complexity of the discrete-sized transcripts. Perhaps they are of limited complexity.

Junghans *et al.* *(145)* reported the synthesis, in Triton X-100-disrupted RSV, of full-length DNA transcripts representative of the entire viral genome. The concentration of Triton X-100 used to disrupt the virus was critical, and it was suggested that, at concentrations exceeding the optimum of 0.02%, nucleases have greater access to the 70 S RNA within the viral core, and consequently DNA is transcribed from partially degraded viral RNA. Likewise, Rothenberg and Baltimore *(146)* reported the synthesis in disrupted Mo-MuLV of large DNA transcripts that may represent full-length copies of viral 35 S and that are competent in transfection experiments (E. Rothenberg and D. Smotkin, personal communication). Only when Mg^{2+} is present in reaction mixtures at a concentration lower than the total nucleotide concentration can these large transcripts be synthesized, and only at this concentration of Mg^{2+} can intact 35 S RNA be recovered. Thus it appears likely that where smaller than full length transcripts appear to represent the entire genome, initiation may have occurred at breaks in the 70 S RNA.

An interesting problem in viral DNA synthesis concerns the mechanism by which synthesis is continued beyond the 5' terminus of the template. Whether the gap is bridged by a tandem alignment of template molecules or by circularization of a single template is not known. Collett and Faras *(147)* reported that initial DNA transcripts as small as the initial heptanucleotide will hybridize to poly(A)-containing viral RNA less than 28 S in size. Since 28 S poly(A)-containing RNA molecules are only about 5000 nucleotides long and should be devoid of sequences near the 5' terminus of intact 35 S RNA molecules, they concluded that sequences complementary to the DNA transcripts initiated at the 5' end of the molecule are also present near the 3' terminus of the RNA. The fact that a 3000-fold excess of RNA to DNA was used to obtain these results make it difficult to rule out contamination by molecules containing 5'-terminal sequences. These results suggest terminally redundant sequences, which could provide a mechanism for circularizing the genome. Kinetic analyses showed *(144)* that the initial 110-nucleotide DNA sequence hybridized several times faster to poly(A)-deficient RNA fragments than to those containing poly(A). Since poly(A)-deficient RNA fragments are enriched for 5'-terminal sequences and poly(A)-containing fragments are enriched for 3'-terminal sequences, they concluded that sequences complementary to the initial 110-nucleotide DNA sequence occur near the 5' terminus. The use of more precisely defined fragments of 35 S RNA should resolve this discrepancy.

It seems reasonably clear, from *in vitro* studies of avian tumor virus systems, that synthesis initiates at a single primer molecule, $tRNA^{Trp}$, located near the 5' terminus of the 35 S RNA template. The mechanism by which continuous transcription from the 5' to the 3' end of the 35 S RNA occurs is unknown. [See Addendum, p. 160.]

In several murine RNA tumor virus systems, $tRNA^{Pro}$ is the tRNA most tenaciously bound to the viral 35 S RNA *(100;* see also Table II). It is also the primer molecule for transcription of Mo-MuLV 70 S RNA *in vitro (147a)*. Mouse cell $tRNA^{Pro}$ when hybridized to AKR-MuLV 35 S RNA, will prime reverse transcription (W.-K. Yang, personal communication). Like tRNA^{Trp} in avian systems, tRNA^{Pro} is located near the 5' terminus of Mo-MuLV 35 S RNA (G. Peters and J. Dahlberg, unpublished data) and the initial DNA transcript d-A-A-T-G-A-A-A-G-A-C is very similar to that determined in avian systems *(143)*. Our finding (see Table II) that tRNA^{Pro} is also the most tenaciously bound tRNA in two strains of FeLV as well as in SSV suggests that tRNA^{Pro} might also function to prime viral DNA synthesis in these viruses. From our analyses of the tRNAs from the feline xenotropic virus, RD-114, (see Section IV and Table II), it appears that tRNA^{Gly} rather than tRNA^{Trp} or tRNA^{Pro} serves as primer in this virus. The primer for visna virus 70 S RNA transcription has been isolated by Faras and Haase and is also different from either tRNA^{Trp} or tRNA^{Pro} (personal communication).

Although the evidence from *in vitro* studies of avian RNA tumor viruses indicating a primer function for tRNA^{Trp} seems most convincing, there is no direct evidence that tRNAs serve as primers *in vivo*. DNA–RNA covalent sequences are found in MuLV *(148)* and in RSV-infected cells, and their appearance upon infection depends upon an active reverse transcriptase *(149)*. However, the size of the RNA portion of the mixed polynucleotide was reported to be considerably larger than 4 S RNA *(149)*.

In addition to the role as primers for DNA synthesis, one could speculate that primer, or other 70 S-associated, tRNAs might play a role in the integration of proviral DNA into the host cell genome. Labeled Friend-MuLV RNA was followed into the nuclei of infected cells, where it was isolated covalently linked to DNA *(150)*. In the presence of host-cell DNA synthesis, virus-specific RNA sequences are found in association with high-M_r nuclear DNA. Even if the association of exogenously supplied RNA with nuclear DNA is not specific to viral RNA, it may nevertheless facilitate integration of viral information into host cell DNA. Primer tRNAs are of cell origin and could be expected to recognize specific DNA sequences in the nucleus. Alternatively it has been suggested that, in complete DNA transcripts of viral RNA, sequences complementary to primer tRNA could determine limited specific sites of integration (W.-K. Yang, personal communication). The nucleotide sequence of the initial DNA transcripts in avian d-A-A-T-G-A-A-G-C and in murine d-A-A-T-G-A-A-A-G-A-C systems are remarkably similar, and it has been suggested that these sequences might be involved in proviral DNA integration *(143)*.

In addition to having a degree of sequence homology with the viral genome, primer tRNAs are recognized by viral reverse transcriptase. The avian reverse transcriptase is composed of an α subunit containing polymerase and RNase H activity *(151, 152)*, and a β subunit enhancing template binding *(152, 153)*. The α subunit is thought to be the product of proteolytic cleavage of the β subunit *(154–157)*. Tryptophan tRNA, the primer in the avian systems, binds to the αβ form of the polymerase under conditions where other tRNAs do not bind *(158)*. The α subunit alone does not bind tRNA^{Trp} but preparations containing predominantly the β subunit do show binding *(159)*. From studies using partial T1 RNase digests of tRNA^{Trp}, it appears that tRNA^{Trp} binding to the avian polymerase *in vitro* requires an essentially intact tRNA^{Trp} molecule (J. M. Bishop, personal communication).

Murine virus reverse transcriptase can be isolated as single polypeptide of M_r 80,000–84,000 that contains polymerase and RNase H activities *(160, 161)*. An 85,000 M_r enzyme from Mo-MuLV efficiently transcribes viral 70 S RNA but does not bind tRNA^{Pro} under conditions that allow binding of tRNA^{Trp} to avian enzyme *(161a)*. A 50,000–80,000 M_r form of R-MuLV transcriptase, active with synthetic templates but relatively inactive with viral RNA, isolated from viral lysates *(162)* does not bind tRNA^{Pro} *(163)*. From the same preparation, a larger form (M_r = 80,000–210,000) that does bind tRNA^{Pro} and transcribes viral RNA much more efficiently has been isolated. Two forms of reverse transcriptase (M_r = 70,000 and 95,000) have been isolated from cells producing RD-114 virus, but only the smaller form was isolated from the virus *(164)*. The relationship of these multiple forms of reverse transcriptases to one another and to other mammalian tumor virus polymerases is not known but is currently being studied.

Although the avian primer is tRNA^{Trp} and the murine primer is tRNA^{Pro}, both polymerases readily utilize the heterologous 70 S RNA as template-primer *(143, 161a)*. The DNA sequences transcribed from 70 S RNA with heterologous polymerase are the same as those transcribed using the homologous enzyme and are unique to the RNA template used *(143)*. Both tRNAs have been sequenced *(129, Harada and Dahlberg, personal communication)* and both contain the unusual ψ-ψ-C-G oligonucleotide sequence, which might be important in enzyme–tRNA interaction. An argument against this possibility comes from studies of chick cell lysine tRNA I, which appears to contain the sequences ψ-ψ-C-G (J. L. Nichols, personal communication) and hybridizes to AMV 35 S RNA, but does not bind to AMV reverse transcriptase under conditions in which tRNA^{Trp} does bind, nor primes reverse transcription (Mullin and Waters, unpublished observation; *110*).

In vitro, the avian polymerase can bind only intact or nearly intact tRNA^{Trp}, yet it can transcribe DNA from a template-primer complex composed of viral 35 S and tRNA^{Trp} fragments containing as little as the 3′ quarter of the molecule (J. M. Bishop, personal communication).

Thus it appears that the ability of an RNA to serve as primer *in vitro* is determined by its ability to form a stable hybrid at its 3' end with the specific template. Although it does not appear to be required *in vitro*, the extent to which reverse transcriptase interacts with the 5' three-quarters of primer tRNAs and whether such interactions might impose any further specificity *in vivo* is not known.

In addition to the specific primer-polymerase binding discussed above, other less specific tRNA-polymerase interactions have been observed. Using a different method of analysis, Cavalieri and Yamaura (165) have shown that most *E. coli* tRNAs interact with AMV reverse transcriptase to the extent that they cosediment in a sucrose gradient. Furthermore, *E. coli* tRNAs in assay mixtures at molar concentrations at least ten times that of the template decreased initial rates of transcription. Although nonpriming tRNAs found in association with viral RNAs do not appear to bind to reverse transcriptase, they may nonetheless affect transcription of 70 S RNA. Viral RNA with primer alone appears to be a better template than RNA with primer plus the less tightly bound 4 S RNAs (166). It is not known whether this effect is incidental or whether nonprimer tRNAs have a regulatory role *in vivo*.

Although a function for nonprimer 70 S-associated tRNAs is not known, they constitute as much as 90% of the associated tRNA and are characterized by consistent and reproducible tRNA composition patterns. Most of the nonprimer tRNAs can be dissociated from 70 S RNA at temperatures below 60°C. It seems of particular interest to note that this fraction of tRNA from a number of different viruses is consistently enriched in the tRNAs accepting basic amino acids, especially tRNA^{Lys} and tRNA^{Arg} (99, 100, and Waters and Mullin, unpublished data). The primer tRNA is not necessarily enriched in this fraction (100). The 70 S-associated tRNAs do not have the same tRNA composition as does the tRNA in the free 4 S RNA fraction, but this does not exclude free 4 S RNA from being the pool from which the associated 4 S RNA is derived.

A possible function for 70 S-associated tRNAs, primers or nonprimers, might be to link 35 S RNA molecules to form the 70 S RNA complex. Electron micrographs of 70 S RNA have shown two 35 S RNA molecules linked together near the 5' end of each (43). Our own data from studies of AMV tRNAs suggest that tRNA^{Lys} might serve such a linker function. Although we have not done a precise dissociation study, our results (99) suggest that tRNA^{Lys} dissociation is coordinated with 70 S to 35 S conversion. Furthermore tRNA^{Lys} hybridizes to AMV 35 S RNA in a 1:1 ratio but does not prime reverse transcription. Determination of the interacting sequences in tRNA^{Lys} should be useful in evaluating such a potential function. To our knowledge, no one has succeeded in reforming the 70 S RNA complex once it has been dissociated.

While considering functions for the viral tRNAs it would be perhaps premature to rule out a purely translational function. Virus-specific proteins are synthesized *in vitro* by isolated AMV cores *(168)*. Several components of a translational apparatus, e.g., tRNA, ribosomes, and aminoacyl-tRNA synthetases, are contained in virus particles. However, some of these components are present in very small quantities in AMV; for instance, tRNA^{Glu} and tRNA^{Val} are present at levels of about one molecule per 25 virus particles (see Section III). The possibility that virus particles, not to mention cores, contain all the components required for protein synthesis is an intriguing one that should be confirmed. Perhaps related to this is the report that a virus-specific protein is synthesized early (within 3 hours) in AMV-infected cells, and that this synthesis is insensitive to pretreatment of the cells with actinomycin D or arabinosylcytosine *(169)*. Whether the translational machinery contained in these viruses is there for the synthesis of such early proteins remains to be determined.

VI. Summary

Clearly, the presence of the majority of the tRNAs inside an RNA tumor virus remains an enigma. The finding that specific tRNA molecules can serve as primers for DNA synthesis is of fundamental importance in the field of molecular biology. The transfer RNAs involved as primers are not restricted to a single tRNA species. The primer tRNA interacts at its 3' end with the template, and this interaction may involve as little as one-fourth of the tRNA molecule. Primer, but not nonprimer, tRNAs interact specifically with RNA-dependent DNA polymerases. On the basis of these properties, a number of techniques have been developed to identify primer tRNAs and to demonstrate their function.

The insight and technology gained from studies of primer structure and function in RNA tumor virus systems should be helpful in the study of DNA replication in other systems. As more complex systems, including nuclei, are probed for the mechanisms by which DNA synthesis is initiated and controlled, we are obliged to consider the possible involvement of tRNAs.

REFERENCES

1. B. N. Ames and P. Hartman, *CSHSQB* **28**, 349 (1963).
2. N. Sueoka and T. Kano-Sueoka, *PNAS* **52**, 1535 (1964).
3. N. Sueoka and T. Kano-Sueoka, This Series **10**, 23 (1970).

4. W. F. Anderson, *PNAS* **62**, 566 (1969).
5. W. F. Anderson and J. M. Gilbert, *BBRC* **36**, 456 (1969).
6. M. Brenner and B. N. Ames, *in* "Metabolic Pathways" (H. J. Vogel, ed.), Vol 5, p. 349. Academic Press, New York, 1971.
7. H. E. Umbarger, *in* "Metabolic Pathways" (H. J. Vogel, ed.), Vol. 5, p. 447. Academic Press, New York, 1971.
8. C. E. Singer, G. R. Smith, R. Cortese and B. N. Ames, *Nature NB* **238**, 72 (1972).
9. G. W. Hatfield and R. O. Burns, *PNAS* **66**, 1027 (1970).
10. D. R. Simpson, S. M. Arnfin and G. W. Hatfield, *FP* **34**, 586 (1975).
11. S. C. Quay, E. L. Kline and D. L. Oxender, *PNAS* **72**, 3921 (1975).
12. K. B. Jacobson, *Nature NB* **231**, 17 (1971).
13. D. Mischke, P. Kloetzel and M. Schwochau, *Nature* **255**, 79 (1975).
14. J. A. Nesbitt and W. J. Lennarz, *JBC* **243**, 3088 (1968).
15. R. M. Gould, M. P. Thornton, V. Liepkalns and W. J. Lennarz, *JBC* **243**, 3096 (1968).
16. A. Kaji, H. Kaji and G. D. Novelli, *JBC* **240**, 1185 (1965).
17. T. S. Stewart, R. J. Roberts and J. L. Strominger, *Nature* **230**, 36 (1971).
18. U. Z. Littauer and H. Inouye, *ARB* **42**, 439 (1973).
19. J. H. Wilson, J. S. Kim and J. N. Abelson, *JMB* **71**, 547 (1972).
20. J. H. Wilson, *JMB* **74**, 753 (1973).
21. A. L. Haenni, A. Prochiantz and P. Yot, *in* "Energy, Regulation and Biosynthesis in Molecular Biology" (D. Richter, ed.), 1974 Lipmann Symp., p. 264. de Gruyter, New York and Berlin, 1974.
22. M. Pinck, P. Yot, F. Chapeville and H. M. Durabon, *Nature* **226**, 954 (1970).
23. T. C. Hall, D. S. Shih and P. Kaesberg, *BJ* **129**, 969 (1972).
24. B. Oberg and L. Philipson, *BBRC* **48**, 927 (1972).
25. R. Salomon and U. Z. Littauer, *Nature* **249**, 32 (1974).
26. A. Prochiantz, C. Benicourt, D. Carre, and A.-L. Haenni, *EJB* **52**, 17 (1975).
27. S. Litvak, A. Tarrago, L. Tarrago-Litvak and J. E. Allende, *Nature NB* **241**, 88 (1973).
28. J. Blumenthal, T. A. Landers and K. Weber, *PNAS* **69**, 1313 (1972).
28a. A. L. Haenni, C. Benicourt, S. Teixeira, A. Prochiantz and F. Chapeville, *Proc. FEBS Meet., 10th* pp. 121–131 (1975).
29. D. Gillespie, W. C. Saxinger and R. C. Gallo, This Series **15**, 1 (1975).
30. W. S. Robinson, A. Pitkanen and H. Rubin, *PNAS* **54**, 137 (1965).
31. F. Galibert, C. Bernard, P. Chenaille and M. Boiron, *Compt. Rend. Acad. Sci.* **261**, 1771 (1965).
32. J. Harel, J. Huppert, F. Lacour and L. Harel, *Compt. Rend. Acad. Sci.* **261**, 2266 (1965).
33. P. H. Duesberg and P. B. Blair, *PNAS* **55**, 1490 (1966).
34. D. Baltimore, *Nature* **226**, 1209 (1970).
35. H. M. Temin and S. Mizutani, *Nature* **226**, 1211 (1970).
36. P. H. Duesberg, *PNAS* **60**, 1511 (1968).
37. R. L. Erikson, *Virology* **37**, 124 (1969).
38. L. Montagnier, A. Golde and P. Vigier, *J. Gen. Virol.* **4**, 449 (1969).
39. A. M. Q. King, *JBC* **251**, 141 (1976).
40. C. H. Riggin, M. Bondurant and W. M. Mitchell, *J. Virol.* **16**, 1528 (1975).
41. H. Delius, P. H. Duesberg and W. F. Mangel, *CSHSQB* **39**, 835 (1974).
42. C. Weissmann, J. T. Parsons, J. W. Coffin, L. Rymo, M. A. Billeter and H. Hofstetter, *CSHSQB* **39**, 1043 (1974).
43. H.-J. Kung, J. M. Bailey, N. Davidson, M. O. Nicolson and R. M. McAllister, *J. Virol.* **16**, 397 (1975).

44. J. M. Taylor, H. E. Varmus, A. J. Faras, W. E. Levinson and J. M. Bishop, *JMB* **84**, 217 (1973).
45. H. Fan and M. Paskind, *J. Virol.* **14**, 421 (1974).
46. M. A. Baluda, M. Shoyab, P. D. Markham, R. M. Evans and W. N. Drohan, *CSHSQB* **39**, 869 (1974).
47. K. Quade, R. E. Smith and J. L. Nichols, *Virology* **61**, 287 (1974).
48. M. A. Billeter, J. T. Parsons and J. M. Coffin, *PNAS* **71**, 3560 (1974).
49. P. Duesberg, P. K. Vogt, K. Beemon and M. Lai, *CSHSQB* **39**, 847 (1974).
50. L.-H. Wang and P. Duesberg, *J. Virol.* **14**, 1515 (1974).
51. L.-H. Wang, P. H. Duesberg, S. Kawai and H. Hanafusa, *PNAS* **73**, 447 (1976).
52. K. Quade, R. E. Smith and J. L. Nichols, *Virology* **62**, 60 (1974).
53. A. M. Q. King and R. D. Wells, *JBC* **251**, 150 (1976).
54. D. Baltimore, *CSHSQB* **39**, 1187 (1974).
55. M. Green and M. Cartas, *PNAS* **69**, 791 (1972).
56. M. C. Lai and P. H. Duesberg, *Nature* **235**, 383 (1972).
57. D. Gillespie, S. Marshall and R. C. Gallo, *Nature NB* **236**, 227 (1972).
58. M. L. Stephenson, J. F. Scott and P. C. Zamecnik, *BBRC* **55**, 8 (1973).
59. A. M. Wu, R. C. Ting, M. Paran and R. C. Gallo, *PNAS* **69**, 3820 (1972).
60. F. Wong-Staal, W. C. Saxinger, R. C. Gallo and D. H. Gillespie, *BBRC* **69**, 599 (1976).
61. Y. Furuichi, A. J. Shatkin, E. Stavnezer and J. M. Bishop, *Nature* **257**, 618 (1975).
62. M. Green, *ARB* **39**, 701 (1970).
63. H. Fan and D. Baltimore, *JMB* **80**, 93 (1973).
64. K. von der Helm and P. H. Duesberg, *PNAS* **72**, 614 (1975).
65. J. G. Levin, P. M. Grimley, J. M. Ramseur and I. K. Berezesky, *J. Virol.* **14**, 152 (1974).
66. E. Canaani, K. V. D. Helm and P. Duesberg, *PNAS* **72**, 401 (1973).
67. K.-S. Cheung, R. E. Smith, M. P. Stone and W. K. Joklik, *Virology* **50**, 851 (1972).
68. J. Robert-Robin, L. d'Auriol and R. Emanoil-Ravicovitch, *J. Gen. Virol.* **30**, 149 (1976).
69. C. M. Stoltzfus and P. N. Snyder, *J. Virol.* **16**, 1161 (1975).
70. R. A. Bonar, L. Sverak, D. P. Bolognesi, A. J. Langlois, D. Beard and J. W. Beard, *Cancer Res.* **27**, 1138 (1967).
71. J. M. Bishop, W. E. Levinson, N. Quintrell, D. Sullivan, L. Fanshier and J. Jackson, *Virology* **42**, 182 (1970).
72. A. J. Faras, A. C. Garapin, W. E. Levinson, J. M. Bishop and H. M. Goodman, *J. Virol.* **12**, 334 (1973).
73. R. C. Sawyer and J. E. Dahlberg, *J. Virol.* **12**, 1226 (1973).
74. R. Emanoil-Ravicovitch, C. J. Larsen, M. Bazilier, J. Robin, J. Peries and M. Boiron, *J. Virol.* **12**, 1625 (1973).
75. J. M. Bishop, W. E. Levinson, D. Sullivan, L. Fanshier, N. Quintrell and J. Jackson, *Virology* **42**, 927 (1970).
76. T. Obara, D. P. Bolognesi and H. Bauer, *Int. J. Cancer* **7**, 535 (1971).
77. C. J. Larsen, R. Emanoil-Ravicovitch, A. Samso, J. Robin, A. Tavitian and M. Boiron *Virology* **54**, 552 (1973).
78. J. L. Nichols and M. Waddell, *Nature NB* **243**, 236 (1973).
79. M. A. Baluda and D. P. Nayok, *PNAS* **66**, 329 (1970).
80. T. A. Walker, N. R. Pace, R. L. Erikson, E. Erikson and F. Behr, *PNAS* **71**, 3390 (1974).
81. J. P. Bader and J. L. Steck, *J. Virol.* **4**, 454 (1969).
82. J. M. Taylor, B. Cordell-Stewart, W. Rohde, H. M. Goodman and J. M. Bishop, *Virology* **65**, 248 (1975).

83. J. Robert-Robin, R. Emanoil-Ravicovitch, M. Bazilier and M. Boiron, *BBRC* **60**, 965 (1974).
84. G. Zieve and S. Penman, *Cell* **8**, 19 (1976).
85. K. Smetana, J. Korb and J. Riman, *Neoplasma* **19**, 437 (1972).
86. W. Levinson, J. M. Bishop, N. Quintrell and J. Jackson, *Nature* **227**, 1023 (1970).
87. J. Riman and G. S. Beaudreau, *Nature* **228**, 427 (1970).
88. N. Biswal, B. McCain and M. Benyesh-Melnick, *Virology* **45**, 697 (1971).
89. W. E. Levinson, H. E. Varmus, A-C. Garapin and J. M. Bishop, *Science* **175**, 76 (1972).
90. A. Kiessling, A. O'C. Deeney and G. S. Beaudreau, *FEBS Lett.* **20**, 57 (1972).
91. A. O'C. Deeney, K. Stromberg and G. S. Beaudreau, *BBA* **432**, 281 (1976).
92. M. Travnicek, *BBA* **182**, 427 (1969).
93. J. W. Carnegie, A. O'C. Deeney, K. C. Olson and G. S. Beaudreau, *BBA* **190**, 274 (1969).
94. K. Stromberg and M. D. Litwack, *BBA* **319**, 140 (1973).
95. D. P. Bolognesi, H. Gelderblom, H. Bauer, K. Molling and G. Huper, *Virology* **47**, 567 (1972).
96. G. S. Beaudreau, L. Sverak, R. Zischka and J. W. Beard, *Natl. Cancer Inst. Monogr.* **17**, 791 (1964).
97. S. Wang, R. M. Kothari, M. Taylor and P. Hung, *Nature NB* **242**, 133 (1973).
98. R. E. Gallagher and R. C. Gallo, *J. Virol.* **12**, 449 (1973).
99. L. C. Waters, B. C. Mullin, E. G. Bailiff and R. A. Popp, *J. Virol.* **16**, 1608 (1975).
100. L. C. Waters, *BBRC* **65**, 1130 (1975).
101. K. Randerath, L. J. Rosenthal and P. C. Zamecnik, *PNAS* **68**, 3233 (1971).
102. E. Erikson and R. L. Erikson, *JMB* **52**, 387 (1970).
103. M. W. Taylor, S. Wang, R. M. Kothari and P. P. Hung, *J. Virol.* **14**, 1092 (1974).
104. S. Wang, R. M. Kothari, M. W. Taylor and P. P. Hung, *BBA* **340**, 52 (1974).
105. K. T. Elder and A. E. Smith, *Nature* **247**, 435 (1974).
106. A. J. Faras, *Virology* **63**, 583 (1975).
107. R. C. Sawyer, F. Harada and J. E. Dahlberg, *J. Virol.* **13**, 1302 (1974).
108. E. Erikson and R. L. Erikson, *J. Virol.* **8**, 254 (1971).
109. L. C. Waters, B. C. Mullin, T. Ho, and W.-K. Yang, *BBRC* **60**, 489 (1974).
110. L. C. Waters, B. C. Mullin, T. Ho and W.-K. Yang, *PNAS* **72**, 2155 (1975).
111. R. A. Davey and A. J. Howells, *Anal. Biochem.* **60**, 469 (1974).
112. L. C. Waters and B. C. Mullin, *FP* **35**, 1736 (1976).
113. P. Duesberg, K. V. D. Helm and E. Canaani, *PNAS* **68**, 2505 (1971).
114. L. J. Rosenthal and P. C. Zamecnik, *PNAS* **70**, 1184 (1973).
115. E. Canaani and P. Duesberg, *J. Virol.* **10**, 23 (1972).
116. D. Baltimore and D. Smoler, *PNAS* **68**, 1507 (1971).
117. J. Hurwitz and J. P. Leis, *J. Virol.* **9**, 116 (1972).
118. I. M. Verma, N. L. Meuth, E. Bromfeld, K. F. Manly and D. Baltimore, *Nature NB* **233**, 131 (1971).
119. J. P. Leis and J. Hurwitz, *J. Virol.* **9**, 130 (1972).
120. R. D. Wells, R. M. Flugel, J. E. Larson, P. F. Schendel and R. W. Sweet, *Bchem.* **11**, 621 (1972).
121. R. M. Flugel and R. D. Wells, *Virology* **48**, 394 (1972).
122. I. M. Verma, N. L. Meuth and D. Baltimore, *J. Virol.* **10**, 622 (1972).
123. H. Okabe, G. G. Lovinger, R. V. Gilden and M. Hatanaka, *Virology* **50**, 935 (1972).
124. R. M. Flugel, U. Rapp and R. D. Wells, *J. Virol.* **12**, 1491 (1973).
125. A. J. Faras, J. M. Taylor, W. E. Levinson, H. M. Goodman and J. M. Bishop, *JMB* **79**, 163 (1973).

126. J. M. Bishop, T.-D. Chun, A. J. Faras, H. M. Goodman, W. E. Levinson, J. M. Taylor and H. E. Varmus, in "Virus Research" (C. F. Fox and W. S. Robinson, eds.), p. 15. Academic Press, New York, 1973.
127. J. E. Dahlberg, R. C. Sawyer, J. M. Taylor, A. J. Faras, W. E. Levinson, H. M. Goodman and J. M. Bishop, J. Virol. 13, 1126 (1974).
128. A. J. Faras, J. E. Dahlberg, R. C. Sawyer, F. Harada, J. M. Taylor, W. E. Levinson, J. M. Bishop and H. M. Goodman, J. Virol. 13, 1134 (1974).
129. F. Harada, R. C. Sawyer, J. E. Dahlberg, JBC 250, 3487 (1975).
130. L. C. Waters, W.-K. Yang, B. C. Mullin and J. L. Nichols, JBC 250, 6627 (1975).
131. A. J. Faras and N. A. Dibble, PNAS 72, 859 (1975).
132. J. J. Eiden, D. P. Bolognesi, A. J. Langlois and J. L. Nichols, Virology 65, 163 (1975).
133. J. J. Eiden, K. Quade and J. L. Nichols, Nature 259, 245 (1976).
134. B. Cordell, E. Stavnezer, R. Friedrich, J. M. Bishop and H. M. Goodman, J. Virol., 19, 548 (1976).
135. J. M. Taylor and R. Illmensee, J. Virol. 16, 553 (1975).
136. K. A. Staskus, M. S. Collett and A. J. Faras, Virology 71, 162 (1976).
137. J. M. Taylor, D. E. Garfin, W. E. Levinson, J. M. Bishop and H. M. Goodman, Bchem. 13, 3159 (1974).
138. P. H. Duesberg and E. Canaani, Virology 42, 783 (1970).
139. H. E. Varmus, W. E. Levinson and J. M. Bishop, Nature NB 233, 19 (1971).
140. L. M. Cashion, R. H. Joho, M. A. Planitz, M. A. Billeter and C. Weissmann, Nature 262, 186 (1976).
141. M. S. Collett and A. J. Faras, J. Virol. 16, 1220 (1975).
142. E. Rothenberg and D. Baltimore, J. Virol. 17, 168 (1976).
143. W. A. Haseltine, D. G. Kleid, A. Panet, E. Rothenberg and D. Baltimore, JMB 106, 109 (1976).
144. J. M. Taylor, R. Illmensee, L. R. Trusal and J. Summers, in "Animal Virology" (D. Baltimore, A. S. Huang and C. F. Fox, eds.), p. 161. Academic Press, New York, 1976.
145. R. P. Junghans, P. H. Duesberg and C. A. Knight, PNAS 72, 4895 (1975).
146. E. Rothenberg and D. Baltimore, J. Virol. 21, 168 (1977).
147. M. S. Collett and A. J. Faras, PNAS 73, 1329 (1976).
147a. G. Peters, F. Harada, J. Dahlberg, A. Panet, W. Haseltine and D. Baltimore, J. Virol. 21, 1031 (1977).
148. T. Takano and M. Hatanaka, CSHSQB 39, 1009 (1974).
149. J. Leis, A. Schincariol, R. Ishizaki and J. Hurwitz, J. Virol. 15, 484 (1975).
150. M. M. Sveda, B. N. Fields and R. Soeiro, J. Virol. 18, 85 (1976).
151. D. P. Grandgenett, G. F. Gerard and M. Green, PNAS 70, 230 (1973).
152. A. Panet, I. M. Verma and D. Baltimore, CSHSQB 39, 919 (1974).
153. D. Grandgenett, in "Viral Nucleic Acid Replication in Eucaryotes" (M. Goulian, P. Hanawalt and C. F. Fox, eds.), p. 760. Benjamin, Menlo Park, California, 1975.
154. K. Moelling, CSHSQB 39, 969 (1974).
155. W. Gibson and I. M. Verma, PNAS 71, 4991 (1974).
156. D. P. Grandgenett and H. M. Rho, J. Virol. 15, 526 (1975).
157. T. S. Papas, D. J. Marciani, K. Samuel and J. G. Chirikjian, J. Virol. 18, 904 (1976).
158. A. Panet, W. A. Haseltine, D. Baltimore, G. Peters, F. Harada and J. E. Dahlberg, PNAS 72, 2535 (1975).
159. D. P. Grandgenett, A. C. Vora and A. J. Faras, Virology 75, 26 (1976).
160. I. M. Verma, J. Virol. 15, 843 (1975).
161. K. Moelling, J. Virol. 18, 418 (1976).

161a. W. Haseltine, A. Panet, D. Smoler, D. Baltimore, G. Peters, F. Harada and J. Dahlberg, in press.
162. A. M. Wu, A. Cetta, M. G. Sarngadharan and R. C. Gallo, *FP* **34**, 1753 (1975).
163. W.-K. Yang, personal communication.
164. B. I. Gerwin, S. G. Smith and P. T. Peebles, *Cell* **6**, 45 (1975).
165. L. F. Cavalieri and I. Yamaura, *NARes.* **2**, 2315 (1975).
166. M. Travnicek, L. Y. Frolova and J. Riman, *J. gen. Virol.* **30**, 187 (1976).
168. I. S. Mendelson, *BBA* **407**, 442 (1975).
169. B. M. Gallis, R. N. Eisenman and H. Diggelmann, *Virology* **74**, 302 (1976).

Addendum

Electron microscopic analysis of hybrids formed between RSV 35 S RNA and large DNA transcripts revealed a significant number of circular structures of 35 S RNA length *(A1)*. This is evidence that such DNA transcripts span the gap between the 5' and 3' ends of 35 S RNA and supports a circularization model of transcription. The terminal redundancy required for circularization of the 35 S RNA template has recently been demonstrated in RSV *(A2, A3)* and in Mo-MuLV (W. A. Haseltine, J. M. Coffin and A. Maxam, personal communication) by nucleotide sequence analysis.

A1. R. P. Junghans, S. Hu, C. A. Knight and N. Davidson, *PNAS* **74**, 477 (1977).
A2. W. A. Haseltine, A. Maxam and W. Gilbert, *PNAS* **74**, 989 (1977).
A3. D. Schwartz, P. Zamecnik and L. Weith, *PNAS* **74**, 994 (1977).

Integration versus
Degradation of Exogenous
DNA in Plants: An Open
Question

> Paul F. Lurquin
>
> Radiobiology Department
> Centre d'Etude de l'Energie
>   Nucléaire
> C.E.N./S.C.K.
> Mol, Belgium

I. Introduction	161
II. Experimental Systems	163
A. Biochemical Evidence	163
B. Biological Evidence	188
III. Attempts to Conclude and Prospectives	196
A. Facts and Fancy	196
B. Missing Links and Further Speculations	198
Notes Added	203
References	204

> "Plusieurs choses certaines sont contredites:
> plusieurs fausses passent sans contradiction.
> Ni la contradiction n'est marque de fausseté,
> ni l'incontradiction n'est marque de vérité."
>
> B. Pascal, *Pensées* (1670)

I. Introduction

The principal aim of this article is to evaluate the relatively numerous and conflicting reports concerning the fate of exogenous DNA applied to plant cells. The literature on the subject falls into two classes, one claiming a functional integration into the host genome, the other denying that such integration occurs.

It may be useful to recall the principal goals of such studies. The general goal is that of "improving" plants in a heritable fashion. Obviously, classical plant-breeding techniques have been quite successful, although they suffer from the disadvantage of being slow. However, the advent of somatic plant cell genetics coupled with protoplast fusion, yielding in some cases complementation between mutant alleles (1–3), may represent a breakthrough in plant breeding, as it bypasses the laborious procedure of field or greenhouse crosses and allows a direct selection of characters under controlled conditions. Although DNA uptake by plants was reported well before the emergence of these still delicate new breeding techniques, the importance is much affected by the success of the other. If exogenous DNA can be taken up by plants and be phenotypically expressed and stably inherited, the consequences for plant breeding would be far more revolutionary than those of somatic cell genetics alone.

Studies of DNA uptake by plant cells began in the early 1960s (4), and the issues raised at that time still remain controversial. The phenomenon is complex, both technically and conceptually.

For comparison, it is of interest to outline briefly what is known about DNA uptake and fate in bacteria and mammalian cells. DNA-mediated genetic transformation is a well-established phenomenon in the genera *Bacillus, Haemophilus* and *Diplococcus*. In all these cases, the homologous donor DNA is taken up by the cells at a definite time in the growth cycle called the "competence phase," and recombination with the host DNA occurs after separation of the donor DNA strands (5–7). DNA extracted from phylogenetically distant bacterial strains is taken up and degraded; no genetic transformation is effected by heterologous DNA (8, 9).

Escherichia coli can be transformed by purified homologous DNA (10, 11) and represents a very unique case, as heterologous DNAs such as *Xenopus laevis* ribosomal DNA (12), *Drosophila* DNA (13), sea urchin DNA (14) and mouse mitochondrial DNA (15), if ligated into plasmids, are taken up and replicated in the bacterial cells. The possible extension of the technique to eukaryotic cells is an obvious target.

Two reports rather convincingly claim biological expression of bacterial genes in human fibroblasts (16, 17). They are emphasized here because they present pertinent analogies with similar work performed in plants. Mutant recipient cell lines were used, and the donor genetic material was either encapsulated in or extracted from transducing coliphages. Human galactosemic cells deficient in α-D-galactose-1-P-uridyltransferase activity were able to survive in the presence of galactose after incubation in the presence of high titers of λpgal bacteriophage carrying the *E. coli* galactose operon (16). Bacteriophage-specific mRNA production in the infected cells was detected by RNA · DNA hybridization, and cell extracts converted UDP-galactose into galactose-1-*P*, whereas control cell extracts were devoid of this activity. Interestingly, purified λ pgal DNA was more efficacious in inducing the appearance of the transferase activity in the treated cells. In appropriate control experiments with phage λ and λ pgal $(K^+T^-E^+)$ containing a mutation that renders the transferase gene inactive, no significant transferase activity was detected. Unfortunately, no direct evidence pleading for the bacterial origin of this enzyme activity was provided.

In conceptually similar experiments involving transfer of the *E. coli* lactose operon to human fibroblasts deficient in β-galactosidase activity, uptake of λplac and λplac DNA was inferred from the induction in the treated cells of a galactosidase activity sharing common chromatographic properties with bona fide *E. coli* β-galactosidase *(17)*. Furthermore, this enzyme activity could be protected from heat denaturation by anti-*E. coli* β-galactosidase antibodies. In this case as well, purified λplac DNA was much more efficient in inducing the phenomenon. Another crucial point is that, for unknown reasons, the mutant fibroblasts responded to λplac in only three out of nineteen experiments, and to λplac DNA in four out of sixteen experiments. Furthermore, induced enzyme levels varied widely among the successful experiments—a feature burdensome for those attempting to duplicate these results. As discussed below (Section II, B, 2; ref. *84*), similar variations are encountered in plant systems. Finally, it must be stressed that nothing is known about the long-term fate of λpgal and λplac DNA taken up by mammalian cells. [Many more studies on the fate of DNA in mammalian cells have been published and have recently been extensively compiled *(18)*.]

In the case of plants, essentially two lines of research have been followed in tracing the fate of exogenous DNA: (i) a biochemical approach in which donor DNA is detected by some property such as an isotopic label or buoyant density in CsCl density gradients; and (ii) biological (or genetic) experiments in which the host plants are examined for the appearance of novel proteins or functions, such as growth under nonpermissive conditions. An analysis of these two categories of experiments furnishes the framework for the discussion that follows.

II. Experimental Systems

Technical details are usually omitted in review articles. However, in this case, the techniques require intensive scrutiny.

A. Biochemical Evidence

Without exception, all claims of successful integration of donor with host DNA, that is, formation of covalent linkages between them, are based only on buoyant density studies of the DNA. In these reports, the radioactive donor DNA was either rich in dG and dC, such as the DNA from *Micrococcus lysodeikticus* or *Streptomyces coelicolor* (containing ca. 70% dG · dC pairs), or was of homologous origin, or contained 5-bromodeoxyuridine, conferring a much higher buoyant density than the recipient DNA.

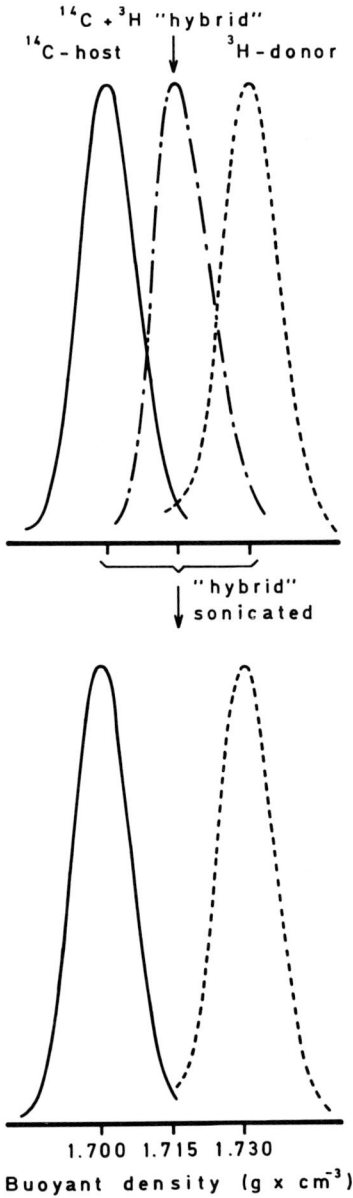

FIG. 1. Idealized representation and behavior of "hybrid" band DNA in a CsCl equilibrium density gradient.

Upper panel: ^{3}H-labeled donor DNA and recipient DNA labeled with ^{14}C form two distinct and well-separated bands in a CsCl gradient. When plant cells are labeled with, e.g., [^{14}C]thymidine, then incubated with nonisopycnic donor [^{3}H]DNA, and finally

Integration of the radioactive donor DNA with host DNA was inferred from the presence of label in the CsCl gradients at a position intermediate between that of the donor DNA and that of the recipient DNA. For example, if the DNA of the recipient cells had a buoyant density of 1.700 g × cm^{-3} and the donor DNA a buoyant density of 1.730 g × cm^{-3}, then, the detection of a "hybrid" band of mean density 1.715 g × cm^{-3} was taken to reflect a covalent bonding of one-half "light" recipient DNA to one-half "heavy" donor DNA. An end-to-end joining of double-stranded DNA molecules was inferred from experiments in which ultrasonication of the "new" DNA fractions yielded DNA molecules displaying the buoyant density values of the original donor and recipient DNAs (see Fig. 1). In turn, incorporation of [^{3}H]- or [^{14}C]thymidine into these "hybrid" molecules was taken to reflect their *de novo* biosynthesis in the donor-DNA-treated plant cells.

Such hybrid DNA molecules were also observed in competent *B. subtilis* cells transformed with density-labeled homologous DNA. In this case, however, only single-stranded donor DNA was integrated into the recipient genome (5). Although the approaches are parallel and similar, it now appears that such analysis of highly complex genomes containing DNA sequences of various degrees of repetition is less conclusive than the analysis of bacterial systems. The recent use of sequence-specific techniques has in fact led to a profound revision of some older observations. Before assessing the current situation, it must be pointed out that the studies devoted to the fate of exogenous DNA in plants have involved a rather scattered array of donor DNA recipient organisms. Therefore, the studies have been arbitrarily grouped into subdivisions to facilitate systematic evaluation.

1. Germinating Seeds and Seedlings

There are at least twenty reports on DNA uptake by seeds or seedlings. Ledoux *et al.* (4) first observed that *E. coli* [^{3}H]DNA could migrate through the endosperm of sectioned barley seeds placed on a drop

---

extracted from the treated tissues and analyzed in a CsCl gradient, a DNA band containing *both* radioisotopes and located at an intermediate position between donor and recipient DNAs is referred to as a "hybrid" band. This "hybrid" band can coexist with residual "free" donor and recipient DNAs.

Lower panel: The "hybrid" band fractions are pooled, possibly purified by rebanding, and submitted to ultrasonic oscillations. After banding of the sonicated "hybrid" band in a CsCl gradient, the donor and recipient DNA bands are regenerated as a consequence of the breaking of phosphodiester bonds originally linking donor and recipient DNAs. This representation does not take into account the band-broadening caused by the depolymerizing effect of sonication.

of the DNA solution. The radioactivity recovered from the roots of the treated seedlings corresponded to donor DNA, judged from CsCl buoyant density analysis, and was said to be located mostly in nuclei undergoing division, as revealed by autoradiography. Furthermore, the preservation of this radioactive donor DNA in the roots was only ensured when the germinating seeds were subsequently transferred to a solution of unlabeled donor DNA for several hours (4, 19). The DNA chase presumably saturated the DNase activity of the cells.

These straightforward conclusions may be challenged. Autoradiographic data do not necessarily indicate the location or the nature of the label. The number of cells thus observed is several orders of magnitude smaller than the number involved in biochemical experiments. The issue of truly representative sampling remains open. So too does the issue of the source of the label. Reutilization of the donor DNA would be expected to yield labeled nuclei. Moreover, donor DNA might remain more or less undegraded in the vascular system of the plant or strongly adsorbed on the outside of the roots despite washing, and contribute to the analytical gradients.

Further studies (20, 21) added the claim that heterologous DNA molecules can become covalently bound *in vivo*. As already indicated, the claim was based on the appearance of fully denaturable "hybrid" density bands in CsCl gradients and their resolution by sonication.

The question now raised concerns the nature of the "hybrid" bands. More than one possibility can be envisaged.

1. "Hybrid" bands are the result of the integration and replication of large pieces of foreign DNA linked to the recipient DNA. The credibility of this claim is questionable on the ground of reproducibility; it has not been observed by others using the same system and similar experimental procedures (22–24).

2. "Hybrid" bands correspond to DNA of contaminating microorganisms. Indeed, barley seeds, especially the covered varieties, are known to be contaminated with bacteria and fungi. To eliminate these, thorough sterilization procedures are necessary (24). Kleinhofs *et al.* (23) and Kleinhofs (24) sterilized their covered barley seeds with 50% H_2SO_4 followed by immersion in 5% $Ca(OCl)_2$, whereas Ledoux and Huart (20, 21) used 3.5% NaOCl only. The former (23, 24) never observed "hybrid" bands displaying the properties described by the latter (20, 21). The more stringent sterilization procedure seems to have eliminated the production of "hybrid" DNA bands from barley. If so, the claim that "hybrid" bands split up into apparently donor and recipient DNA bands upon sonication is unexplained. Moreover, plants grown in the presence of [^{3}H]thymidine alone do not show any un-

usual DNA bands. The answer may lie in the character of the profiles of the barley DNA in CsCl density gradients; they are very broad and thus indicative of a low degree of polymerization. In contrast, the profiles of the "hybrid" bands in the same gradients are always narrow and suggest a preferential extractability of these molecules. Moreover, the incorporation of [^{3}H]thymidine is always much higher in the "hybrid" bands as compared to the barley DNA. This suggests an exogenous origin of these unusual DNA species. It is indeed difficult to understand how the integration of a segment of bacterial DNA could promote a preferential synthesis of the site of integration and its surroundings unless the specific mechanisms of DNA replication are fundamentally altered as a consequence of the presence of large amounts of unrelated DNA.

3. The "hybrid" band and its sonic dissociation are due to impurities present in the barley DNA preparations. This was shown to occur by Kleinhofs *et al.* (23), but their data are not really representative of what was published by Ledoux and Huart (20, 21). On the other hand, the "hybrid" band is present even in preparations further purified by gel filtration on agarose gels (L. Ledoux, personal communication).

4. The "hybrid" band is an amplified preexisting plant satellite DNA. Barley cells contain two satellite DNA species separable from main-band DNA by centrifugation in Ag^+/Cs_2SO_4 gradients (25). They are not good candidates for the "hybrid" band since their buoyant densities in CsCl are 1.702 and 1.698 $g \times cm^{-3}$. However, barley main-band DNA contains slowly reassociating DNA fractions displaying heterogeneous banding patterns in CsCl gradients with buoyant density values distinctly higher than 1.711 $g \times cm^{-3}$. Furthermore, very rapidly reassociating fractions exhibit sharp bands at a density of 1.707 $g \times cm^{-3}$. The amplification of these DNA fractions as the result of unusual physiological conditions may give results similar to those of Ledoux and Huart (20, 21), but this remains to be proved.

5. The "hybrid" band is a complex mixture of amplified barley DNA and contaminant DNA. In such a case, it may very well be that a definitive answer will never be obtained owing to the variations in the local microbial flora and barley varieties.

Essentially two groups have challenged the possibility that bacterial DNA might integrate within barley DNA. The work of one of them (23, 24) is considered above. Hotta and Stern (22) also oppose the integration hypothesis as an explanation of "hybrid" bands. In their hands, barley seeds allowed to germinate under physiological conditions do not integrate *M. lysodeikticus* DNA into their own DNA so as to yield a "hybrid" density band in CsCl gradients. In this, they are in agreement with Kleinhofs et al. (23). Interestingly, they detected a DNA band reminiscent of the "hybrid" band when the barley seeds were X-irradiated or kept at low humidity after exposure to bacterial DNA. Furthermore, a large amount of the bacterial and "hybrid" DNA appeared to be present in the cytoplasmic fraction of the barley homogenate. It was concluded that no covalently linked hybrid material was present inside the nuclei and that such a binding was the result of a pathological process.

It is rather clear from the above that no consensus regarding the fate of bacterial DNA in barley has been reached. These results also point out the failure of the CsCl density gradient centrifugation technique to yield by itself a straightforward answer to all the questions raised.

The claim of heterologous DNA integration and replication in barley ultimately relies on a single physical parameter: resolution of a hybrid band by sonication into two bands possessing buoyant density values close to those of donor and recipient DNAs. Quite evidently, the question of "hybrid" bands must be subjected to critical evaluation by techniques based on other parameters, such as base-pairing properties. The reannealing of single-stranded DNA is known to be very specific. Hence, a new molecule containing both plant and bacterial DNAs should hybridize with both donor and recipient DNAs. Positive preliminary results of DNA of hybridization experiments (26). cannot be regarded as final since the reassociation of DNA was done in solution using high-M_r bacterial DNA as a driver; high-M_r DNA does not allow specific reassociation of the DNA strands (27).

A new insight into the controversy can be found in the work of Charles (P. Charles, personal communication), which confirms that covered barley seeds sterilized with 3.5% NaOCl and further incubated for 24 hours in the presence of [^{3}H]thymidine contain only one labeled DNA species of buoyant density 1.702 g $\times$ cm^{-3}, corresponding to barley DNA synthesized in the embryo. However, if the source of radioactive label is *M. lysodeikticus* [^{3}H]DNA extensively degraded by incubation with an acellular extract of germinating barley, there is preferential incorporation in the endosperm of radioactivity into DNA molecules displaying buoyant density values between 1.709 and 1.715 g $\times$ cm^{-3}. Thus, incubation in the presence of polymerized donor DNA is not a prerequisite to observe "hybrid" bands in CsCl density gradients. That these bands consist of DNA molecules is shown by their sensitivity to pancreatic deoxyribonuclease and their denaturation at alkaline pH.

The mechanism by which "hybrid" band DNA synthesis occurs is still obscure, and it still has to be demonstrated whether Charles' "hybrid" band DNA corresponds or not to similar DNA found by others (20, 21). In any case, these experiments show that that incubation of seeds with [^{3}H]thymidine alone is not the best available control in DNA uptake experiments. It may be that donor DNA degradation products stimulate the growth of microorganisms not killed by the sterilization procedure or else that these unusual conditions perturb DNA synthesis in the developing embryo.

Subsequent to the studies on barley, it was claimed that the crucifer *Arabidopsis thaliana* is also able to integrate high-density bacterial DNA into its genome (28–30). In this case, radioactive donor DNA was supplied to swelling seeds and was later traced in the growing plants. The DNA extracted from young plants treated as seeds with *S. coelicolor* [^{3}H]DNA ($d = 1.730$ g $\times$ cm^{-3}) displayed two radioactive bands in CsCl gradients. The main band corresponded to labeled endogenous DNA ($d = 1.698$ g $\times$ cm^{-3}), and the second band had a buoyant density ($d = 1.720$ g $\times$ cm^{-3}) intermediate between that of donor and recipient DNAs. The latter was thus a "hybrid" band covalently linking donor and recipient DNAs. As before, sonication split this "hybrid" band. The response of *Arabidopsis* to heteropycnic DNA resembled that of barley. Despite the identity in both the experimental approach and techniques involved, however, the study with *Arabidopsis* was much more elaborate than that with barley inasmuch as *Arabidopsis* can be grown to maturity under laboratory conditions. On analysis of the mature plant, molecules of hybrid density were found mainly in cotyledons, whereas roots, leaves and flowers contained mixtures of donor and recipient radioactive DNA molecules. Control experiments with [^{3}H]thymidine as the sole source of radioactivity never yielded "hybrid" bands; only labeled endogenous DNA was found under those conditions.

Experiments with *E. coli* [^{3}H]DNA ($d = 1.710$ g $\times$ cm^{-3}) gave results similar to those reported for the uptake of *S. coelicolor* DNA except that the "hybrid" band had a buoyant density of 1.708 g $\times$ cm^{-3}. However, owing to the small difference in density between *E. coli* DNA and the recipient DNA, it seems perilous to draw a parallel between experiments performed with *E. coli* DNA and those done with donor DNA richer in G · C pairs. Indeed, artifactual "hybrid" bands can result from the combination of the Gaussian distributions of two DNA species, depending on their degree of polymerization and the difference in buoyant density (31). Experiments with *S. coelicolor* DNA seem more appropriate for analysis. It was considered that the latter spontaneously breaks up its covalent association with the recipient DNA at flowering time and migrates toward the flowers, where it is stored mainly as free exogenous DNA. It then reappears as a "hybrid" band in the seeds derived from the [^{3}H]DNA-treated plants. The fate of foreign DNA in *Arabidopsis* as described is complicated, since it requires specific integration and release mechanisms to effect a translocation of the donor DNA molecules and to preserve them through meiosis. Moreover, recognition of donor DNA seems to depend only on its being foreign, not, for instance, on its base sequence, since the above pattern is said to hold true for *E. coli, B. subtilis* and *S. coelicolor* DNAs.

Another intriguing observation is that sonication of the "hybrid" band or of the recipient DNA released radioactive molecules said to be of donor nature. This observation implies either two different mechanisms for integration of exogenous DNA, or a preferential breaking of DNA molecules during the isolation procedure. Indeed, a simple calculation *(32)* shows that a "hybrid" band of density 1.720 g × cm^{-3} assumed to contain donor DNA of density 1.730 g × cm^{-3} and recipient DNA of density 1.698 g × cm^{-3} contains ca. 70% donor DNA and 30% recipient DNA. Since the declared M_r of DNA extracted from *Arabidopsis* is 10^7 *(30)*, the "hybrid" band DNA must contain stretches of *S. coelicolor* DNA having an M_r around 7 × 10^6. On the other hand, endogenous material is said to contain integrated pieces of foreign DNA with M_r no higher than 5 × 10^5. Such putative integration of foreign DNA requires either two homogeneous populations of donor DNA fragments, or that integration of foreign DNA causes a weakening of the DNA backbone at some specific sites. If stretches of donor DNA were integrated irrespective of their length or without causing site-specific DNA fragility, one would expect to obtain broad banding patterns with extreme buoyant density values ranging from that of donor DNA to that of recipient DNA depending on the respective proportions of these DNA stretches in the "hybrid" molecules.

The replication of bacterial DNA supplied to *Arabidopsis* was traced by applying unlabeled *M. lysodiekticus* DNA followed by [^{3}H]thymidine to swelling seeds *(33, 34)*. No unusual DNA band was found in plants growing from the DNA-treated seeds, but "loading" successive generations with constant amounts of *M. lysodiekticus* DNA caused the appearance of a more and more prominent satellite band in CsCl gradients. The proportion of this satellite DNA relative to main-band DNA as well as its buoyant density increased as a function of the number of treatments with foreign DNA. It was concluded that these "hybrid" bands reflected increased integration and replication of foreign DNA. Here also, sonication of the "hybrid" bands was used to infer the exogenous nature of part of the constituents of this band.

From these observations, it can be conservatively estimated that ca. 10% of the DNA present in the plants after four successive treatments is of foreign origin. Owing to the low DNA content per nucleus of *Arabidopsis*, this large amount of supposedly integrated and replicated foreign DNA would represent the equivalent of about 100 bacterial genomes per plant cell genome. It may be wondered whether plant cells can actually accommodate such a great proportion of unrelated DNA. The implications of such a phenomenon are profound, and it would therefore be useful to take a close look at DNA metabolism in senescing cotyledons of *Arabidopsis*, since they play such an important role in foreign DNA translocation.

Studies by others do not support the formation of "hybrid" band DNA in *Arabidopsis* seedlings treated with exogenous DNA in liquid medium. Figure 2 shows some typical CsCl equilibrium density gradients of DNA extracted from *Arabidopsis* seedlings treated for various periods of time with *Agrobacterium tumefaciens* [^{3}H]DNA ($d = 1.718$ g × cm^{-3}). The plants were grown for 7 days until expansion of the cotyledons occurred and then treated with radioactive donor DNA. It can be seen that some heavily degraded bacterial DNA appears after 8 hours of contact in close association with the seedlings (deoxyribonuclease-resistant). However, this broad band subsequently decreased and vanished at the expense of the endogenous DNA band, which became prominently labeled after 192 hours of exposure to radioactive donor DNA. No high-M_r donor DNA was thus found associated with the treated plants. On the contrary, it seems that rapid breakdown of donor DNA occurred and that incorporation of label into plant DNA corresponded to the reutilization of the breakdown products. Degradation of the donor DNA in the culture medium is indeed shown in Fig. 3, where it can be seen that the sedimentation velocity of donor DNA present in the medium drastically decreased with time. A similar drop in sedimentation velocity in neutral sucrose gradients (not shown) ruled out the production of single-strand breaks only. Thus, *Arabidopsis* seedlings grown in liquid medium do not preserve the structure of bacterial DNA.

In another experiment, donor [^{3}H]DNA was mixed in liquid medium together with ungerminated seeds, and growth was allowed for 9 days. Again, only labeled endogenous DNA was found as in Fig. 2, meaning that seeds swelling in liquid medium and growing in the presence of the donor DNA do not maintain the high-M_r donor DNA

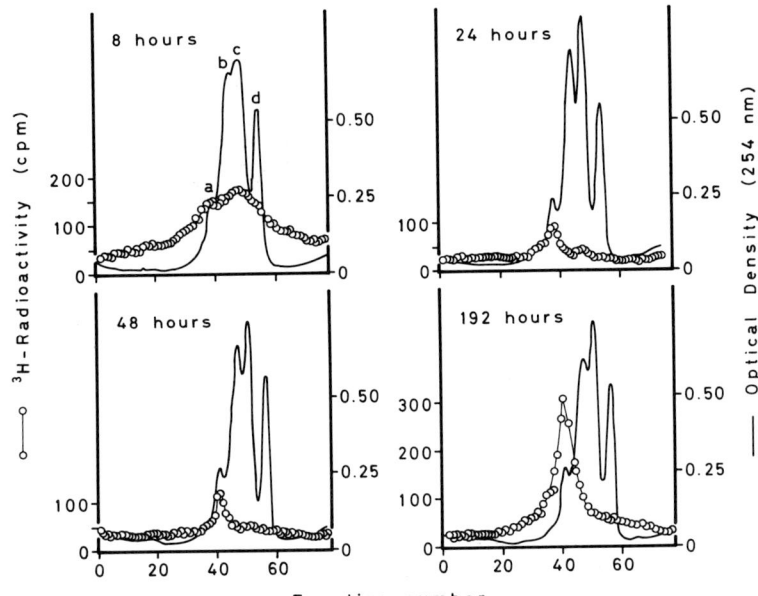

FIG. 2. Uptake of bacterial DNA by *Arabidopsis* seedlings. Two hundred *Arabidopsis thaliana* seeds were allowed to germinate in 20 ml of liquid medium containing, per liter: 20 g of glucose, 0.1 g of $MgSO_4 \cdot 7H_2O$, 0.1 g of KH_2PO_4, 0.05 g of K_2HPO_4, 0.1 g of $CaH_4(PO_4)_2$, 0.4 g of $Mg(NO_3)_2 \cdot 6H_2O$, 0.4 g of KNO_3, 0.026 g of Na_2EDTA, 0.025 g of $FeSO_4 \cdot 7H_2O$, 0.0015 mg of MoO_3, 1.8 mg of $MnCl_2 \cdot 4H_2O$, 0.22 mg of $ZnSO_4 \cdot 7H_2O$, 7.9 μg of $CuSO_4 \cdot 5H_2O$, and 2.9 mg of H_3BO_3, pH 6.0. Growth was allowed for 8 days under continuous illumination. *Agrobacterium tumefaciens* [^{3}H]DNA (100,000 cpm/μg) was then added to a concentration of 10 μg/ml. After the indicated incubation times, plantlets were recovered, washed, treated with pancreatic DNase, and processed for DNA extraction (53). The *Arabidopsis* DNA (a) ($d = 1.696$ g × cm^{-3}) was analyzed in CsCl density gradients in the presence of (b) *Escherichia coli* ($d = 1.710$ g × cm^{-3}), (c) *A. tumefaciens* ($d = 1.717$ g × cm^{-3}) and (d) *Micrococcus lysodeikticus* ($d = 1.731$ g × cm^{-3}) reference DNAs. Symbols: ———, optical density at 254 nm; ○——○, ^{3}H.

either. No experiments were done to detect the possible presence of small pieces of donor DNA in the recipient DNA.

When *Arabidopsis* seeds were incubated for 48 hours in the presence of *M. lysodeikticus* [^{3}H]DNA as described by Ledoux et al. (30), or in the presence of an equivalent amount of the same DNA extensively degraded by pancreatic deoxyribonuclease (Fig. 4), the DNA extracted from callus tissue derived from these treated seeds did not contain labeled DNA species other than the endogenous one. In addition, it is clear that previously degraded donor DNA is a much better substrate for endogenous DNA synthesis than high-M_r donor DNA. It is possible that uptake of low-M_r compounds is more efficient than uptake of polymerized DNA. Thus, it seems that bacterial DNA taken up by *Arabidopsis* seeds is not preserved or integrated to yield a "hybrid" band if the seeds are grown on medium allowing callus formation.

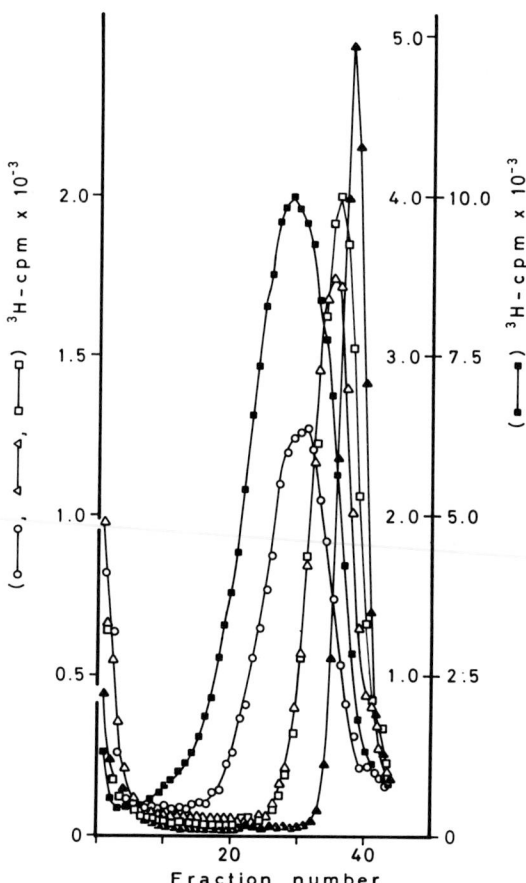

FIG. 3. Analysis in 5 ml of 5 to 20% alkaline sucrose gradients of aliquots of the culture medium from Fig. 2 containing the donor [^{3}H]DNA. Centrifugation was for 2 hours at 45,000 rpm and at 20°C in a Martin-Christ SW52 rotor. Symbols: ■——■, donor DNA at zero time; ○——○, donor DNA after 8 hours of incubation with the plantlets; △——△, donor DNA after 24 hours; □——□, donor DNA after 48 hours; ▲——▲, donor DNA after 192 hours.

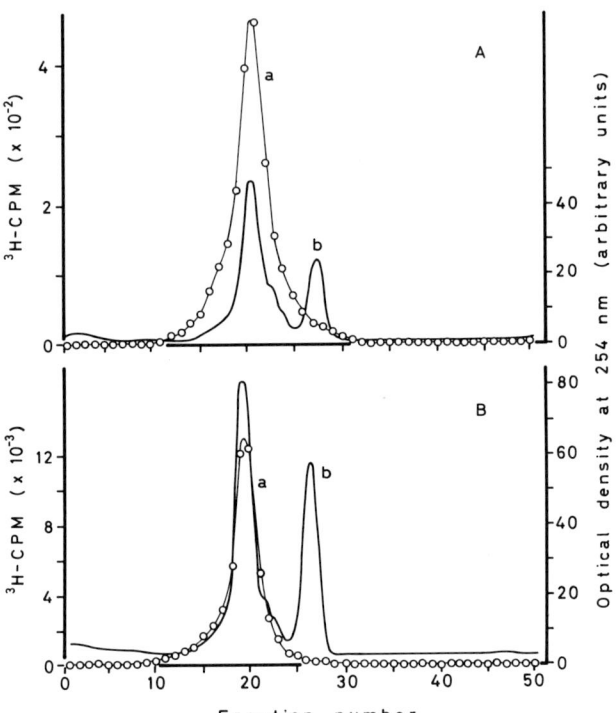

FIG. 4. Fate of bacterial DNA taken up by *Arabidopsis* seeds. One hundred *Arabidopsis thaliana* seeds were incubated for 48 hours either in the presence (A) of [^{3}H]DNA from *Micrococcus lysodeikticus* according to (30) or (B) in the presence of an equivalent amount of *M. lysodeikticus* [^{3}H]DNA previously degraded with pancreatic deoxyribonuclease. Seeds were then plated on B$_5$-agar, and callus growth was allowed for 2 months at 27°C in the dark. DNA was then extracted and analyzed by CsCl density gradient centrifugation (53). *Arabidopsis* DNA ($d = 1.695$ g $\times$ cm^{-3}) is at (a). The reference DNA (b) is from *Streptomyces coelicolor* ($d = 1.730$ g $\times$ cm^{-3}). Note the difference in scale. Symbols: ——, optical density at 254 nm; O——O, ^{3}H.

It may be that absorption of donor DNA by dry seeds followed by growth into fully differentiated plants is a prerequisite for long-lasting preservation of the foreign DNA.

The heterogeneity seen on the dense side of *Arabidopsis* mainband DNA is reproducibly observed and may represent satellite or organelle DNA. Figure 5 shows that rebanding of *Arabidopsis* callus DNA labeled with [^{3}H]thymidine causes a definite heterogeneity in the CsCl banding patterns. The nature and location of the DNA species of buoyant density 1.706 g $\times$ cm^{-3} is still unknown, but its existence cannot be ignored when interpreting results of heteropycnic DNA uptake experiments.

Unusual DNA species appearing after treatment of plants with exogenous DNA have also been observed by others. Rebel et al. (35) discovered that the crucifer *Matthiola incana* apparently took up ^{32}P-labeled bacteriophage T4 DNA whose buoyant density ($d = 1.694$ g × cm^{-3}) is 0.004 g × cm^{-3} less than that of the recipient DNA. Nevertheless, most of the radioactivity extracted from the DNA-treated seedlings was found to band in CsCl gradients at a buoyant density of 1.724 g × cm^{-3}. These molecules were fully denaturable and were said to correspond to the integration of fragments of T4 DNA within a recipient DNA fraction of very high buoyant density. Indeed, labeling of *Matthiola* seedlings with [^{32}P]orthophosphate or with T4 [^{32}P]DNA previously degraded with DNase produced a shoulder on the heavy side of main-band DNA. This time, however, sonication of the prominently labeled DNA band observed after incubation of seedlings with polymerized T4 [^{32}P]DNA caused no shift or splitting of the DNA distributions in a CsCl gradient. It seems that concluding integration of foreign DNA at this stage may be somewhat premature.

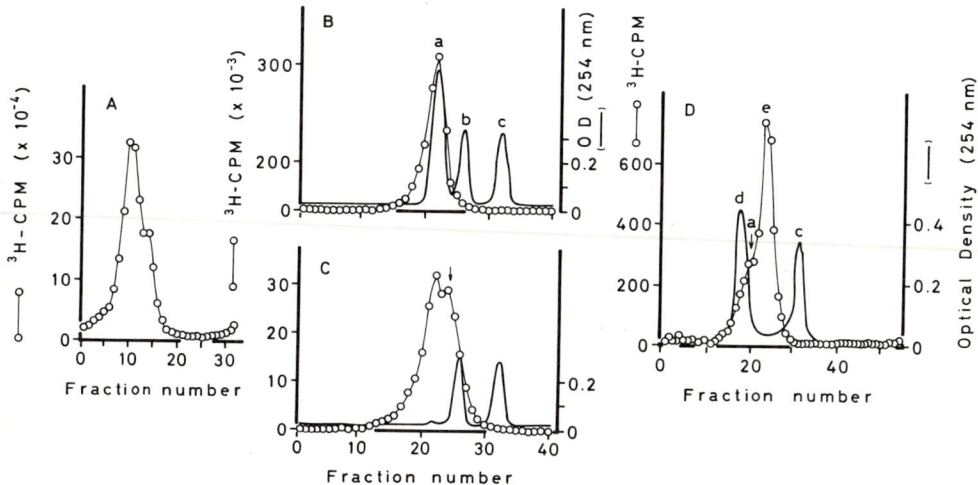

FIG. 5. Detection of satellite DNA in *Arbidopsis* cells growing in tissue culture. *Arabidopsis thaliana* cells in callus tissue culture were labeled with [6-^{3}H]thymidine (18 Ci/mmol) for a period of 6 days and then processed for DNA extraction (53). (A) CsCl banding pattern of total DNA in a preparative Spinco 40 rotor; (B) Rebanding of the left-hand part of the distribution shown in (A); (C) Rebanding of the right-hand part of (A); (D) Rebanding of the dense fractions (starting from the arrow) shown in (C). Gradients in (B), (C) and (D) were spun in a Martin-Christ SW60 rotor. Reference DNAs are from (d) *Clostridium perfringens* ($d = 1.691$ g × cm^{-3}), (b) *Escherichia coli* ($d = 1.710$ g × cm^{-3}), and (c) *Micrococcus lysodeikticus* ($d = 1.731$ g × cm^{-3}); (e) denotes the *Arabidopsis* satellite-DNA band ($d = 1.706$ g × cm^{-3}). Symbols: ———, optical density at 254 nm; ○———○, ^{3}H.

These studies were further extended to the uptake of other foreign DNAs (36). The incubation of *Matthiola* seedlings with labeled DNA extracted from *Streptomyces griseus, Diplococcus pneumoniae* and *B. subtilis* invariably gave rise to the incorporation of label into endogenous DNA only. Moreover, transformation experiments involving *D. pneumoniae* cells and DNA extracted from seedlings inclubated with *D. pneumoniae* DNA were all negative. In other words, integration within a dense *Matthiola* satellite DNA would seem to be a particular property of bacteriophage T4 DNA. Furthermore, the 1.724 g × cm^{-3} DNA band resulting from the incubation with T4 [^{32}P]DNA was found to be preferentially lost upon transfer of the treated seedlings to water and did not replicate to any extent. Filter hybridization of DNA to DNA, performed with filter-bound T4 DNA and the labeled dense satellite gave a positive response, but unfortunately there was no attempt to determine the melting temperature of the hybrid formed. In any case, the effect induced by T4 DNA seems to be purely transient.

It is recalled (36) that T4 DNA is highly glucosylated and might therefore be more resistant to plant deoxyribonucleases than other heterologous DNAs. This suggestion could easily be checked, as nonglucosylated T-even bacteriophage DNA can be obtained by growing the phage on a suitable *E. coli* strain. The possible existence in plants of a mechanism similar to the DNA restriction and modification systems known in bacteria is worth mentioning and should be subjected to some experimental work.

Following the idea that restriction and modification of foreign DNA might lead to its rapid destruction by recipient plant cells, Hemleben et al. (37) fed homologous density-labeled radioactive DNA to *Matthiola* seedlings. In these experiments, density-labeled homologous DNA essentially replaced the high-density G · C-rich bacterial DNA used by others. Another interesting difference is that excess unlabeled competitor (adenine) was present in the incubation medium to decrease the reutilization by the plant cells of the breakdown products of radioactive donor DNA. Otherwise, the sequence of experiments closely followed the others (21, 30), including the use of buoyant density determinations in CsCl gradients. The radioactive donor DNA, density-labeled with 5-bromodeoxyuridine, had a buoyant density of 1.741 g × cm^{-3} in CsCl density gradients as compared to 1.698 g × cm^{-3} for the recipient DNA. When sonicated, it looked rather homogeneous as judged from the banding profile in a neutral CsCl gradient but seemed fairly heterogeneous when banded in an alkaline CsCl gradient.

Upon incubation of *Matthiola* seedlings with this DNA, a prominent "hybrid" band of density 1.717 g × cm^{-3} was observed in CsCl gradients together with a less important band of donor DNA. Virtually no radioactivity was located at the level of the recipient DNA. Conversely, in a novel procedure, when radiolabeled "light" *Matthiola* DNA was fed to density-labeled *plants,* the result almost mirrored the previous one in that a "hybrid" band of density 1.719 g × cm^{-3} coexisted with a band of "light" donor DNA. Appropriate controls on the incorporation of 5-bromodeoxyuridine and [^{3}H]adenine in the presence or in the absence of unlabeled competitors eliminated a gross artifactual origin of these "hybrid" bands, but one wonders if these are the right controls for this kind of experiment (see above).

Here also, sonication of the "hybrid" band caused peak splitting, reminiscent of what was observed by others *(21, 30),* except that in the case of *Matthiola* sonication causes about only one-third of the radioactivity associated with the "hybrid" band to shift toward a buoyant density corresponding to that of the donor DNA, the remainder still banding at the position of the "hybrid" band, not at that of recipient DNA. This result is not easy to understand if one assumes that depolymerization by ultrasonication yields DNA fragments of rather homogeneous length.

Again, integration of exogenous DNA was claimed on the basis of the presence of a "hybrid" band and its dissociation after sonic treatment. However, in view of the possible presence of higher-density satellite DNA in *Matthiola* in an earlier study *(35),* one would like to see additional evidence for integration. Hybridization techniques are obviously not applicable in this case, but the analysis of nearest-neighbors in this "hybrid" band DNA should be feasible and should give an identical result when performed on recipient main-band DNA and "hybrid"-band DNA. A significant deviation of the nearest-neighbor frequency in the "hybrid" band could indicate the presence of DNA sequences not originating from main-band DNA.

Other reports do not favor the integration of exogenous DNA in young plants. Bendich and Filner *(38)* reported the failure of *Pseudomonas aeruginosa* [^{32}P]DNA to be preserved in pea seedlings. Even if low amounts of *P. aeruginosa* DNA could be detected in DNA extracted from treated seedlings, it was obvious that partial degradation and reutilization of the donor DNA occurred within 3 days after incubation. However, no trace of donor DNA could be found inside the seedlings upon analysis of a crude nuclear fraction and the corresponding cytoplasmic extract separately. The interpretation was that the fractionation of the plant tissue into subcellular components favored the degradation of unbound bacterial DNA taken up. Although this is plausible, an alternative explanation can be offered, taking into account the fact that the seedlings were grown for another 27 hours in the presence of unlabeled donor DNA after incubation with radioactive donor DNA. Indeed, excess unlabeled donor DNA competes with radioactive donor DNA loosely bound to tomato tissues and causes the disappearance of a radioactive donor DNA band originally associated with isolated tomato nuclei (22). Thus, Bendich and Filner (38) did not detect unusual DNA species after incubation of pea seedlings with heteropycnic bacterial DNA. In a later study (23), replication as such or as a "hybrid" band of *P. aeruginosa* DNA administered to pea seeds, seedlings and shoots was completely ruled out on the basis of buoyant density analysis.

A similar conclusion was drawn by Kado and Lurquin (39), who showed that *A. tumefaciens* DNA supplied to swelling mung bean seeds did not replicate in the growing plants to any extent detectable by DNA · DNA filter hybridization. This absence of foreign DNA replication could not be attributed to a very rapid degradation of the donor DNA since Kado and Mojica-a (40) showed that transforming activity could be recovered from the leaves of young plants derived from mung bean seeds incubated with *B. subtilis* transforming DNA. Since no transforming activity was found associated with isolated cells obtained from plants otherwise treated in the same way, these authors concluded that transforming DNA was present in the extracellular spaces and was not inside the cells. Finally, Luyindula and Lurquin (unpublished) observed a very rapid breakdown of *A. tumefaciens* [^{3}H]DNA by germinating soybean seeds. Only labeled endogenous DNA was found in plants grown from these treated seeds.

As a partial conclusion, it can be said that attempts to duplicate results claimed to reflect integration of exogenous DNA within plant genomes have failed. One reason for this may be that some unrecognized subtle experimental conditions are stringently required to obtain a positive response from the plants. Even so, it would be of interest to submit these putative hybrid DNA molecules to sensitive and specific DNA · DNA hybridization techniques. Unfortunately, the latter are also not without limitations, as is discussed in Section III, B.

2. WHOLE PLANTS AND PLANT PARTS

After showing that *A. tumefaciens* [^{3}H]DNA could be taken up by tomato shoots where it was partially preserved and partially degraded and reutilized (41), Anker and Stroun (42) claimed that this donor DNA could become integrated within the plant DNA. Their observations can be summarized as follows. The DNA isolated from tomato shoots dipped into a solution of *A. tumefaciens* [³H]DNA showed two bands when analyzed by CsCl density gradient centrifugation. One band corresponded to donor DNA ($d = 1.722$ g × cm^{-3}) and the other one had the density of tomato DNA ($d = 1.692$ g × cm^{-3}). Thus, no "hybrid" band was observed, but donor DNA sequences were said to be released from bulk plant DNA by sonication as shown by the appearance of a band at the buoyant density of donor DNA. In view of the considerable overlapping of the two DNA bands (donor and recipient) detected in the original extract, it may be wondered whether the tomato DNA molecules were actually obtained free of contaminating donor DNA before sonication. Since no rebanding of tomato DNA was done, it seems unlikely that at least part of the donor DNA sequences said to be released by sonication were not already present as such as the result of peak cross-contamination.

Shortly afterward, Stroun *et al. (43)* studied the possible replication of bacterial DNA in tomato shoots. A classical experimental procedure *(21)* was followed and led to the detection of "hybrid" bands in CsCl gradients when the shoots were given unlabeled DNA from *A. tumefaciens, E. coli* and *M. lysodeikticus* followed by [³H]thymidine. In each of these cases, the buoyant density of the "hybrid" band was a function of the buoyant density of the donor DNA. For instance, incubation with *Clostridium perfringens* DNA, which has a buoyant density very close to that of tomato DNA, yielded no "hybrid" band. Surprisingly, and reminiscent of the results obtained with barley *(21)*, virtually all the [³H]thymidine was incorporated into the "hybrid" band whereas almost no label was found at the position of tomato DNA. In this study also, the "hybrid" band was denaturable and split upon sonication in such a way as to give two DNA bands in CsCl gradients, one corresponding to donor DNA and the other to recipient DNA. Thus, integration and replication of bacterial DNA was concluded as before. Again, one deplores the absence of more specific techniques (including bacteriological tests) owing to the revolutionary implications of the conclusions drawn from these results.

Hotta and Stern *(22)* attempted to reproduce the observations described above. They fed tomato shoots with *M. lysodeikticus* [³²P]DNA and subsequently isolated nuclei from the treated tissues. They found it necessary to do so since removal of donor DNA from intercellular spaces and from vascular tissues was extremely difficult to achieve. DNA purified from these isolated nuclei contained, in addition to labeled tomato DNA, large amounts of donor DNA, which could be completely eliminated provided the treated plants were homogenized in the presence of excess calf-thymus DNA. These authors concluded that the label present in tomato DNA was due to the reutilization of donor DNA breakdown products, as more label became incorporated into host DNA along with the production of radioactive acid-soluble compounds in the tissues. Moreover, sonication of the recipient DNA peak did not modify its unimodal distribution in a CsCl gradient. The displacement of donor DNA from the nuclear fraction also eliminated any presumptive chemical association of this DNA with nuclear structures. Along the same line, Kleinhofs et al. (23) could not demonstrate incorporation of [^{3}H]thymidine into a "hybrid" band after feeding tomato shoots with A. tumefaciens DNA. All the incorporated radioactivity, as in the control plants, corresponded to the host DNA band with a slight trailing of radioactivity toward higher density ascribed to the nuclear satellite present in taomato (44).

A critical step in the study of "hybrid" bands was later taken by Hanson and Chilton (45), who attempted to duplicate the results of Stroun et al. (46) according to which the incubation of tomato shoots in the presence of A. tumefaciens DNA or living A. tumefaciens cells led to the incorporation of [^{3}H]thymidine not only into plant DNA, but also into a "hybrid" band of intermediate buoyant density. This time, Hanson and Chilton (45) were able to detect this new DNA peak, which was most prominent when shoots were incubated in the presence of living A. tumefaciens. Incidentally, a shoulder on the dense side of main band DNA was also detected in control plants. Only the "hybrid" band appearing in the plants upon incubation with A. tumefaciens was studied in detail by DNA-filter hybridization, CsCl density gradient analysis, and DNA · DNA reassociation measurements.

These results showed that purified "hybrid" fractions did indeed separate into two components (d = 1.711 and 1.699 g × cm^{-3}) after sonication or shearing in a French pressure cell. The buoyant density of the less dense component was somewhat variable. This was in fact the first time a "hybrid" band behaving like those reported by others (21, 43, 46) could be reproduced and submitted to a detailed analysis. It turned out that this "hybrid" band DNA did not significantly bind to filter-bound A. tumefaciens DNA, but did bind to filter-bound tomato DNA extracted from whole shoots or nuclei. Residual duplexes formed with A. tumefaciens filter-bound DNA exhibited melting profiles incompatible with the thermal stability of authentic A. tumefaciens DNA duplexes. Moreover, the rate of reassociation of "hybrid" band DNA was not accelerated by driver A. tumefaciens DNA, but was considerably increased by either total or nuclear tomato DNA. Thus, the DNA molecules from the "hybrid" band contained less than 3% (limit of detection) bacterial DNA and preexisted in untreated tomato plants. These authors (45) concluded that there is at present no evidence for the integration and replication of *A. tumefaciens* DNA in tomato shoots and that CsCl buoyant density characterization of plant DNA is not without pitfalls. The location as well as the apparent amplification of this new tomato satellite DNA are not yet understood.

Although this latter study seems to terminate the controversy about "hybrid" bands in tomato, it should be kept in mind that Hanson and Chilton (45) focused their attention on the phenomenon induced by *A. tumefaciens* cells and that it is not certain whether all other "hybrid" bands found in tomato and elsewhere have the same endogenous origin.

Similar experiments were performed by us (Luyindula and Lurquin, unpublished) with soybean shoots as a recipient system. [^{3}H]labeled DNA from *M. lysodeikticus* was supplied to soybean shoots incubated in nutrient medium together with 1 mM unlabeled thymidine as a competitor (75-fold excess as compared to the [^{3}H]thymine content of the donor DNA). Figure 6 shows that considerable amounts of donor DNA could be detected in DNA extracted from the stems of the treated shoots even after pancreatic deoxyribonuclease treatment. Little endogenous DNA became labeled despite the presence of even higher amounts of alcohol-soluble donor DNA breakdown products, because of the competing effect of exogenous thymidine. A 30-hour chase period of the treated shoots with unlabeled *M. lysodeikticus* DNA caused a large decrease in the radioactivity associated with the donor DNA band together with a concomitant increase in low-M_r radioactive compounds, poorly reutilized as expected. Thus, even if the donor DNA could migrate in the plant stems, it was not sufficiently stably bound to escape plant nuclease attack. Autoradiography showed that donor DNA was located in extracellular spaces in close association with cell walls. Moreover, hardly any donor DNA could be detected in the leaves, where a prominent peak of labeled endogenous DNA was found. No "hybrid" band was seen in experiments with either [^{3}H]labeled *A. tumefaciens* DNA or unlabeled DNA followed by [^{3}H]thymidine.

3. Cells in Suspension and in Tissue Culture

Plant cells grown in liquid medium or in tissue culture offer a much more homogeneous cell population than differentiated plants. They therefore constitute a material of choice for the study of the

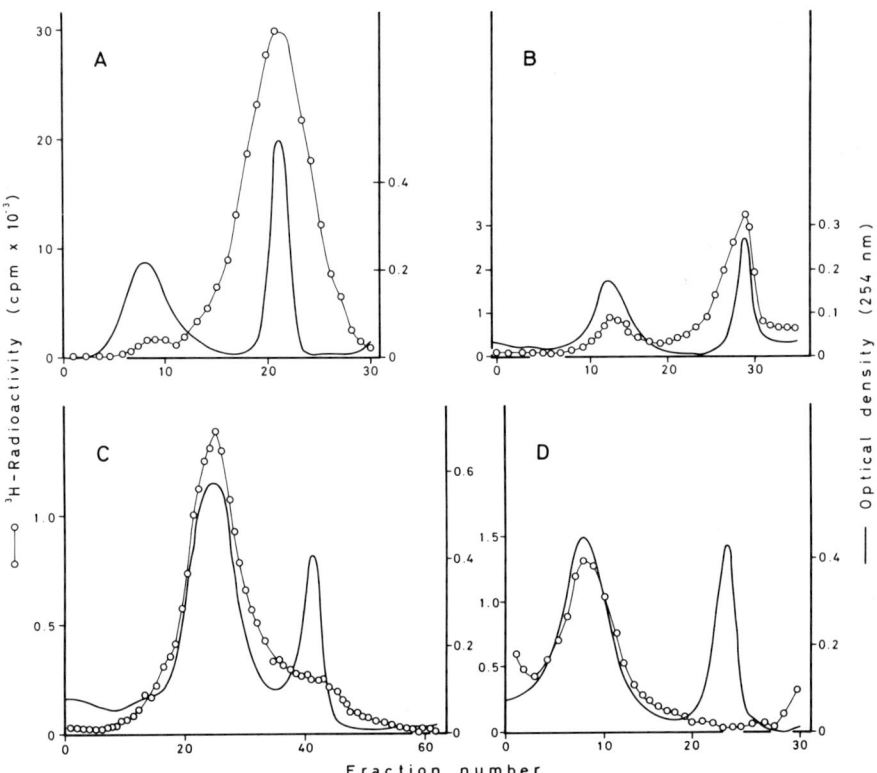

FIG. 6. Translocation and fate of bacterial DNA in soybean shoots. Soybean seeds var. Japan Black were allowed to germinate and grow in medium containing per liter: 0.1 g of $CaCl_2 \cdot H_2O$, 0.12 g of $MgSO_4 \cdot 7H_2O$, 0.15 g of $Na_2HPO_4 \cdot 2H_2O$, and 0.05 g of $FeSO_4$. The plants were harvested after 10 days of growth, their roots were amputated, and the stems were dipped into individual test tubes containing 2 ml of nutrient medium plus 50 µg of [^{3}H]DNA from *Micrococcus lysodeikticus* (950,000 cpm/µg). Unlabeled 1 mM thymidine was present in order to minimize reutilization of donor DNA breakdown products. Leaves were exposed to facilitate transpiration, and plants were collected after most of the DNA solution had been taken up (usually 120 hours during which rooting occurred). Stems and leaves were dissected, washed, and treated with pancreatic DNase and DNA was extracted and analyzed in CsCl density gradients. The chase experiment consisted in transferring the [^{3}H]DNA-treated plants to fresh medium containing 100 µg of unlabeled donor DNA per milliliter for 48 hours. The reference DNA is from *M. lysodeikticus* ($d = 1.731$ g $\times$ cm^{-3}). (A) DNA extracted from the stems; (B) DNA from the stems after the chase period (note the difference in scale); (C) DNA from the leaves; (D) DNA from the leaves after the chase period. Symbols: ——, optical density at 254 nm; O———O, ^{3}H.

uptake, fate and expression of exogenous DNA. In addition, regeneration into normal plants does not represent an unsurmountable obstacle. Comparatively fewer studies have been done with these systems. They were all either negative or inconclusive regarding the integration of exogenous DNA.

Hotta and Stern (22) first published DNA uptake studies on lily meiotic cells cultured at various stages of meiosis in the presence of either *M. lysodeikticus* [^{32}P]DNA or lily [^{32}P]DNA, density labeled with 5-bromodeoxyuridine. They found a very poor incorporation of exogenous DNA at all meiotic stages despite intense incorporation of [^{3}H]thymidine. Moreover, no donor DNA persisted intracellularly after 48 hours of incubation, and all the radioactivity was present in recipient DNA as indicated by CsCl density gradient analysis. This recipient DNA was not analyzed for the presence of small amounts of donor DNA.

Bendich and Filner (38) measured the uptake by tobacco cells in suspension culture of radiolabeled DNA extracted from *P. aeruginosa* and from tobacco cells grown in the presence of D_2O and $K^{15}NO_3$ as the sole nitrogen source. The fate of these heteropycnic DNAs was followed by CsCl density gradient centrifugation. They first found that the cells apparently excreted an enzyme into the medium that rapidly degraded the donor DNAs. This difficulty could be circumvented by washing the cells with fresh culture medium and treating them with Pronase. Under such conditions, an important fraction of the donor DNA remained acid-precipitable after 22 hours of contact with the culture medium in which tobacco cells had been growing for 7 days. After the cells were subjected to the washing/Pronase procedure, they were found to take up about 20% of the radioactivity available as donor DNA. Approximately 0.5% of the input radioactivity was found in recipient DNA isolated from the nuclear fraction, and this could be partly attributed to the presence of depolymerized donor DNA as indicated by the broad banding pattern of the high-density donor DNA peak in CsCl gradients. Furthermore, the transfer of the cells treated with radioactive DNA to fresh medium for another 35 hours led to the disappearance of the donor DNA band, suggesting that the band broadening observed at an earlier time was a sign of progressive donor DNA degradation. Thus, homologous DNA taken up by tobacco cells did not retain its integrity and was broken down and reutilized by the recipient cells.

Experiments with *P. aeruginosa* [^{32}P]DNA gave essentially identical results; that is, 0.1% of the input donor DNA was recovered from a nuclear fraction after a 12-hour incubation. Again, this donor DNA associated with tobacco nuclei had an M_r roughly equivalent to 5–10% that of the input bacterial DNA. The authors concluded that donor DNA (homologous and heterologous) was not integrated within the recipient genomes, but was progressively degraded and reutilized for endogenous DNA synthesis.

In order to study a hypothetical transfer of DNA from the plant tumor inducing *Agrobacterium tumefaciens* to plant cells *(47)*, Heyn and Schilperoort *(48)* investigated the uptake of radioactive *A. tumefaciens* DNA by tobacco cells in suspension culture. Using another variety of tobacco cells, they found very little deoxyribonuclease activity released into the culture medium but suggested that this activity could be induced by the addition of donor DNA. Their donor DNA was labeled with [^{3}H]thymidine and [^{32}P]orthophosphate and was incubated with tobacco cells in the presence of a large excess of unlabeled thymidine in order to minimize reutilization. After a 24-hour incubation, small amounts (0.025% of input) of donor DNA were associated with protoplasts derived from the treated tobacco cells. Isolation of protoplasts after treatment with DNA was performed by these authors in order to decrease the amount of donor DNA adsorbed on the cell walls. Indeed, the treatment of the tobacco cells previously incubated in the presence of donor DNA with Macerozyme and cellulase resulted in the loss of most of the radioactivity initially bound to the cells. Unfortunately, neither Bendich and Filner *(38)* nor Heyn and Shilperoort *(48)* treated their cells with, for example, pancreatic deoxyribonculease prior to DNA isolation. Therefore, it is not certain whether the low amounts of donor DNA recovered from the tobacco cells were truly taken up or merely exchanged either between the cell walls and the nuclear membranes *(38)* or between the cell walls and the cell membranes *(48)*. Heyn and Schilperoort *(48)* also found that pretreatment of the cells with 100 μg/ml DEAE-dextran significantly increased the amount of *A. tumefaciens* DNA bound to the cells. This pretreatment interestingly decreased (almost suppressed) the extent of donor DNA reutilization, but it may be wondered whether this high concentration of polycations did not alter the metabolism of the cells as was found with *Chlamydomonas (49)*. Finally, Heyn and Schilperoort *(48)* considered their results as preliminary and neither proving nor disproving integration of *A. tumefaciens* DNA within tobacco recipient DNA.

Lurquin and Behki *(50)* made a study of the fate of bacterial DNA supplied to cultures of the lower eukaryotic alga *Chlamydomonas reinhardi*. This study, like previous ones, was based on DNA buoyant density analyses in CsCl gradients. However, an attempt was made to determine the heterogeneity in M_r of the donor DNA molecules irreversibly absorbed by the *Chlamydomonas* cells. Indeed, in other studies dealing with DNA uptake by plant cells, only two criteria were used to determine the fate of the donor DNA, namely, acid-precipitability and banding properties in CsCl gradients. Precipitation with trichloroacetic acid may yield an overestimate of DNA uptake because of inclusion of fragments below 5×10^5 daltons (hence not biologically "meaningful") whereas banding in CsCl gradients will obscure the presence of large oligonucleotides of about $1-2 \times 10^5$ daltons, especially if other DNA species with a higher degree of polymerization are present. Therefore, it is desirable to estimate DNA molecular weight heterogeneity, as, for instance, by molecular sieving of DNA on agarose gels (51). Using this procedure, *E. coli* [^{3}H]DNA taken up by *Chlamydomonas* cells and insensitive to pancreatic DNase appeared extremely heterogeneous in size. No labeled recipient DNA was found after incubation of the cells with *E. coli* radiolabeled DNA in exponential phase of growth. In the stationary phase, low-M_r radioactive compounds together with radioactive recipient DNA were detected. The labeling pattern of the endogenous DNA was such that only reutilization could be concluded, since under these conditions the specific radioactivity of chloroplast DNA was much higher than that of nuclear DNA. This pattern is characteristic of the use of [^{3}H]thymidine as a precursor for *Chlamydomonas* DNA synthesis (49, 52).

The same paper (50) discussed for the first time the usefulness of CsCl gradients in the study of a putative integration of foreign DNA within a recipient genome. It was calculated that if the equivalent of 10 bacterial genes were integrated in the genomes of each of 5×10^7 *Chlamydomonas* cells (in a total of 10^8 cells, easily handled in one DNA uptake experiment), then, even if the specific radioactivity of the donor DNA was of the order of 10^5 cpm/μg, *no* radioactivity at all would have been detected in a CsCl gradient because of the low though biologically very "meaningful" amount of donor DNA involved. It was concluded that sensitive DNA · DNA hybridization techniques should rather be used, as was done for the first time by Kleinhofs (24). DNA · DNA reassociation kinetics were used by Lurquin and Hotta (53) to determine the long-term fate of [^{3}H]DNA from *E. coli* and *M. lysodeikticus* supplied to *Arabidopsis* cells growing in callus tissue culture. These studies revealed the complete absence of donor DNA after 7 days of incubation although high amounts of radioactivity were incorporated into the recipient DNA. The latter was analyzed for the presence of donor DNA sequences, considering that integration of heteropycnic DNA does not necessarily lead to the production of a "hybrid" band in CsCl gradients (see Section III). Using a sensitive DNA · DNA reassociation kinetic analysis, it was concluded that, if exogenous DNA sequences were indeed present within the recipient DNA molecules, it must have represented no more than 2–3% of the radioactivity recovered in the host DNA.

Essentially the same results were obtained by Kado and Lurquin (54) in their study involving tobacco cells in tissue culture and *A. tumefaciens* [^{3}H]DNA. Although Kovoor (55) previously claimed that purified *A. tumefaciens* DNA was able to induce tumor formation in axenic plant cells in culture, integration and replication of this DNA could not be detected in treated tobacco callus cells. The techniques used in these two studies (53, 54) were at least 100-fold more sensitive than the generally used CsCl gradient analyses, yet they failed to demonstrate integration (and in one case also replication) of the donor DNA within the host genome. Results obtained with soybean cells growing in tissue culture also indicate rapid degradation of bacterial DNA and extensive reutilization (Luyindula and Lurquin, unpublished).

In conclusion, it appears that, at least in the case of tobacco, lily, soybean and *Arabidopsis*, plant cells in suspension and in tissue culture are refractory to the preservation of exogenous DNA. This is very unfortunate since there was some hope of applying selective cloning techniques in order to check for a biological expression of the genes taken up. However, as discussed in Section III, it may very well be that the biochemical approach (the search for the presence of donor DNA within plant cells) of assessing the "transformability" of plant cells is unproductive.

4. PROTOPLASTS

Plant protoplasts appear to be excellently suited for DNA uptake studies. They are devoid of the thick plant cell wall, can be prepared in rather large amounts and can be plated on agar and cloned almost as easily as bacterial cells. Screening methods can thus be applied to very large numbers of identical and, in principle, totipotent cells. Surprisingly, the information about exogenous DNA uptake and fate in protoplasts is scarce and preliminary. It is hoped that some efforts will be devoted to these systems since other nucleic acids, namely RNA extracted from tobacco mosaic virus and from cowpea chlorotic mottle virus, are not only taken up by protoplasts, but are also fully biologically active in the sense that viral RNA and viral protein synthesis occurs (56, 57). Similarly, the application of the immunofluorescence technique to the detection of donor gene products as used by plant virologists might be useful in the case of protoplasts fed with small-sized DNA coding for a limited number of functions, such as plasmid and bacteriophage DNAs.

Ohyama et al. (58) were the first to report on DNA uptake by plant protoplasts. Their data indicate that after feeding Ammi visnaga, carrot and soybean protoplasts with E. coli [^{14}C]DNA and then treating them with deoxyribonuclease, some radioactivity remained bound to protoplasts, of which approximately 20% was acid-precipitable. Polycations were found to stimulate deoxyribonuclease-resistant binding of donor DNA but turned out to be toxic to the protoplasts (59, 60). The fact that 80% of the radioactivity taken up was acid-soluble seems to favor external or internal degradation of donor DNA. It was concluded that the 20% acid-precipitable radioactivity could not correspond to the reutilization of donor DNA breakdown products since no endogenous DNA synthesis was detected during the period of incubation with DNA. In the light of the recent results of Howland and Yette (61) showing that DNA synthesis in freshly prepared protoplasts is readily measurable provided 5-fluorodeoxyuridine is present together with [^{3}H]thymidine, it may be argued that Ohyama et al. (58) were not using the best conditions to measure endogenous DNA synthesis, especially if donor DNA breakdown occurred primarily inside the protoplasts, not outside. More convincing is Ohyama's observation (59) that donor DNA could be reextracted from incubated protoplasts and characterized by its buoyant density in a CsCl gradient. However, it appears from the broad banding pattern of this DNA compared to that of the host DNA that significant depolymerization had taken place. Furthermore, the donor DNA band is so broad that it considerably overlaps with the host DNA band, making it impossible to rule out reutilization. Nevertheless, if the latter has occurred, its extent must not exceed about 20% of the total radioactivity recovered from the gradient.

Hoffman and Hess (62) studied the uptake of homologous [^{3}H]DNA in Petunia hybrida protoplasts by autoradiography. Most of the radioactivity was located in the nuclei, and the apparent absence of deoxyribonuclease activity in the incubation medium led these authors to conclude that the exogenous DNA was actually taken up by the nuclei. As was said above, autoradiography alone cannot discriminate among the kinds of DNA present, and the experiments did not rule out internal nuclease activity.

Somewhat more convincing are the results published by Hoffman (60) showing that double-labeled (^{3}H/^{14}C) homologous DNA could be recovered from the nuclear fraction of treated Petunia protoplasts.

This recovered DNA represented 0.04% of input donor DNA and displayed the original ^{3}H to ^{14}C ratio of the donor DNA. Whether the latter observation really rules out reutilization is also an open question. Nevertheless, it was concluded that donor DNA was present in the nuclei and, to a much lesser extent, in the cytoplasm of the protoplasts. These results support those of Ohyama et al. (58), but these experiments, to be credible, need to be repeated with much higher radioactivity inputs and under conditions allowing a direct comparison of the banding patterns of the donor and recipient DNAs before and after uptake. This was not possible in Hoffman's study (60) where DNA was banded in CsCl-ethidium bromide gradients.

Finally, protoplasts as recipients for exogenous DNA deserve serious attention, perhaps because they have been less used than other plant systems. It must be stressed that, if little is known about DNA uptake in protoplasts, nothing is known about the fate of donor DNA taken up when protoplasts form their cell walls and resume division.

5. POLLEN GRAINS

This is a novel recipient system inaugurated by Hess et al. (63). The idea of using pollen grains as a vehicle to transfer foreign genetic information into plants is interesting. Hess et al. incubated squashed anthers from *Nicotiana glauca* and from *Petunia hybrida* in pollen germination medium containing [^{14}C]DNA from *Rhizobium leguminosarum* or [^{3}H]DNA from *E. coli* R1 drd 16 carrying a plasmid coding for the resistance to kanamycin. The pollen was incubated with donor DNA, washed, treated with deoxyribonuclease and processed for DNA isolation. This DNA was banded in CsCl density gradients and revealed the presence of labeled DNA presumably banding at the position of donor DNA. When *E. coli* DNA was supplied, the radioactivity values recovered from the gradients were too low for unambiguous interpretation of the data. Nevertheless, these preliminary results are encouraging, and, if the DNA can be shown to be truly intracellular, this sytem could be of considerable importance, since little if any reutilization was observed.

B. Biological Evidence

This section discusses the biological effects of exogenous DNA supplied either free or encapsidated in bacteriophages.

1. EFFECTS OF PURIFIED DNA

Hess (64, 65) first claimed "transformation" of *Petunia hybrida* for anthocyanin synthesis. Mutant seedlings defective in flower pigment production were treated with DNA from wild-type plants. Up to 15% of the mutant plants treated in such a way produced red to reddish flowers reminiscent of the wild type (64). No segregation of flower color was observed upon selfing of the so-called transformed plants, meaning that both alleles had been effectively transformed. This high frequency of "transformation" was then reduced to 0.06% in a study published somewhat later (65). The genetic data published by Hess were scrutinized by Bianchi and Walet-Foederer (66), who concluded from morphological observations of *Petunia* flowers that somatic mutations of epidermal cells unrelated to exogenous DNA could very well explain Hess' results. Flower color may not be a reliable genetic marker in such experiments, as it is difficult to imagine how selection pressure could be exercised in order to favor exogenous DNA expression.

A related study was undertaken by Wilcockson and Harrison (67) using color-deficient mutants of *Antirrhinum*. Three homozygous strains blocked for anthocyanin production were treated with DNA extracted from the leaves of the full-red-flowering dominant strain. All the plants growing from the DNA-treated seeds or seedlings showed the mutant phenotype. However, these authors based their experimental system on the high yield of "transformation" published previously (64) and used no more than 50 to 200 plants per group. If the presumptive "transformation" frequency had been close to that reported later for *Petunia* (65), then no positive result could be expected from their experiments.

The second case of successful genetic correction of plant material by exogenous DNA was reported by Ledoux *et al.* in 1971 (26). Mutant seeds of *Arabidopsis thaliana* blocked either in the thiamin pathway or in the tryptophan pathway were treated with DNA from various sources, with the exception of wild-type *Arabidopsis* DNA. Viable and fertile plants were recovered after treatment of both types of mutants. The correction frequency was as high as 12% with thiamin mutants and 16% with tryptophan mutants. A detailed genetic analysis was performed with corrected thiamin mutants only as it appeared that the tryptophan mutants were not conditional lethals. The genetic study revealed no segregation of the corrected phenotype after selfing, and this up to the third generation issued from the corrected plants. Here also, correction frequencies were reduced to about 0.7% as reported in a subsequent paper (68), and this probably as a result of a large increase in the number of seeds used in the correction experiments. Nevertheless, suitable control experiments consisting in treating mutant seeds with bacterial DNA extracted from a thiamin A mutant of *E.*

coli or from bacteriophages T7 and 2C, which do not contain the thiamin genes, all failed to promote plant growth in the absence of added thiamin. In one case, it was shown that DNA extracted from *E. coli thi* A carrying the F110 *thi*$^+$ episome and previously degraded with pancreatic DNase had no effect on the absence of growth of *Arabidopsis* mutant plants whereas intact DNA did exercise a correcting effect *(68a)*. The fact that the corrected plants behaved as homozygotes was considered to reflect the selection of gametes having incorporated the donor thiamin genes at the expense of uncorrected gametes. Furthermore, by crossing the corrected types with the wild type or the mutant types, a resurgence of the original mutant phenotype did occur in the second generation. In other words, it seems that an extragenic suppressor is present in the corrected types, an interpretation that was extended to the notion of donor genes being added to the recipient genomes and not substituting the mutant genes. The authors further hypothesized that donor DNA could be integrated within the recipient genomes, but not at the mutant locus, and was sometimes lost, as during chromosome pairing. Thus, the donor genes in this foreign environment would share some of the properties displayed by bacterial episomes. Whatever the interpretation of these genetic experiments may be, it must be realized that there is at present no physical evidence of the presence of donor genes in the corrected plants, nor is there any biochemical evidence that bacterial enzymes of the thiamin pathway are actually present and functional.

Finally, there is the fact that gene expression could be changed as a result of epigenetic modification without change in genetic information content. The epigenetic control mechanisms in plants are not known, as molecular genetics of plant cells is still in its infancy. Thus the more general questions asked concern the nature of mutations. For instance, pyrimidine, thiazole and thiamin mutants of *Arabidopsis* specifically respond to the substrate corresponding to the genetic block when grown in the light on minimal medium. Surprisingly, this is not true when the seeds are plated on B5-agar medium allowing callus formation. Wild-type callus proliferation does not occur in the absence of added thiamin, but rapid cell division takes place when either thiamin *or* 5-β-hydroxymethyl-4-methyl thiazole *or* 4-amino-5-aminomethyl-2-methyl pyrimidine at a concentration of 0.1 mM are present in the medium. Thus, *Arabidopsis* cells in tissue culture on B5 medium present an auxotrophy for thiamin or its moieties. This means that cultural conditions can deeply modify gene expression and that growth can be promoted by any one of the two precursors of the required metabolite. Obviously, the control of thiamin synthesis in these cells is not a simple one.

Moreover, mutant seeds respond in much the same way as wild-type seeds in the sense that the thiazole moiety of thiamin promotes growth of the pyrimidine mutant and, similarly, the pyrimidine moiety of thiamin is able to promote growth of the thiazole mutant. Strangely enough, the thiamin-requiring mutant specifically responds to thiamin only, and not to its precursors (Lurquin, unpublished). Clearly, the genetic control of thiamin production in wild-type and mutant cells of *Arabidopsis* deserves further attention. To the elaborate genetic study of the correction phenomenon by Ledoux *et al. (68)* one can oppose a short negative report from Feenstra *et al. (69)*, who failed to reproduce correction of *Arabidopsis* mutants with bacterial DNA. They treated two batches of 1000 thiamin-mutant seeds (blocked in the synthesis of the pyrimidine moiety of this compound) with *E. coli* or *B. subtilis* DNA and were unable to obtain a single green plant growing in the absence of thiamin. The number of treated seeds should have allowed detection of the average 0.7% correction frequency reported by others *(68)*. The reason for this discrepancy is not known, but it points out the fact that this kind of experiment may not be easily repeatable.

A recent genetic study by Rédei *et al. (69a)* seems to rule out any participation of foreign genetic information in the correction phenomenon described by Ledoux *et al. (68)*. This study consisted of a genetic analysis of several corrected seed stocks provided by L. Ledoux. Allelism tests and linkage studies both revealed that the corrected plants analyzed so far cannot be distinguished from plants mutated in the th-2 locus and able to survive in the absence of added thiamin. Furthermore, the corrected plants display physiological properties similar to those of the th-2 viable mutants used by Rédei *et al. (69a)* in their crosses with the corrected types. It is concluded from this genetic analysis that the apparent correction is not due to the expression of bacterial genes within the plant cells, but rather to a mechanical contamination by either the th-2 viable mutant, other mutants, or the wild type *(69a)*. It remains to be ascertained whether all the corrected types described by Ledoux *et al. (68)* correspond to this viable th-2 mutant.

Two kinds of biological experiments involving DNA feeding were undertaken by Holl *et al. (70)*. The first consisted in incubating soybean protoplasts in the presence of DNA extracted from *Azotobacter vinelandii*. This bacterium is able to utilize mannitol as a carbon source whereas soybean cells do not normally grow in its presence. Cells regenerated from the DNA-treated protoplasts grew on mannitol *plus* lactose (frequency not given), but callus growth on mannitol alone did not occur. Mannitol uptake was slow, which may explain the absence of callus growth on this medium. Since soybean cells do not grow in the presence of lactose either, it seems difficult to attribute the observed effect to a simple transfer of the bacterial genes coding for the biosynthesis of D-mannitol dehydrogenase (EC 1.1.1.67) alone. The preliminary character of these observations was fully recognized by the authors *(70, 71)*.

The second experiment involved the genetic control of nitrogen fixation exercised by the plant cells. When seeds of mutant field peas unable to nodulate in the presence of *R. leguminosarum* were incubated with DNA from the wild-type plants and then inoculated with a commercial culture of *R. leguminosarum*, it was found that 1–2% of the plants derived from the DNA-treated seeds showed nodule formation. Control experiments with unrelated DNA or with DNA extracted from the mutant seedlings were all negative. Here also, the authors avoided "premature interpretation and gross speculation" about the molecular mechanism of this phenomenon *(70, 71)*.

The third well-documented study reporting effective DNA-mediated "transformation" of plants is that of Turbin *et al.* *(72)*. These authors examined the rate of reversion to wild-type of barley *waxy* mutants whose grains at milk maturity stage were injected with DNA from wild-type barley. The main advantage of their approach was that it enabled them to analyze the effect of exogenous DNA at the cellular level. *Waxy* mutants do not synthesize amylose, which is thus absent from the pollen grains, which in turn do not stain blue when incubated with I_2 dissolved in KI. Back mutation was thus easily detectable and could be investigated on a very large scale. It was found that injection of water or *E. coli* DNA into the grains approximately doubled the frequency of appearance of wild-type pollen grains in the offspring. This frequency was 230-fold higher (equal to 5×10^{-3}) than control values in pollen grains derived from plants themselves grown from grains injected with wild-type barley DNA of high M_r and slightly deproteinized. Strangely enough, extensively deproteinized high-M_r barley DNA increased the reversion frequency only by a factor of three. The interpretation of this small effect is a hypothetical one involving an increased resistance of crude DNA to nuclease attack. This hypothesis is in fact no more plausible than an alternative explanation invoking a biological effect of contaminants of crude barley DNA rather than that of DNA itself. It remains that their system allowed a good screening of individual pollen grains (they scored more than 4×10^6 of them) and that artificial DNA · protein or DNA · polycation complexes could help solve the question of resistance to nucleases.

Kovoor *(55)* reported that purified *A. tumefaciens* DNA was able to induce the formation of crown gall (a plant tumor growing in tissue culture in the absence of added auxin) in plant tissue culture. This biological effect of DNA extracted from the causative bacterial agent could not be reproduced by others (for review see 73) and no auxin-autotrophic plant cells could be recovered from protoplasts treated with *A. tumefaciens* DNA *(74)*. Wild-type tobacco DNA seemingly has no effect on haploid protoplasts derived from tobacco cells unable to grow without lysine or hypoxanthine *(75)*. A total of 5.2×10^7 hypoxanthine-requiring protoplasts showed no significant increase in the frequency of colonies recovered in the absence of the substrate. Similar results were obtained with lysine-requiring cells and protoplasts. One obvious drawback in these preliminary experiments *(75)* is the very low concentration of donor DNA in the medium (0.01 µg/ml). Such a low concentration is not routinely used in the well-characterized and efficient *B. subtilis* transformation systems.

M. lysodeikticus and *E. coli* DNAs have no effect on several arginine mutants of *Chlamydomonas reinhardi* with or without cell wall *(76)*. In some cases, a reversion frequency of 10^{-7} would have been detected. This negative result can be understood in terms of donor DNA degradation and reutilization after uptake *(50)*.

Increased resistance to the pathogen *Phytophthora infestans* of potato leaves painted with DNA extracted from *Nicotiana tabacum*, *Datura abla* and from a resistant potato hybrid has been observed, and painting leaves with exogenous DNA resulted in profound modifications in protein biosynthesis *(77)*. What appears to be the most striking observation made in the field of heterologous DNA uptake and biological expression in plant cells is that of Sander *(78)*, showing that purified single-stranded DNA from bacteriophage fd induces complete phage particle synthesis in tobacco leaves. Inoculation with intact phage particles gave rise to no phage progeny. Moreover, the phage particles recovered after rubbing the leaves with fd DNA were identical to *bona fide* fd particles in host specificity, plaque morphology, serological behavior and buoyant density in CsCl gradient. In a later paper *(79)*, it was reported that fd particles recovered from tobacco leaves had lost their typical specificity, and it was suggested that tobacco cells possess DNA modification systems somewhat comparable to the restriction-modification controls of *E. coli* B. Finally, it was shown that fd DNA was not able to direct mature bacteriophage formation in bacteria commonly found to live on tobacco leaves under the experimental conditions used.

Despite its originality and its deeply unorthodox nature, this research has not been pursued nor contradicted. It would be of tremendous interest to find out whether or not this phenomenon is easily reproduced, since many of the questions raised by foreign DNA uptake, transcription, translation and replication might be resolved with this experimental system.

2. Effects of Bacteriophages

Bacteriophages possess several advantages over "naked" DNA: (a) the gene dosage per DNA molecule is high; (b) bacteriophages are rather easily prepared in large amounts; (c) many transducing bacteriophages are known; (d) phage DNA is relatively small and homogeneous, helpful in DNA reassociation experiments to detect the intracellular presence of phage particles or DNA; and (e) their genetic material is efficiently protected. This last point may also be disadvantageous, since it is hard to imagine how coated DNA can be fully expressed. Hence, it must be assumed that phage DNA has to be released from the particles in a "safe" place inside the cells in order to be biologically active at all. Nothing is known about this presumptive bacteriophage stripping, nor is anything known about phage uptake itself. Studies dealing with bacteriophage–plant cell interactions have been limited to the detection of specific enzymic functions coded for by the phage DNA. No attempts have been made to obtain physical evidence for the presence of phage particles or phage genes inside the treated plant cells.

Carlson (80) claimed that *Nicotiana tabacum* protoplasts start synthesizing bacteriophage T3 RNA polymerase and S-adenosylmethionine-cleaving enzyme several hours after inoculation with high titers of bacteriophage T3. A control experiment conducted with an amber mutant of the phage gave a negative result consistent with the location of the amber mutation in the structural gene coding for the phage RNA polymerase activity. The latter dropped to virtually zero 96 hours after inoculation with normal phage whereas the S-adenosylmethionine-cleaving activity remained fairly constant up to 120 hours after inoculation. It is also difficult to explain the differential effect of cycloheximide and chloramphenicol on both enzyme activities apparently expressed in the infected protoplasts. The RNA polymerase was more sensitive to cycloheximide, the S-adenosylmethionine-cleaving function being more sensitive to chloramphenicol. It can be assumed that the aim of this experiment was to locate the site of synthesis (organelle or cytoplasm) of these enzymes. Clearly, this result is puzzling and does not simplify the models one could make to rationalize the biological mechanisms responsible for these observations.

More detailed studies on the effects of bacteriophages on plant cells were published by Doy et al. (81), who used the transducing phages λpgal and ϕ80plac to treat haploid tomato and *Arabidopsis thaliana* callus cells. These were then allowed to grow on nutrient medium containing either galactose or lactose as sole carbon source, conditions not allowing growth of these plant tissues. Callus tissue treated with λpgal⁻ or ϕ80plac⁻ died on these media, as did the untreated tissues, but tissues inoculated with high titers of λpgal⁺ or ϕ80plac⁺ grew slowly but significantly on the same media. In these experiments, bacteriophage ϕ80plac⁺ contained the gene z coding for the *E. coli* β-galactosidase. Treatment of the calluses with ϕ80plac⁺ increased the endogenous β-galactosidase activity by a factor of about 25, and this enzyme turned out to be protected against heat denaturation by a rabbit antiserum specific for the bacterial enzyme and known to protect *E. coli* β-galactosidase against heat denaturation.

Callus growth on lactose was promoted by treatment of the cells with λpgal⁺ but not by λpgal⁻ containing a nonsense mutation in the transferase gene (81). Unfortunately, no selective test was done to ascertain the presence of the three enzymes (epimerase, kinase and transferase) coded for by the galactose operon carried by the transducing phage. Such experiments are certainly feasible, since these three *E. coli* enzymes have been fully characterized.

Finally, Doy et al. (81) observed cell death when callus tissue was exposed to phage ϕ80supF⁺, which specifies insertion of tyrosine at amber codons. They concluded that UAG nonsense codons are essential to ensure correct functioning and viability of plant cells. This observation leads one to think that all cells in the calluses took up phage particles or that cell-to-cell transmission of phage occurs. However, this would somewhat be in contrast to the reported sectorial growth of callus tissue treated with ϕ80plac⁺ (82).

The results of Doy et al. (81) were grossly confirmed by Johnson et al. (83), who reported continued growth of diploid sycamore cells in suspension cultures containing 2% lactose as the sole carbon source, provided that the cells were incubated in the presence of high titers of bacteriophage λplac5. Treatment with λ⁺ was completely ineffective, and peaking of galactosidase activity was noticed as in the work of Doy et al. (84).

Nevertheless, Johnson et al. noted a fundamental difference between their system and that of Doy et al. Indeed, they were unable to detect *E. coli*-specific β-galactosidase in the growing plant cells inoculated with bacteriophages containing the relevant genes. Both untreated and treated cells contained enzymes hydrolyzing o-nitrophenyl galactopyranoside, but none of these activities responded to the expected protective effect of rabbit anti-β-galactosidase serum. This serous discrepancy was further aggravated by Smith et al. (85), who showed that a specific inhibitor of β-galactosidase, such as β-phenyl thiogalactoside, was without effect on the galactosidase activity found in extracts from plant cells treated with λplac5. They consider that there is at present no conclusive evidence that E. coli β-galactosidase is actually synthesized in the sycamore cells.

III. Attempts to Conclude and Prospectives

This critical review of this field leaves one with mixed feelings. The numerous discrepancies and contradictions are rather confusing, but at the same time, poorly explored areas seem to leave the door open to some more comforting possibilities. As the title indicates, it is not a simple task to draw definite conclusions or to build models from scattered, and often contradictory, data. This is why this section is limited to a comparison between natural systems in which heterologous genome interactions occur and artificial systems involving purified DNA. Finally, additional speculations based on modern techniques are here presented.

A. Facts and Fancy

The field as a whole presents an image of extraordinary confusion despite the existence of several reviews (18, 82, 86–92). However, these reviews have not given the necessary detailed attention to the present subject because DNA uptake was only one of several topics covered. Not until recently has the integration and replication of exogenous DNA been submitted to specific and sensitive molecular hybridization techniques. The results so obtained invariably showed that if such a phenomenon as integration and replication of bacterial DNA in plant cells did occur, it would have been undetectable by present techniques.

The controversy about integration of foreign DNA in plants has made it clear that a combination of several stringent techniques will be required to substantiate the claims of heterologous DNA integration and replication in eukaryotic cells. It is essential to eliminate such gross artifacts as microbial contamination, peak overlapping, or viscosity effects in CsCl gradients. Moreover, genetic or biological effects attributed to foreign DNA can gain wide acceptance only when they are independently confirmed in several laboratories. Also, a more rigorous detailing than has been customary of experimental procedures that have led to claims of success is in order. This would make possible an objective assessment of the applicability of the procedures.

We now consider some *natural* systems where our notions on genome–genome interactions had to be dramatically revised in the last years. It was formerly believed that crown-gall-tumor DNA contained appreciable amounts of DNA from the inciting bacterium *A. tumefaciens* and its bacteriophage PS8. These conclusions were subsequently challenged in several laboratories (review in 73 and 94), where it was found that if *A. tumefaciens* or PS8 DNA were present within plant tumor cells, it must have been to an undetectable extent. Moreover, these tumor cells were even found to contain less than one copy, if any, of the *A. tumefaciens* plasmid DNA present in the inciting bacteria (95). In other words, bacterial DNA initially enclosed in living bacteria does not seem to be detectably replicated in axenic plant tissues whose biological properties obviously originated from the invasion of the tissues by bacteria known to undergo lysis during the tumorigenic process. Of course, if only a few bacterial genes were present in each tumor cell, these would not readily be detectable using the present techniques of DNA hybridization, although they might be responsible for profound alterations of the cells' physiology.

In turn, nitrogen fixation by legume root nodules had been thought for a long time to be triggered by the interactions between plant and *Rhizobium* genomes established upon symbiotic association (96). Indeed, it was then believed that free-living *Rhizobium* cells were completely unable to fix N_2. Recent discoveries (discussed in 97) ruled out this hypothesis by demonstrating that free *Rhizobium* cells can indeed fix N_2 in the complete absence of plant cells or plant cell extracts provided they are grown in synthetic medium under certain well-defined conditions. Furthermore, Rake (98) demonstrated the absence of detectable DNA homology between soybean and its symbiotic *Rhizobium* bacteria, thus rendering unlikely a hypothetical transfer of genes from one type of cell to the other that might explain the specificity of *Rhizobium* cells.

This little digression was made because the crown-gall and root-nodule systems were occasionally put forward in order to support conclusions drawn from experiments in which purified DNA was supplied to plants. The recent findings quoted above clearly modify our views on the mechanistic aspects of these genomic interactions.

Shanmungan and Valentine (99) speculated on the transfer of the amplified *nif* genes from *Klebsiella pneumoniae* to plant protoplasts in order to obtain whole plants constitutively possessing the capacity of fixing N_2. Although insertion of the *nif* genes in a fast replicating plasmid is technically possible (see Introduction), the uptake of this hybrid plasmid into plant protoplasts followed by its replication and biological expression in the regenerated plants would meet formidable obstacles, as discussed by Postgate (100). Nevertheless, such an experimental "scenario" is certainly worth a trial. Indeed, Dixon *et al.* (101) recently constructed a recombinant plasmid RP41 starting from the well-known P-group R factor RP4 and the FN68 plasmid that carries the *nif* genes from *K. pneumoniae*. This plasmid, despite the fact that its synthesis is not of the relaxed type, could be used in DNA feeding experiments, and would even allow a double selection test, since it also carries the genes determining resistance to kanamycin, an antibiotic to which several plant species are susceptible (see below).

B. Missing Links and Further Speculation

Clearly, the ideal study reporting successful transformation of plant cells should demonstrate (a) the existence of a DNA transport mechanism, (b) the intracellular persistence of donor DNA sequences, (c) their replication in tandem with host DNA, (d) their transcription, (e) the translation of the mRNA of donor type into specific proteins, and (f) the same events (except for condition a) in the progeny. These six conditions have not been simultaneously fulfilled in papers claiming "transformation" of eukaryotic cells. Some aspects of these various steps are described here.

1. UPTAKE

Cell permeability problems are among the most difficult to solve. Moreover, it is hard, from an anthropomorphic viewpoint, to conceive how a macromolecule such as DNA can cross the thick cell walls of plants. However, this is probably a misconception, since uptake of purified cauliflower mosaic virus DNA, for instance, must occur to account for its infectivity in whole plants (102). Nevertheless, the uptake process is very poorly understood, although it is a prerequisite for studying the fate of exogenous DNA in plant cells under reproducible conditions. It seems that protoplasts are well suited for uptake studies, since they engulf various sorts of inert particles, bacteria, and chloroplasts (103). Moreover, protoplasts take up and allow expression of tobacco mosaic virus, cowpea chlorotic mottle virus, and their respective RNAs (56, 57). It follows that protoplasts might as well take up DNA and bacteriophages, especially under conditions allowing protoplast fusion (104). Indeed, part of the foreign elements adsorbed on the cell membrane would eventually penetrate the fused protoplasts.

2. Intracellular Persistence and Replication

If replication of exogenous DNA can be demonstrated, it must have remained long enough to allow at least partial *de novo* synthesis. In other words, condition (c) partially renders condition (b) obsolete and somewhat simplifies the experimental approach. Actually, preservation of donor DNA has usually been studied by feeding plants with radioactive exogenous DNA and analyzing the nature of DNA extracted from the treated plants. Such a procedure may lead to erroneous conclusions, as it is extremely difficult to distinguish intracellular foreign DNA from its extracellular counterpart. On top of that, some reutilization of radioactive donor DNA breakdown products is expected to occur in most cases and further complicates the picture.

A different approach consists in feeding the plants with unlabeled donor DNA and using the total DNA from the treated plants as a driver to measure its effect on the rate of reassociation of a probe DNA corresponding to the DNA sequences supplied to the plants. In this case also, nonabsorbed donor DNA will give a positive response.

Alternatively, plants incubated with unlabeled donor DNA can further be labeled with radioactive precursors of DNA synthesis and the extracted DNA then used as a probe in DNA reassociation experiments in the presence of a large excess of unlabeled donor DNA. Here, a positive response will be obtained only if donor DNA sequences became labeled *in vivo*, but this approach has more limitations than the previous one, since presumably replicated donor DNA sequences might represent an extremely low percentage of plant DNA and could therefore be totally masked by the endogenous DNA.

Nevertheless, DNA reassociation experiments are several orders of magnitude more sensitive than the CsCl density gradient technique regarding the detection of low amounts of unrelated DNA sequences (Table I). Persistence of bacterial DNA in barley tissues was demonstrated by Kleinhofs *(24)* using this sensitive method. However, the author considered that this result in itself did not prove the intracellular presence of exogenous DNA nor its replication, and he suggested the use of the more selective hybridization technique developed by Varmus *et al. (105)*. In brief, this technique relies on the fact that integrated oncogenic viral DNA cosediments with high-M_r cellular DNA previously denatured and renatured under conditions allowing network formation via reassociation of repetitious DNA sequences. Unintegrated viral DNA is not found in these networks and is thus easily separated from cellular DNA containing integrated viral sequences by centrifugation. DNA reassociation experiments performed with network DNA should give a clear-cut answer regarding

TABLE I

CORRELATION BETWEEN THE AMOUNT OF PRESUMABLY INTEGRATED DONOR
DNA AND ITS DETECTION BY MEANS OF DNA REASSOCIATION KINETICS
(HYBRIDIZATION) OR CsCl DENSITY ANALYSIS[a]

Kinetic complexity of bacterial DNA presumably integrated	Bacterial genome[b]/ plant genome[c]	Percent bacterial DNA versus plant DNA	Expected density shift[d] ($g \times cm^{-3}$)
ca. one gene $\begin{bmatrix} 10^6 \\ 2 \times 10^6 \end{bmatrix}$	0.0005	0.00004	NO[e]
	0.001	0.00008	NO
2×10^7	0.01	0.0008	NO
2×10^8	0.1	0.008 hybrid-	NO
ca. one genome $\begin{bmatrix} 2 \times 10^9 \end{bmatrix}$	1	0.08 ization	NO
2×10^{10}	10	0.8	NO
2×10^{11}	100	8	0.002 to 0.003
4×10^{11}	200	16 density	0.004 to 0.006
2×10^{12}	1000	80 analysis	0.012 to 0.014

[a] Hybridization refers to the measurement of the reassociation kinetics of trace amounts of labeled donor DNA (probe DNA) in the presence of a very large excess of DNA extracted from plant tissues assumed to contain donor DNA sequences. The expected density shift is calculated assuming a linear relationship between DNA buoyant density in CsCl and the respective proportions of donor and recipient DNAs linked by end-to-end joining. These values are based on the production of discrete bands in CsCl gradients. The arrows indicate the limits of detection of each method. In fact, the lower limit of the hybridization procedure is determined only by the specific radioactivity of the probe DNA or, alternatively, by the solubility of the experimental DNA. Obviously, the ratios between bacterial and plant genomes vary according to their respective sizes. The values reported here correspond to idealized conditions, but should not vary by more than one order of magnitude in actual experiments. Lower proportions of foreign DNA can be detected, provided the donor sequences correspond to low-complexity genomes, such as those from bacteriophages or plasmid DNA.

[b] Assuming an M_r of 2×10^9.

[c] Assuming an M_r of 2.5×10^{12}.

[d] Assuming a difference of density between donor and recipient DNA of 0.030 $g \times cm^{-3}$.

[e] NO = not observable.

the state of exogenous and "hybrid"-band DNA found in association with plant tissues.

3. TRANSCRIPTION

It is not clear whether heterologous DNA can be effectively transcribed in eukaryotic systems. A *fortiori*, nothing is known about the regulation of foreign gene transcription in eukaryotic cells. However, the work of Gurdon et al. (106), Laskey and Gurdon (107), and Colman (108) convincingly showed that unrelated DNA is not only replicated, but also transcribed in amphibian eggs. Although such organisms bear little resemblance to plant cells, it is encouraging to note that replication and at least partial transcription of the donor DNA did occur in a eukaryotic system. Also encouraging are the results reported by Gardner et al. (109), who demonstrated the transcription of single-stranded circular DNA of bacteriophage ϕX174 by purified maize RNA polymerase II.

The work of Goebel and Schiess (110) clearly showed that Col E1 covalently-closed-circular DNA taken up by mammalian cells undergoes temporary replication but is eventually degraded, and that Col E1 complementary RNA is synthesized in the cells during the persistence of the donor DNA. These results should prompt those interested in DNA uptake by plants to perform similar experiments.

4. Translation

This is of course another crucial step in successful plant genetic engineering. Any modification of the genetic content of plant cells is subordinate to the biosynthesis of active enzyme molecules compatible with the cellular functions of the recipient organism. Here also, very little is known about the translation of heterologous messenger RNA (bacterial) in eukaryotic cells. At least one study (111) showed that bacteriophage Qβ RNA is efficiently transcribed in a Krebs-ascites cell-free system. Suitable controls performed with Qβ RNA containing an amber mutation demonstrated, by gel electrophoresis of the polypeptides synthesized in vitro, that the eukaryotic system respects the specificity of translation. This also means that initiation and termination of protein synthesis does not necessarily constitute an unsurmountable barrier between widely separated species. Similarly, wheat germ cell-free systems are able to carry out meaningful translation of heterologous (eukaryotic and prokaryotic) mRNAs (112–113b). Of course, some of the theoretical problems raised by the hypothetical interactions between heterologous genomes as well as interactions between donor DNA and RNA and the machinery subsequently synthesizing protein in the recipient cells might not be so acute if more closely related DNA were used in uptake experiments.

5. Transmission of the Added Trait

If we assume that all the questions raised above are answered in a positive way, then, the most economical and time-saving aspect of plant-cell genetic engineering is the stable inheritance of the added genetic trait. This may also be the most demanding event in the sequence. Indeed, the foreign genetic information supposed to be expressed in its new environment must be stably stored and transmitted through the complicated and sensitive mechanisms of differentiation and meiosis. It will also have to resist DNA repair mechanisms known to exist in plants *(114)*. In view of the many and partly unpredictable difficulties awaiting foreign DNA in plant cells, it may be justifiable to wonder whether the current biochemical approach should be continued as such. As noted above, most workers studying the fate of foreign DNA in plants have tackled the problem by investigating its behavior via buoyant density and DNA hybridization analyses, but both techniques suffer from severe limitations; for instance, it can be calculated that even if each cell of a plant tissue has incorporated the equivalent of one foreign gene (that is, a DNA stretch of M_r of the order of 10^6), this amount of donor DNA would not normally be detected using the present DNA hybridization procedures. Thus, actual transformation might occur without being noticed if only physical methods of detection are utilized.

One can imagine that the approach could be the converse of what it has been so far; that is, once a DNA-mediated biological phenomenon is reproducibly established *(115)*, then, and preferably only then, can its molecular mechanisms be seriously envisaged. In other words, it could be more rewarding to invest some effort in developing methods of selection allowing the detection of biological effects of exogenous DNA. Since the latter may be expected to occur at rather low frequencies, it would be necessary to treat very large numbers of cells. Once again, cells in suspension cultures or, better, protoplasts appear to be excellently suited for this kind of investigation. Unfortunately, the plant mutants required to test transformability are scarce and one will have to await a somewhat better understanding of somatic plant-cell genetics before genetic modifications can actually be attempted in a controlled manner.

In the meantime, slightly different biological experiments can be performed with, for instance, homogeneous DNA molecules coding for a limited number of functions, such as drug-resistance plasmids either free or recombined with transducing bacteriophages. Plant cells are sensitive to certain antibiotics *(116)*. Kanamycin, streptomycin and chloramphenicol, among others, are inactivated by enzymes coded for by several bacterial plasmids *(117)*, thus allowing direct selection of the transformed cells., It may even be speculated that some day, a molecular vehicle such as cauliflower-mosaic-virus DNA will be utilized for the transfer of exogenous genetic information to plants as was so successfully done in the case of *E. coli* with Col E1 and pSC101 plasmids. Many more systems of molecular cloning in plant cells can be imagined. Only the future will tell whether these are realistic or not.

In conclusion, it can be said that research on DNA uptake and fate in plant cells has not yet reached a mature stage. More and more difficulties of interpretation have emerged since the first claims were made, and it is now clear that no simple model can be built on the basis of the present experimental evidence. Fortunately, many problems have been recognized and can now be dealt with. The near future should tell us whether the hypothesis of DNA-mediated transformation of plant cells was of any heuristic value as compared to other means of modifying the genetic characteristics of plants.

Notes Added

1. After submission of this paper it was shown that DNA infection of monkey kidney cells with fragments of SV40 DNA and phage λDNA generated by restriction endonuclease cleavage followed by *in vitro* ligation led to the propagation, and in one case to the encapsidation, of λDNA sequences into progeny SV40 virions. [D. Ganem, A. L. Nussbaum, D. Davoli and G. C. Fareed, *Cell* **7**, 349 (1976) and A. L. Nussbaum, D. Davoli, D. Ganem and G. C. Fareed, *PNAS* **73**, 1068 (1976).]

Thus, it seems that molecular cloning techniques are also applicable to the case of eukaryotic cells susceptible to infection by purified viral DNA.

2. In more recent experiments, using *M. lysodeikticus* DNA of M_r about 10^7, Ledoux and Huart were again able to obtain DNA satellites in barley of "hybrid" nature as defined in the text and appearing to reassociate with the donor DNA. Very recently, reassociation kinetics and thermal denaturation have been applied to this material with results seemingly favoring the integration and replication of the foreign DNA (Ledoux, Huart and Janowski, unpublished).

3. It now seems that bacterial DNA can be preserved and replicated in plant cells under a very specific set of conditions. Crown gall tumor cell DNA contains a 3×10^6 dalton piece of the plasmid DNA initially present in the inciting bacterium *A. tumefaciens*. This was demonstrated by reassociation kinetics studies involving unlabeled driver tumor DNA and highly radioactive fragments of plasmid DNA generated by digestion with the Sma restriction endonuclease [E. W. Nester, M.-D. Chilton, M. Drummond, D. Merlo, A. Montoye,

D. Sciaky and M. P. Gordon, *in* "Impact of Recombinant Molecules on Science and Society," 10th Annu. Miles Symp.(F. Young, ed.), Raven press, New York, in press]. This discovery opens new possibilities for genetic engineering in plants as these inserted bacterial DNA sequences, if manipulated, could act as vectors for the transfer of foreign genetic material from *A. tumefaciens* to plant cells susceptible to infection.

The opinions expressed in this article arose gradually during personal experience in the field and also after spending many hours of sometimes passionate discussion with a large number of people involved or interested in these questions. I am aware that this area is now progressing faster than in the past five years. Therefore, experiments suggested here may already have been performed and some of the hypotheses may either have been demonstrated or eliminated by the time this paper appears.

Acknowledgments

I am particularly indebted to Drs. R. M. Behki, P. Charles, Y. Hotta, C. I. Kado, A. Kleinhofs, L. Ledoux and M. Mergeay for expressing their own ideas about genetic engineering in plants. I am also grateful to Drs. F. Cannon, J. Postgate, H. Smith and H. Stern for constructive criticism and suggestions. Special thanks are due to Dr. P. Charles for communicating his results before publication.

References

1. P. S. Carlson, *PNAS* **70**, 598 (1973).
2. O. Schieder, *Z. Pflanzenphysiol.* **74**, 357 (1974).
3. G. Melchers and G. Labib, *Mol. Gen. Genet.* **135**, 277 (1974).
4. L. Ledoux, R. Huart and W. Baeyens, *Arch. Int. Physiol. Biochim.* **73**, 563 (1965).
5. W. F. Bodmer and A. T. Ganesan, *Genetics* **50**, 717 (1964).
6. N. K. Notani and S. H. Goodgal, *J. Gen. Physiol.* **49**, 197 (1966).
7. M. S. Fox and M. K. Allen, *PNAS* **52**, 412 (1964).
8. E. M. Silverstein and B. M. Mehta, *Bchem* **10**, 683 (1971).
9. M. Piechowska, A. Soltyk and D. Shugar, *J. Bact.* **122**, 610 (1975).
10. S. D. Cosloy and M. Oishi, *PNAS* **70**, 84 (1973).
11. S. D. Cosloy and M. Oishi, *Mol. Gen. Genet.* **124**, 1 (1973).
12. J. F. Morrow, S. N. Cohen, A. C. Y. Chang, H. W. Boyer, H. M. Goodman and R. B. Helling, *PNAS* **71**, 1743 (1974).
13. M. Thomas, J. R. Cameron and R. W. Davis, *PNAS* **71**, 4579 (1974).
14. L. H. Kedes, A. C. Y. Chang, D. Houseman and S. N. Cohen, *Nature* **255**, 533 (1975).
15. A. C. Y. Chang, R. A. Lansman, D. A. Clayton and S. N. Cohen, *Cell* **6**, 231 (1975).
16. C. R. Merril, M. R. Geier and J. C. Petricciani, *Nature* **233**, 398 (1971).
17. J. Horst, F. Kluge, K. Beyreuther and W. Gerok, *PNAS* **72**, 3531 (1975).
18. P. M. Bhargava and G. Shanmugam, This series **11**, 103 (1971).
19. L. Ledoux and R. Huart, *BBA* **134**, 209 (1967).

20. L. Ledoux and R. Huart, *Nature* **218**, 1256 (1968).
21. L. Ledoux and R. Huart, *JMB* **43**, 243 (1969).
22. Y. Hotta and H. Stern, *in* "Informative Molecules in Biological Systems" (L. Ledoux, ed.), p. 176. North-Holland Publ., Amsterdam, 1971.
23. A. Kleinhofs, F. C. Eden, M.-D. Chilton and A. J. Bendich, *PNAS* **72**, 2748 (1975).
24. A. Kleinhofs, *in* "Genetic Manipulations with Plant Material" (L. Ledoux, ed.), p. 461. Plenum, New York, 1975.
25. P. K. Ranjekar, D. Pallotta and J. G. Lafontaine, *BBA* **425**, 30 (1976).
26. L. Ledoux, *in* "Genetic Manipulations with Plant Material" (L. Ledoux, ed.), p. 479. Plenum, New York, 1975.
27. R. J. Britten, D. E. Graham and B. R. Neufeld, *in* "Methods in Enzymology," Vol. XXIX, E (L. Grossman and K. Moldave, eds.), p. 363. Academic Press, New York, 1974.
28. L. Ledoux and M. Jacobs, *Arabidopsis Inf. Serv.* **6**, 5 (1969).
29. L. Ledoux and M. Jacobs, *Arabidopsis Inf. Serv.* **6**, 7 (1969).
30. L. Ledoux, R. Huart and M. Jacobs, *EJB* **23**, 96 (1971).
31. P. F. Lurquin, M. Mergeay and J. Van der Parren, *in* "Uptake of Informative Molecules in Living Cells" (L. Ledoux, ed.), p. 254. North-Holland Publ., Amsterdam, 1972.
32. L. Ledoux, "L'absorption des acides désoxyribonucléiques par les tissus vivants" (Vaillant-Carmanne, ed.). Liège, 1968.
33. R. Huart, J. Swinnen-Vranckx, M. Jacobs and L. Ledoux, *Arch. Int. Physiol. Biochim.* **78**, 591 (1970).
34. L. Ledoux, J. Brown, P. Charles, R. Huart, M. Jacobs, J. Remy and C. Watters, *in* "Advances in the Biosciences" (G. Raspé, ed.), p. 347. Pergamon, Oxford, 1972.
35. W. Rebel, V. Hemleben and W. Seyffert, *Z. Naturforsch.* **28**, 473 (1973).
36. W. Gradmann-Rebel, Ph. D. Thesis, Karl Eberhard University, Tübingen, German Federal Republic, 1975.
37. V. Hemleben, N. Ermisch, D. Kimmich, B. Leber and G. Peters, *EJB* **56**, 403 (1975).
38. A. Bendich and P. Filner, *Mutat. Res.* **13**, 199 (1971).
39. C. I. Kado and P. F. Lurquin, *BBRC* **64**, 175 (1975).
40. C. I. Kado and T. Mojica-a, *Arch. Int. Physiol. Biochim.* **83**, 190 (1975).
41. M. Stroun, P. Anker, P. Charles and L. Ledoux, *Nature* **215**, 975 (1967).
42. P. Anker and M. Stroung, *Nature* **219**, 932 (1968).
43. M. Stroun, P. Anker and L. Ledoux, *Curr. Modern Biol.* **1**, 231 (1967).
44. M.-D. Chilton, *Genetics* **81**, 469 (1975).
45. R. S. Hanson and M.-D. Chilton, *J. Bact.* **124**, 1220 (1975).
46. M. Stroun, P. Anker, P. Gahan, A. Rossier and H. Greppin, *J. Bact.* **106**, 634 (1971).
47. D. M. Yajko and G. D. Hegeman, *J. Bact.* **108**, 973 (1971).
48. R. F. Heyn and R. A. Schilperoort, *in* "Protoplastes et fusion de cellules somatiques végétales," Colloq. int. CNRS p. 385. Inst. Natl. Recherche Agron., Paris, 1973.
49. P. F. Lurquin and R. M. Behki, *in* "Genetic Manipulations with Plant Material" (L. Ledoux, ed.), p. 429. Plenum, New York, 1975.
50. P. F. Lurquin and R. M. Behki, *Mutat. Res.* **29**, 35 (1975).
51. A. J. Faras and R. L. Erikson, *BBA* **182**, 583 (1969).
52. D. C. Swinton and P. C. Hanawalt, *J. Cell Biol.* **54**, 592 (1972).
53. P. F. Lurquin and Y. Hotta, *Plant Sci. Lett.* **5**, 103 (1975).
54. C. I. Kado and P. F. Lurquin, *Physiol. Plant Pathol.* **8**, 73 (1976).

55. A. Kovoor, *C. R. Hebd. Seances Acad. Sci.* **265**, 1623 (1967).
56. S. Aoki and I. Takebe, *Virology* **39**, 439 (1969).
57. F. Motoyoshi, J. B. Bancroft, J. W. Watts and J. Burgess, *J. Gen. Virol.* **20**, 177 (1973).
58. K. Ohyama, O. L. Gamborg and R. A. Miller, *Can. J. Bot.* **50**, 2077 (1972).
59. K. Ohyama, *in* "Plant Tissue Culture Methods" (O. L. Gamborg and L. R. Wetter, eds.), p. 28. N.R.C. Publ., Ottawa, 1975).
60. F. Hoffman, *Z. Pflanzenphysiol.* **69**, 249 (1973).
61. G. P. Howland and M. L. Yette, *Plant Sci. Lett.* **5**, 157 (1975).
62. F. Hoffmann and D. Hess, *Z. Pflanzenphysiol.* **69**, 81 (1973).
63. D. Hess, H. Lörz and E.-M. Weissert, *Z. Pflanzenphysiol.* **74**, 52 (1974).
64. D. Hess, *Z. Pflanzenphysiol.* **60**,348 (1969).
65. D. Hess, *Z. Pflanzenphysiol.* **66**, 155 (1972).
66. F. Bianchi and H. G. Walet-Foedere, *Acta Bot. Neerl.* **23**, 1 (1974).
67. J. Wilcockson and B. J. Harrison, *Annu. Rep. John Innes Inst.* p. 61 (1972).
68. L. Ledoux, R. Huart and M. Jacobs, *Nature* **249**, 17 (1974).
68a. L. Ledoux, R. Huart, M. Mergeay, P. Charles and M. Jacobs, *in* "Genetic Manipulations with Plant Materials" (L. Ledoux, ed.), p. 499. Plenum, New York, 1975.
69. W. J. Feenstra, D. L. De Heer and F. J. Oostindier-Braaksma, *Arabidopsis Inf. Serv.* **10**, 33 (1973).
69a. G. P. Rédei, G. Acedo, H. Weingarten and L. D. Kier, *in* "Cell Genetics in Higher Plants" (D. Dudits, G. L. Farkas and P. Maliga, eds.), Publ. House Hung. Acad. Sci. In press.
70. F. B. Holl, O. L. Gamborg, K. Ohyama and L. Pelcher, *in* "Tissue Culture in Plant Science" (H. E. Street, ed.), p. 301. Academic Press, New York, 1974.
71. F. B. Holl, *Can. J. Genet. Cytol.* **17**, 517 (1975).
72. N. V. Turbin, V. N. Soyfer, N. A. Kartel, N. M. Chekalin, Y. L. Dorohov, Y. B. Titov and K. K. Cieminis, *Mutat. Res.* **27**, 59 (1975).
73. K. A. Drlica and C. I. Kado, *Bacteriological Rev.* **39**, 186 (1975).
74. J. W. Watts, D. Cooper and J. M. King, *in* "Modification of the Information Content of Plant Cells" (R. Markham, D. R. Davies, D. A. Hopwood and R. W. Horne, eds.), p. 119. North-Holland Pub., Amsterdam, 1975.
75. P. S. Carlson, *Genetics* **71**, Abstr. s9 (1972).
76. V. Konvalinkova, Mémoire de Licence, Université de Liège, Belgium (1973).
77. M. Yamamoto and K. Matsuo, *Nature* **259**, 63 (1976).
78. E. Sander, *Virology* **24**, 545 (1964).
79. E. Sander, *Virology* **33**, 121 (1967).
80. P. S. Carlson, *PNAS* **70**, 598 (1973).
81. C. H. Doy, P. M. Gresshoff and B. G. Rolfe, *PNAS* **70**, 723 (1973).
82. P. M. Gresshoff, *in* "Genetic Manipulations with Plant Material" (L. Ledoux, ed.), p. 539. Plenum, New York, 1975.
83. H. Smith, R. A. McKee, T. H. Attridge and D. Grierson, *in* "Genetic Manipulations with Plant Material" (L. Ledoux, ed.), p. 551. Plenum, New York, 1975.
84. C. H. Doy, P. M. Gresshoff and B. G. Rolfe, *Nature NB* **244**, 90 (1973).
85. C. B. Johnson, D. Grierson and H. Smith, *Nature NB* **244**, 105 (1973).
86. L. Ledoux, This series **4**, 231 (1965).
87. R. F. Heyn, A. Rörsch and R. A. Schilperoort, *Q. Rev. Biophys.* **1**, 35 (1974).
88. D. Hess, *Biol. Rundschau* **12**, 297 (1974).
89. C. R. Merril and H. Stanbro, *Z. Pflanzenphysiol.* **72**, 371 (1974).

90. C. B. Johnson and D. Grierson, *Curr. Advances Plant Sci.* **9**, 1 (1974).
91. P. J. Bottino, *Radiat. Bot.* **15**, 1 (1975).
92. R. S. Chaleff and P. S. Carlson, *Annu. Rev. Genet.* **8**, 267 (1974).
93. Reference deleted.
94. J. D. Kemp and D. J. Merlo, *BBRC* **67**, 1522 (1975).
95. M.-D. Chilton, S. K. Farrand, F. Eden, T. Currier, A. J. Bendich, M. P. Gordon and E. W. Nester, *in* "Modification of the Information Content of Plant Cells" (R. Markham, D. R. Davies, D. A. Hopwood and R. W. Horne, eds.), p. 297. North-Holland Publ., Amsterdam, 1975.
96. M. J. Dilworth and C. A. Parker, *J. Theor. Biol.* **25**, 208 (1969).
97. J. Postgate, *Nature* **256**, 363 (1975).
98. A. V. Rake, *Genetics* **71**, 19 (1972).
99. K. T. Shanmugam and R. C. Valentine, *Science* **187**, 919 (1975).
100. J. Postgate, *in* "Genetic Manipulations with Plant Materials" (L. Ledoux, ed.), p. 123. Plenum, New York, 1975.
101. R. Dixon, F. Cannon and A. Kondorosi, *Nature* **260**, 268 (1976).
102. R. J. Shepherd, G. E. Bruening and R. J. Wakeman, *Virology* **41**, 339 (1970).
103. E. C. Cocking, *in* "Genetic Manipulations with Plant Materials" (L. Ledoux, ed.), p. 311. Plenum, New York, 1975.
104. S. Sarkar, M. Upadhya and G. Melchers, *Mol. Gen. Genet.* **135**, 1 (1974).
105. H. E. Varmus, P. K. Vogt and J. M. Bishop, *PNAS* **70**, 3067 (1973).
106. J. B. Gurdon, M. L. Birnstiel and V. A. Speight, *BBA* **174**, 614 (1969).
107. R. A. Laskey and J. B. Gurdon, *EJB* **37**, 467 (1973).
108. A. Colman, *EJB* **57**, 85 (1975).
109. C. O. Gardner, Jr., P. Achey and R. J. Mans, *FEBS Lett.* **63**, 205 (1976).
110. W. Goebel and W. Schiess, *Mol. Gen. Genet.* **138**, 213 (1975).
111. T. G. Morrison and H. F. Lodish, *PNAS* **70**, 315 (1973).
112. B. E. Roberts and B. M. Paterson, *PNAS* **70**, 2330 (1973).
113. K. Benveniste, J. Wilczek, A. Ruggieri and R. Stern, *Bchem* **15**, 830 (1976).
113a. J. W. Davies and P. Kaesberg, *J. Virol.* **12**, 1434 (1973).
113b. S. Wang, K. B. Marcu and M. Inouye, *BBRC* **68**, 1194 (1976).
114. G. P. Howland, R. W. Hart and M. L. Yette, *Mutat. Res.* **27**, 81 (1975).
115. G. P. Redei, *Annu. Rev. Genet.* **9**, 111 (1975).
116. J. W. Watts and J. M. King, *Planta* **113**, 271 (1973).
117. M. H. Richmond, This series **13**, 191 (1973).

Initiation Mechanisms of Protein Synthesis

MARIANNE
GRUNBERG-MANAGO

*Institut de Biologie
Physico-Chimique
Paris, France*

FRANÇOIS GROS

*Institut Pasteur
Paris, France*

I. Introduction	209
II. Components Involved in the Initiation Step	210
A. Initiator tRNA	210
B. Initiation Factors	215
III. Total Reaction Sequence	220
A. Prokaryotes	220
B. Eukaryotes	226
IV. Reversible Association of Ribosomal Subunits	227
V. Role of Ribosomes and Initiation Factors in Recognizing the Initiation Signals	235
A. Requirement for IF3 with Synthetic Messengers	235
B. Structure of Initiation Signals	237
C. Mechanism of Cistron Recognition in Prokaryotes	247
D. Mechanism of Cistron Recognition in Eukaryotes	263
VI. Pending Questions	268
A. Precise Operational Sequence of Initiation Complex Formation in Prokaryotes	268
B. Number of tRNA Binding Sites on 30 S Subunits	270
C. ATP Requirement	271
D. Role of the Formyl Group in Prokaryote Initiator tRNA	272
VII. Conclusion	274
References	276

I. Introduction

Initiation, the act preceding the start of polypeptide chain synthesis, consists in a series of events ensuring the "recognition" of particular regions of the messenger RNA chain, called the "initiation signals," that specify the place where decoding of the template RNA begins. During this step, the tRNA providing the N-terminal aminoacyl residue of the nascent chain is positioned opposite the initiation triplet, present within the initiation signal.

It is the formation of such an initiation complex that is critical in phasing the "readout" of the messenger RNA and that is probably the rate-limiting step in protein synthesis (1). Initiation also represents an obvious and important control point in translation. For instance, alteration of any one of the components involved in the formation of the initiation complex could theoretically eliminate recognition of a particular initiator region, cause a latent site to become active, or alter the relative efficiency of ribosome binding to various initiation triplets in a messenger RNA. Although results obtained from several systems hint that such phenomena may play a role in biological regulations, the evidence presently available is far from conclusive.

In this review, the emphasis is placed on the molecular involvement of the initiation components, as currently conceived.

II. Components Involved in the Initiation Step

A. Initiator tRNA

The first clue to the existence of a special mechanism for initiation of protein synthesis arose from the finding that methionine occurs as the N-terminal amino acid in 40% of the ribosomal proteins, although it constitutes only 2.5% of the total amino acids in proteins (2); the remaining 60% of the N-terminal amino acids are either alanine or serine. It is now established that in prokaryotes as well as in eukaryotes, the first residue incorporated into a peptide linkage at the NH_2 terminus in a growing protein chain is always the same, a methionine residue, whatever the nature of the protein. This residue is inserted on the messenger–ribosome complex as a thermodynamically reactive ester, methionyl-tRNA, the tRNA involved in this step being different from the tRNA positioning methionine internally in the protein chain.

1. PROKARYOTES

Two species of methionine-accepting tRNAs are found in prokaryotes: an initiator tRNA (tRNA^{fMet}) and an elongation tRNA (tRNA^{Met}); both are acylated by the same synthetase, and both recognize the triplet AUG, but initiator tRNA recognizes GUG *in vitro* as well; priming of the translation process by methionyl-tRNA^{fMet} with 70 S ribosomes requires the aminoacyl residue to be N-substituted by a formyl group (3, 4). The discovery of an N-substituted form of methionine was of cardinal importance for the discovery of initiator tRNA. It was made during the course of a search for sequencing techniques (3); therefore Met-tRNA that could be heavily labeled *in vivo* was chosen. Two methionine-containing products, adenosyl-methionine and adenosyl-formylmethionine, were isolated after RNase hydrolysis. Subsequently, it was established that formylation is catalyzed by a special transformylase (5) that can transfer a formyl residue from an N-formyltetrahydrofolate donor to Met-tRNA^{fMet}, but that cannot formylate the methionine of Met-tRNA^{Met}.

Shortly after the discovery of fMet-tRNA, it was found that the coat protein of phage Qβ, when synthesized *in vitro* from viral RNA, has the formylmethionine residue at the N-terminal position instead of alanine, the amino acid present at this terminal when the protein is synthesized *in vivo*. Subsequently, formylmethionine was found at the N-terminal position of two of the proteins synthesized *in vitro* from the cistrons of small RNA phages, and later it was identified as the N-terminal residue of three of these proteins (6–8), namely, the maturation protein, the coat protein, and the synthetase. These observations could thus be considered as definite proof that fMet-tRNA is *the* protein chain initiator for each of the products that are coded for by a polycistronic mRNA, such as phage RNA.

It should be noted that, *in vivo*, the formyl group, initially present on the growing chain, is rapidly removed by a "deformylase." An enzyme activity that removes the formyl group from proteins synthesized *in vitro* has been identified in *Escherichia coli* extracts; it readily uses fMet-peptides, fMet-Phe-tRNA, or fMet-puromycin as substrate, but not fMet-tRNA (9–11), it is very unstable *in vitro*, and it is inhibited by -SH reagents. Once the methionine from the growing polypeptide is deformylated, other enzymes split off the methionine residue itself. Indeed, both *E. coli* and *Bacillus subtilis* contain an aminopeptidase that removes the methionine residue from Met-puromycin (but not from fMet-puromycin) or from the N-terminal position of the hexapeptide of phage f2 coat protein (after treatment by a deformylase) (9, 12). This explains what appeared at first as a paradox, that *in vivo* the coat protein appears to "start" not with methionine, but with alanine.

2. EUKARYOTES

In eukaryotes, the situation is more complex, and a distinction should be made between the translation system that operates on the true cytoplasmic ribosomes (the 80 S) and those functioning on the ribosomes originating from cytoplasmic organelles (mitochondria or chloroplasts). For translation on the 80 S, the initiator is a special Met-tRNA (often designated tRNA$_i^{Met}$) responding to AUG as well as

GUG *(4, 13–16)*. Although formylation does not occur naturally on such a species, it can be formylated artificially by an *E. coli* formylase *(14–16)*; it can also be acylated by the *E. coli* synthetase, although the *E. coli* enzyme does not recognize the tRNA^{Met} serving as adaptor for internal residues in eukaryotes. The internal methionine residues arise, as in prokaryotes, from the Met-tRNA^{Met}, distinct from initiator tRNA *(17–19)*. An aminopeptidase is also present in the reticulocyte system: it removes methionine from the terminal position of the hemoglobin chain, but does not remove incorporated formyl-methionine *(19)*.

When translation proceeds on organelle-derived ribosomes, the initiator is, as in bacteria, an fMet-tRNA^{fMet} whose RNA moiety differs from that of the cytoplasmic initiator; moreover, and as expected, there exists a transformylase in mitochondria and chloroplasts *(20)*. This should be related to the fact that mitochondria and chloroplast ribosomes show many analogies to ribosomes from bacteria.

It is therefore interesting, for the sake of comparative biochemistry, to find the extent to which prokaryotic and eukaryotic initiation systems can be crossed. When artificially formylated by *E. coli* transformylase, Met-tRNA^{fMet} from guinea pig can be recognized as an initiator by *E. coli* 70 S ribosomes (+ AUG and factors), whereas when not formylated it cannot be recognized *(21)*. The same applies to initiator tRNA from rabbit reticulocytes. On the contrary, when the "internal" aminoacyl adaptors, *E. coli* Met-tRNA^{Met} or eukaryotic Met-tRNA^{Met} are artificially formylated, they are not used as initiators by the 70 S ribosomes *(21)*. Conversely, the eukaryotic initiation factors form an initiation complex preferentially with nonformylated eukaryotic Met-tRNA. They discriminate also against prokaryotic initiator tRNA whether formylated or not *(22)*. Therefore, in prokaryotes, recognition of an aminoacyl-tRNA as initiator by the 70 S ribosomes implies (i) an N-substitution of the aminoacyl moiety by the formyl group, and (ii) a particular, presumably steric, structure for the tRNA. Both prokaryotic and eukaryotic initiator tRNA possess that structure. (We comment further, in Section VI, D, on the importance of the formyl group for the correct positioning of initiator tRNA).

The nucleotide sequence of both *E. coli* tRNA^{fMet} *(23)* and tRNA^{Met} *(24)* have been determined, as has the sequence of yeast *(25)*, rabbit liver, myeloma initiator tRNA, and sheep mammary gland *(26)* initiator tRNAs (Fig. 1). Comparison of these various initiator tRNAs reveals no unusual characteristics that might account for the uniqueness of their role. However, it should be noted that, in the case of eukaryotic

FIG. 1. Sequences of various initiator tRNAs. (a) *Escherichia coli* CA 265 tRNA[fMet] *(23)*. (b) *Escherichia coli* CA 265 tRNA[Met] *(24)*. (c) Yeast tRNA[fMet] *(25)*. (d) Myeloma, rabbit liver, sheep mammary gland tRNA[fMet] *(26)*.

initiator tRNA, the G-T-Ψ-C sequence is replaced by a G-A-U-C sequence. It is also striking that the last three eukaryotic initiator tRNAs mentioned above have very similar sequences. It should also be noticed that the prokaryote initiator tRNA has one base-pair missing at the -C-C-A extremity. This may possibly account for the fact that it is not recognized by elongation factor EF-T (27) but, even in the nonenzymic binding reaction, initiator tRNA behaves differently from Met-tRNA^{Met} (see Section VI, D).

While formylation of Met-tRNA^{fMet} is generally required for recognition by the complex of 70 S ribosomes and messenger RNA, an interesting exception has been established, in *Streptococcus faecalis*, that sheds some light on the role of this N-formylation in constructing the normal complex. *S. faecalis* cannot synthesize folic acid or its derivatives *de novo* (28). However, the organism can grow essentially at the normal rate in the absence of any of these compounds, provided the medium contains serine, methionine, thymine, a purine base, and pantothenate, compounds whose synthesis involves a folate derivative. When grown in a medium containing folate or its derivatives, the organism readily synthesizes fMet-tRNA^{fMet}, and this takes part in the process of chain initiation. But when the bacteria are grown in a medium lacking folic acid, the consequences are 2-fold: (i) uracil is no longer methylated to yield the thymine occurring in the G-T-Ψ-C sequence (the explanation is that in this bacterium the thymine of tRNA is synthesized not through S-adenosylmethionine but through folic acid) (28); (ii) the methionine of initiator tRNA is not formylated. These two changes enable the bacteria to use an unformylated initiator for protein synthesis, suggesting that a change in the G-T-Ψ-C loop somehow compensates for the steric modification of the initiator molecule as a whole, caused by the lack of formylation. The ability to use an unformylated initiator is not due to the modification of a component of the protein-synthesizing machinery other than that occurring within the sequence of this tRNA (29). A similar situation was observed in two other gram-positive bacteria, *B. subtilis* and *Micrococcus luteus* (30); here, trimethoprim (an inhibitor of dihydrofolate reductase) caused the same modifications as those described above: in initiator tRNA, no formylated Met-tRNA could be detected. Nevertheless, both organisms synthesize proteins at normal rates under these conditions, provided that the low-M_r[1] products of folate metabolism are supplied in the medium.

---

[1] M_r is the IUPAC-recommended symbol for "molecular-weight ratio," commonly termed "molecular weight" [Ed.].

B. Initiation Factors

1. PROKARYOTES

Positioning of initiator tRNA on the messenger-ribosome complex involves, as the first step, the small 30 S ribosomal subunit *(31, 32)*. The complex thus achieved is usually designated the "preinitiation complex"; its formation involves the cooperative interaction of a variety of ligands, among which are the initiation factors. These proteins, isolated independently by three groups of workers *(33–35)*, exist in a state of loose association with the 30 S subunits, from which they can be detached by washing with a high concentration of salt (such as 1 M NH_4Cl).

E. coli initiation factors are now reasonably well characterized. Three general types are known: IF1, IF2 and IF3. They have been purified to homogeneity *(36a–41)*.[2] IF1 has even been crystallized *(38)*. The general requirement for these factors can be shown operationally, and in a first approximation, by the use of *in vitro* systems including the RNA from an RNA phage as messenger, NH_4Cl-washed 70 S ribosomes, GTP, the elongation factors EF-Tu, EF-Ts, and EF-G, the termination factors, aminoacyl-tRNAs, an energy-yielding system, and the appropriate ions (Mg^{2+}, K^+, etc. . . .). In such a reconstituted system, no protein synthesis occurs unless the three initiation factors are present. By contrast, when a synthetic messenger, such as poly(U), is used, and provided that appropriate ionic conditions are met (high Mg^{2+} concentrations), translation occurs at maximal rate in the absence of initiation factors. This type of experiment indicates that the step at which factors from a ribosomal wash are involved is not the elongation process per se, but rather that natural messengers carry particular signals, the recognition of which requires the presence of both a special tRNA and initiation factors.

IF1 is a basic protein with an M_r of 9500 *(38)*; it is not retained on DEAE-cellulose. When bacteria are harvested at the beginning of their exponential growth, IF1 is usually found in tight association with IF3, and the complex cannot be separated into its constituents by filtration on a Sephadex–urea column (J. Dondon and M. Grunberg-Manago, unpublished); the amino-acid composition of IF1 is similar to that of most ribosomal proteins *(38)*.[3]

---

[2] A very reproducible procedure to simultaneously prepare the three factors with a good yield is that of Hershey *et al. (287)*.

[3] However, IF1 prepared by Hershey *et al. (287)* and by Gualerzi (personal communication) shows an amino acid composition quite different from that reported by Lee-Huang *et al. (38)*.

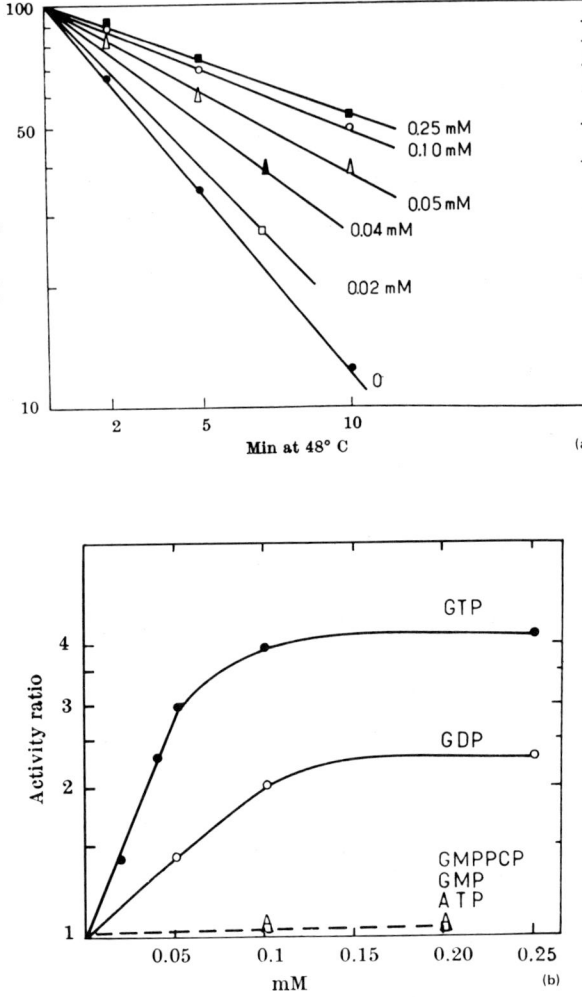

FIG. 2. (a) GTP protection of initiation factor IF2. From a mixture (100 μl) containing 10 μg of factor in 10 mM TrisCl, (pH 7.5), 0.5 mM dithiothreitol, 200 μg/ml bovine serum albumin, and various concentrations of GTP from 0 to 0.25 mM, 30-μl aliquots were heated at 48°C for the indicated times. From each tube, 2 μg of IF2 was used to measure poly(A,U,G)-dependent [^{3}H]fMet-tRNA binding to 150 μg of ribosomes in the presence of 10 μg of initiation factor IF1 and GTP (final concentrations, 1 mM). (b) Compared affinities of guanosine derivatives and ATP for initiation factor IF2. Nucleotides were substituted for GTP as indicated, and all tubes were heated at 48°C for 10 minutes. The results are expressed, on a log scale, as the ratio of factor activity, heated in the presence and in the absence of protector. (a) and (b) from Lelong *et al.* (42).

TABLE I
MOLECULAR WEIGHT OF PROKARYOTE IF2

Method used	Factors	
	IF2a	IF2b
Polyacrylamide/sodium dodecyl sulfate gel electrophoresis	107,000	84,000
Varying gel concentration[a]	98,000	250,000
		160,000
		85,000[b]
Sephadex G-200	95,000	—
Sedimentation constant (sucrose)	3.5 S[c]	—

[a] Method of Hedrick and Smith *(44)*.
[b] From Gros *et al. (39)*.
[c] From Lelong *(40)*.

IF2 exists in two forms, IF2a and IF2b, IF2b being slightly smaller in size with an M_r of 85,000 while IF2a has an M_r estimated between 95,000 and 118,000, depending on the technique used *(36a,b, 39–43, 287)* (Table I). Denaturation of IF2a does not seem to affect its M_r, whereas native IF2b appears as an aggregate with a majority of trimers, but the chromatographic properties of both factors are very similar. IF2b is less thermostable than IF2a (Fig. 2). The sedimentation constant of IF2a is about 3.5 S, a value suggesting that the relatively high-M_r compound has an elongated shape. Both IF2 forms exhibit the same specific activity in catalyzing phage RNA or poly(U,G)-directed fMet-tRNA binding *(40, 41, 43)*. *E. coli* contains twice as much IF2a as IF2b, and it has been suggested (Traut and Hershey, personal communication; *41*) that IF2b results from a proteolytic cleavage of IF2a.

IF3 has an M_r of 22,000 *(37, 45–50)* and is a thermostable protein. The sedimentation constant reported, 2.2 S *(50)*, is in accordance with the M_r determined by acrylamide/dodecylsulfate gel electrophoresis. However, the M_r determined by gel filtration on Sephadex was estimated to be approximately 32,000 *(49)*, which may indicate that this factor does not behave as a typical globular protein.

It has been reported *(45, 46)*, but this is not universally found *(47, 50, 287)*, that *E. coli* contains at least two forms of IF3 with almost identical M_r's: IF3α (22,500) and IF3β (21,000). IF3α is supposed to catalyze the translation of MS2 RNA preferentially to late T4 messenger; IF3β shows the inverse specificity. However, the difference is not striking. It is not known whether these two proteins are products from different genes. It is possible that IF3β results from a proteolytic cleavage of IF3α *(51b)* as suggested for IF2b.

The mapping of a thermosensitive mutant strain characterized by thermolabile IF3 activity *(51a,b,c)* localized on the *E. coli* chromosome the structural gene for this factor and should eventually reveal whether or not there is only one gene for IF3. It is not clear yet if the thermosensitivity is directly related to the thermolability of IF3, as the mutant strain has actually two defects: one in the IF3 activity and the other in the phenylalanyl-tRNA synthetase activity.

In *Bacillus stearothermophilus (52)* and *Caulobacter crescentis (53, 54)*, only two initiation factors have been found and purified; they correspond to IF2 and IF3. This situation might reveal a somewhat dispensable nature of IF1 in the other prokaryotic systems.

2. EUKARYOTES

In eukaryotes, the picture is much more complex, and the exact number of translation initiation factors is still a matter of debate. At least six different components have been described, the presence of which would be essential for the initiation step to occur at maximum rate *(55–62)*. Quite recently a seventh has been described; it was contaminating one of the others (see Table II). Unfortunately, complete agreement has not yet been reached as to nomenclature, but at a Symposium on Protein Synthesis, held at the National Institutes of Health and cosponsored by the John E. Fogarty International Center and the National Heart, Lung, and Blood Institute, an effort was made to arrive at a single nomenclature for the eukaryotic initiation factors; the system recommended is set out in Table II together with the two main nomenclatures used so far.[4] This multiplicity of factors, most of which have been isolated from liver, reticulocytes, or muscle cells, probably corresponds to the reality and most of these factors are not ribosomal proteins having the property of potentiating the action of true initiation factors. There are some differences: M1, for instance, and perhaps M2Bα, which, according to the NIH group are true eukaryotic factors, are not necessary for the translation of the globin messenger.

A number of these factors have been purified to homogeneity. Four of them are single polypeptide chains with approximate M_r's of 15,000 (eIF-1), 50,000 (eIF-4A), 160,000 (eIF-5), and 80,000 (eIF-4B). By contrast, eIF-2 appears to be composed of three subunits of about 35,000, 50,000, and 55,000 *(55, 57b)*. The M_r of native eIF-2, as estimated from sedimentation analysis, is (within the limit of error) compatible with the sum of these three subunits, or possibly with two of the larger and one each of the others *(55)*. eIF-3 is a large protein complex that, in its

---

[4] The new system is used in the present chapter.

TABLE II
NOMENCLATURE FOR THE INITIATION FACTORS INVOLVED IN
EUKARYOTIC PROTEIN SYNTHESIS[a]

New[b,c]	Basel[d]	NIH[e]
eIF-1	IF-E1	—
eIF-2[c]	IF-E2	IF-MP
eIF-3[f]	IF-E3	IF-M5
eIF-4A	IF-E4	IF-M4
eIF-4B	IF-E6	IF-M3
eIF-4C	IF-E7	IF-M2B$_\beta$
eIF-4D	—	IF-M2B$_\alpha$
eIF-5	IF-E5	IF-M2A

[a] Developed at an International Symposium on Protein Synthesis held at the U.S.A. National Institutes of Health, October 18, 1976 [W. F. Anderson et al., FEBS Lett. 76 (1),1–10 (1977).]

[b] eIF stands for eukaryotic initiation factor. The "e" can be replaced by: a = animal; p = plant; v = viral; y = yeast; etc.

[c] Special case: M1^e (which appears to be an evolutionary remnant) is eIF-2A.

[d] Staehelin et al. (55).

[e] Shafritz et al. (57b).

[f] Subunits of eIF-3 (when purified to homogeneity) should be named eIF-3a, eIF-3b, etc. (lower case letters)

native state, sediments at 17 S and that behaves during electrophoresis as a single and well-defined physical unit. Upon dissociation in dodecyl sulfate, 9 or 10 subunits, some apparently not in stoichiometric amounts, are revealed. It is probable that initiation factor eIF-3 possesses some kind of organelle-like structure. By and large, the eukaryotic factors have rather high M_r's and show a tendency to form homo- or heteroaggregates. Furthermore, as discussed below (Section III), while some prokaryote factors exhibit several different activities, this is apparently not the case in eukaryotes.

III. Total Reaction Sequence

In order to discuss more easily the part played by factors in the translation process, it may be convenient at this point to describe in general terms what is likely to be the temporal sequence of the initiation events. It must be borne in mind that the very high degree of cooperativity implied in the formation of the preinitiation complex makes it quite difficult to study the binary interactions between the various ligands involved. Also, the cases of eukaryotes and prokaryotes must be distinguished.

A. Prokaryotes

In a cell growing normally, the factors are not present in stoichiometric amounts with respect to ribosomes: one 30 S ribosomal subunit out of 20, for example, is equipped with a molecule of IF3, one out of 30 with a molecule of IF1 (63). The sequence of events, illustrated in Fig. 3, can be viewed approximately as follows:

When RNA translation is terminated, the 70 S ribosomes are released from the polysomal complex and are then in equilibrium with their subunits. But, as seen in (Section IV), the dissociation rate in 5 mM Mg^{2+} is slow, and the equilibrium tips toward the 70 S (Fig. 2). One of the major roles of IF1 is to activate this rate precisely; 70 S dissociation would then unmask a site on the free 30 S particle for stoichiometric binding of IF3, thus shifting the equilibrium in favor of the free subunits. Beyond that stage, the sequence of events is not firmly established. A possible route is that, under the influence of IF3, the 30 S subunit interacts with the messenger initiation signals *before* attachment of the initiator tRNA. Another possibility is that initiator tRNA would *first* bind to the ribosome under the influence of IF2, this attachment being a prerequisite to the correct binding of messenger RNA. It has also been suggested that, depending on the mRNA structure, either the first or the second route can be taken. (This is discussed in further detail in Section VI, A.) A number of workers regard IF2 as "fMet-tRNA recognition or carrier protein." It is tempting to establish a parallel with EF-Tu, which is responsible for the recognition and positioning of internal aminoacyl-tRNAs in the presence of GTP. However, whereas EF-Tu forms, with GTP and aminoacyl-tRNA, a stable ternary complex in the absence of ribosomes, interactions between IF2 and GTP, or IF2 and fMet-tRNA, or IF2, GTP and fMet-tRNA, are not stable enough in the absence of ribosomes to allow separation of the corresponding complexes. The situation is different in the case of eukaryotes (Section III, B).

FIG. 3. Ribosomal equilibria in the formation of the initiation complex. ⓘ ② ③ ─•─•─•─•─•─ indicates mRNA; ∪ indicates fMet-tRNA^{fMet}; I = initiation complex; PI = preinitiation complex, for equilibria in which ribosomes with two initiation factors are involved; tRNA is optional when one of the factors is IF2, and mRNA is optional when one of the factors is IF3.

INITIATION MECHANISMS OF PROTEIN SYNTHESIS 221

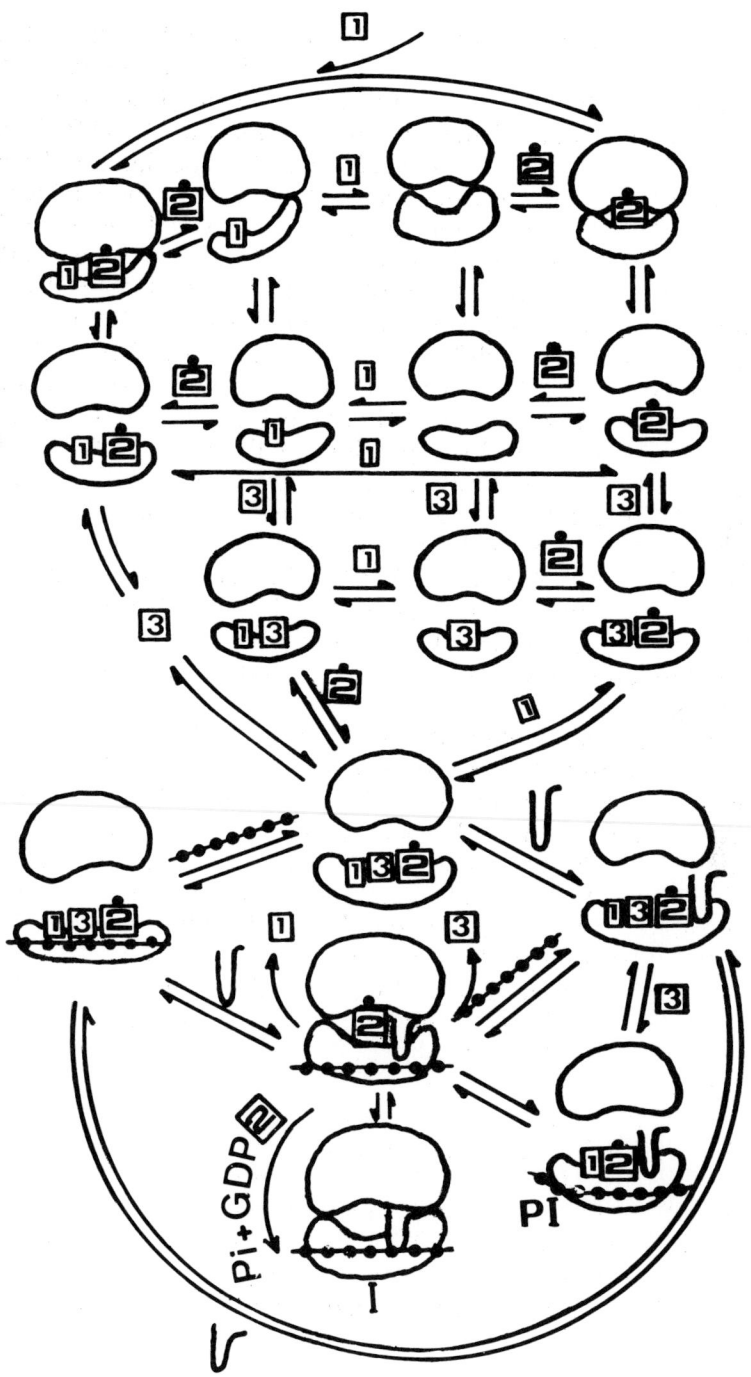

Nevertheless, the existence of weak sites for GTP can be indirectly demonstrated: GTP protects IF2 against the denaturing effects of -SH substitution *(64)*; furthermore, GTP, and GDP to a lesser extent, protects IF2 against thermal inactivation (around 50°C); ATP, GMP, or GMP-P[CH$_2$]P are without effect *(42)* (Fig. 2). Calculations show that GTP and IF2 associate with an apparent K_A of $10^4 M^{-1}$, which indicates, as was stated, that the interaction is very weak. However, GTP is required for the correct binding of IF2 to ribosomes (Section IV).

Concerning the ternary complex (IF2 · GTP · fMet-tRNA), the melting denaturation profile spectrum of the initiator tRNA as a function of temperature is modified in the presence of IF2 plus GTP, the other initiation factors having no effect *(65)*. Also, a ternary complex, IF2 · GTP · fMet-tRNA (not Met-tRNA), can apparently be separated from the free components by Sephadex gel filtration on a G-50 column *(65–67a)* or by glycerol gradient centrifugation *(67b)*. However, since highly purified IF2 does not give rise to such a complex, it is possible that some other ribosomal proteins are required. Recently, with purified IF2 in the absence of Mg^{2+}, a binary complex, IF2 · fMet-tRNA has been detected on nitrocellulose filters *(68)*. However, only a small fraction of input IF2 becomes bound to fMet-tRNA^{fMet}. The complex formation is GTP-independent, and dissociates in the presence of Mg^{2+}.

According to a number of authors *(69–71)*, IF3 and IF1 are detached from the initiation complex before the joining of the 50 S particle, IF3 being displaced by fMet-tRNA *(72)*, and IF1 by the junction of the large subunit in the absence of GTP hydrolysis *(69)* (IF1 has, however, some affinity for the 70 S, which is discussed in Section IV). The recycling of IF2, in contrast, requires both the joining of the 50 S particles to the 30 S initiation complex, and GTP hydrolysis *(73)*. Indeed, after the 70 S couple is reformed, a GTPase is activated by the combination of 50 S particle and IF2, and the hydrolysis of GTP into GDP + P$_i$ is probably what causes a change in ribosome or IF2 conformation, allowing IF2 release *(42, 74, 75)*. However, in the absence of GTP, this can occur, albeit less efficiently *(75)*, but in the presence of GMP-P(CH$_2$)P, IF2 is not released from the 70 S complex *(73–77)*. IF1 and IF3 markedly stimulate the recycling of IF2. Nevertheless, attachment of fMet-tRNA to 70 S ribosomes can occur in the absence of IF1 and IF3, but it is then stoichiometric to the fMet bound. The reaction is catalytically stimulated by IF2 only in the presence of either one of the other two factors *(37, 78–80)*. As expected, IF1 increases IF2-dependent GTPase activity, while IF3 does so only

TABLE III
IF2-DEPENDENT GTPASE[a]

Additions to incubation mixture	Initial rate of P_i liberated (pmol/min)
None	<4
Ribosomes	<4
Ribosomes + IF1 + IF3	4
Ribosome + IF2	39.5
Ribosomes + IF2 + IF1	145
Ribosomes + IF2 + IF3	52
Ribosomes + IF2 + IF3 + IF1	86

[a] The incubation mixture contained, in a total volume of 50 μl: 10 mM TrisCl, pH 7.5; 100 mM NH_4Cl; 5 mM Mg acetate, 2.5 mM phosphate and 0.11 mM GDP (the latter two added to inhibit endogenous ribosomal GTPase); 50 μM [^{32}P]GTP (150 cpm/pmol); plus ribosomes 11.02 A_{260}/ml and, when indicated, IF1 (520 pmol/ml), IF2 (140 pmol/ml), IF3 (158 pmol/ml). Phosphate liberation was tested at 37°C after 5 and 10 minutes of incubation. The 70 S ribosomes (56 A_{260} units/ml) were first incubated in 30 mM Tris, 15 mM Mg acetate, 300 mM NH_4Cl, for 20 minutes at 37°C (Godefroy-Colburn, unpublished).

slightly, probably because its recycling effect is antagonized by preventing the junction of the 50 S subunits necessary for IF2 recycling (Table III). The GTPase activity is stimulated by poly(A,U,G), but not by fMet-tRNA unless poly(A,U,G) is also present. Maximal activity occurs in the presence of all the components of initiation complex (Th.Godefroy-Colburn, unpublished).

The rationale for the GTP-dependent release of factor IF2 to ensure a catalytic positioning of initiator tRNA can be understood in the following terms. Initiator tRNA(fMet-tRNA^{fMet}) has the same general conformation as a polypeptidyl-tRNA. It therefore is located at the P site, a specific 70 S territory made up, when the 50 S joins the 30 S subunit on the messenger, by the overlapping of certain regions of both particles (81). Cross-linking experiments show that proteins L2 and L27[5] are actually the attachment site of the methionyl residue (82). That L2 and L27 also belong to site P can be deduced from experiments with N-substituted Phe-tRNA^{Phe} (83, 84). For the N-substituted methionyl radical of the donor to establish peptide linkage with the

[5] The nomenclature of Kaltsmidt and Wittmann is used for the designation of ribosomal proteins. See article "The Ribosomes of Escherichia coli" by Brimacombe et al. in Vol. 18 of this series.

amino-acid radical corresponding to the second residue of the chain, the latter must be positioned at a site sufficiently adjacent to site P, called site A. Moreover, a series of observations (examined below) support the notion that when IF2 interacts with the ribosome and positions initiator tRNA, the factor itself occupies the A site. Since IF2 occupies the A site, it must be driven out for the elongation step to proceed. A complex and very precise mechanism must come into play to ensure a sequential order for the initiation and elongation reactions.

Evidence that IF2 occupies the A site on the ribosomal complex is based on the following observations.

1. IF2 prevents the EF-Tu-dependent positioning of Phe-tRNA on a 70 S · poly(U) complex *(77, 85, 86a)*. When the initiation complex is "frozen," thus preventing the recycling of IF2 [omitting IF1 or replacing GTP by its nonhydrolyzable analog, GMP-$P(CH_2)P$], the positioning of the second aminoacyl-tRNA, a process catalyzed by EF-Tu, is blocked. Furthermore, puromycin, an analog of the internal aminoacyl-tRNAs, no longer has access to site A to form the peptide link with initiator tRNA, leading to fMet-puromycin *(77, 86b)*. When IF2 is eliminated from the complex (by passing the latter through a Sephadex column, a treatment that does not eliminate initiator tRNA), the fMet residue can react freely with puromycin *(77)*.

2. Competition experiments between IF2 and EF-G suggest that the ribosomal site involved in the IF2-dependent GTPase is very close to that of EF-G-dependent GTPase *(86a)*.

3. Thiostrepton, an inhibitor of the elongation reaction (EF-Tu and EF-G-dependent) does not inhibit the positioning of fMet-tRNA in the presence of IF2, but blocks the recycling of this factor by inhibiting the IF2-dependent GTPase *(86a, 87)*.

4. Proteins L7 and L12 are involved in the positioning of IF2 on the 50 S subunits *(87–90a)*; moreover, specific cross-linking of factor IF2 to proteins L7/L12 occurs *(90b)*. These 50 S proteins also serve as attachment sites for the elongation factors EF-T and EF-G *(91–93)*.

A well-set chronology must therefore guide the attachment of factors since they all appear to bind to the same site, equivalent to or part of the ribosome acceptor site.

One of the key steps in the release of translation factors from ribosomes thus appears to be a factor- and ribosome-dependent hydrolysis of GTP. This general multicomponent reaction ensures IF2 release during initiation and EF-Tu release during elongation. Moreover, GTP hydrolysis is necessary for the liberation of EF-G, involved in translocation.

The reason for the difference in nucleotide requirement probably results from conformational changes occurring when the factors bind to either GTP or GDP; in the first case, binding results in a conformation with a high affinity for ribosomes while GDP binding results in a low affinity *(94–96)*. Hence the necessity for GTP hydrolysis. At the present time, the location of the catalytic site for the GTP hydrolysis on the ribosome or the factor is still a matter under investigation. One should emphasize the difficulty of determining whether a definite protein is or is not part of a specific site; a large number of parameters are inherent to a system of such great complexity, and any change in a protein could alter the structure of the whole ribosome, affecting either the catalytic site or the ability of the ribosome to bring about the correct conformation of the factors. The evidence presented is therefore only tentative.

Carefully prepared ribosomes show no GTPase activity, yet this activity has been assigned to a complex of 5 S RNA and proteins L18 and L25 *(97, 98)*. One explanation is that, in whole ribosomes, these proteins have a structure that prevents them from either attaching or hydrolyzing GTP. However, the binding of factors to proteins L7 and L12 would cause a long-range modification in the structure of L18 and L25 in such a way as to elicit a GTPase activity. Furthermore, thiostrepton, an antibiotic that binds irreversibly to ribosomes, inhibits factor binding and hence GTP hydrolysis. This antibiotic does not interact with proteins L7 and L12, as was previously believed, but with protein L11 *(99)*. The latter is also involved in the interaction of aminoacyl-tRNAs at the A site *(100)* as well as in the binding of EF-G *(101)*.

The sum of events involved in a factor- and ribosome-dependent GTPase activity and consequently in the process of recycling is probably even more complex if one realizes that the 30 S subunit is also necessary to bring about the appropriate conformational changes in the 50 S *(102, 103a)*[6] and that depletion of L18 and L25 does not completely abolish GTPase activity *(104)*.

It is also possible that the catalytic site for GTP hydrolysis is located on the factor and that ribosomes bring about conformational changes of the protein factor required for GTP hydrolysis. The evidence for the GTP hydrolysis catalytic site being located on the factor

---

[6] At least, at physiological ionic concentrations. At low ionic concentrations, the 30 S subunit is not necessary *(103b)*.

is best demonstrated for EF-Tu-dependent GTPase. It has been shown that this factor is able to hydrolyze GTP in the absence of ribosome but in the presence of an antibiotic, kirromycin *(105a)*. It appears therefore that the catalytic site for GTP hydrolysis is located primarily on the elongation factor and that the ribosome or the antibiotic affect the conformation of the EF-Tu molecule. [Direct evidence confirming this statement has been obtained by nuclear magnetic resonance (NMR) *(105b)*.]

In brief, IF2 positions initiator tRNA at site P, the factor itself occupying site A; when the 70 S couple is reformed, the factor is ejected, and this requires the presence of IF1 and the hydrolysis of GTP. As for IF1, its release probably occurs at the same time, although some experiments *(69)* suggest that its liberation, concomitant with the formation of the 70 S couple, can nonetheless take place in the presence of GMP-$P(CH_2)P$ and could therefore precede the ejection of IF2; IF3, which is initially attached to the 30 S subunit, would be detached before the 70 S couple is formed.

B. Eukaryotes

The sequence of events in eukaryotes is probably still more complicated because of the existence of (at least) seven factors. Nevertheless, some of the steps are better understood, and the scheme in Fig. 4 summarizes our current view in the assembly process of initiation *(55)*.

In a first step, a stable complex forms between factor eIF-2, GTP and tRNA^{fMet} (initiator tRNA).

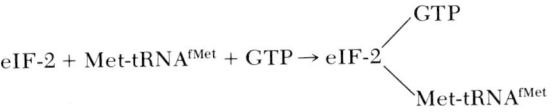

(this reaction bears resemblance to that catalyzed by the prokaryotic factor EF-Tu). The ternary complex thus formed would be transferred to the 40 S subunit, in the absence of mRNA, this interaction being stabilized by eIF-1 and -3. Binding of initiator tRNA to the 40 S in the absence of mRNA occurs *in vitro*. More convincing is the observation *(106)* that there exists in the cell a (Met-tRNA^{fMet}) · 40 S intermediate that can be converted into a preinitiation complex.

eIF-3 is absolutely required for the next step: i.e., the attachment of the messenger RNA, a reaction that also depends on eIF-4A, -4B, and -4C and that, according to Staehelin *(55)*, requires ATP.

The reassociation of the 40 S · (initiator tRNA) · (mRNA) complex with the 60 S subunit would be catalyzed by eIF-5; this reaction also

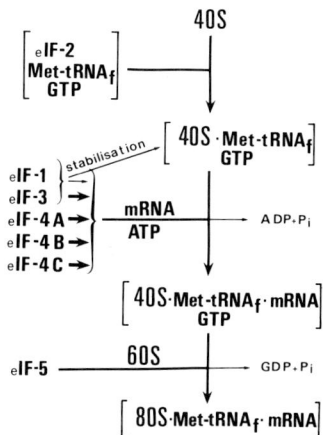

FIG. 4. Sequential steps of initiation complex formation with 80 S ribosomes. From Staehelin et al. (55).

requires the hydrolysis of GTP into GDP + P_i. eIF-5 can also promote mRNA-independent junction of 60 S subunits to the 40 S · Met-tRNA^{fMet} complex; this reaction is thus completely independent of ATP.

eIF-1 (Staehelin's "policeman" factor) would have two functions: it would block formation of a "forbidden" complex, (Met-tRNA^{fMet}) · 80 S, in the *absence* of messenger, but would stimulate formation of this complex when the messenger is *present*. Thus, with an excess of eIF-1, formation of the (initiator-tRNA) · 80 S complex is strongly dependent, not only upon the presence of factor eIF-5 (sub-unit junction factor), but also upon the presence of messenger, ATP, and eIF-4. *In vivo*, it can be imagined that binding of messenger RNA and junction with the 60 S subunits are directed by a strong coopera-tive action of factors eIF-1, -4, and -5, as well as of ATP. The exact role of ATP is not clear at present. A requirement for ATP has also been found in a plant system (107).

We discuss in more detail below (Section V, C, and D) the specific-ity of the factors in the translation process and also envisage the possi-ble function of additional proteins, believed to modulate the recogni-tion of specific messengers.

IV. Reversible Association of Ribosomal Subunits

The interaction between the two ribosomal subunits is a key step in protein synthesis. As already stated, initiation in prokaryotes starts on the 30 S particle, which subsequently joins the 50 S. Moreover, analysis of the interaction between free subunits (in the absence of the other protein synthesis components) affords the most sensitive and critical method to detect the conformational differences of ribosomes.

Although it is known since 1971 *(108, 109)* that ribosomes are in dynamic equilibrium with their subunits, quantitative data are still very scarce. The nature of the inter-subunit bond is an important question, which can be approached by quantitative studies and thermodynamic analysis of the reversible association of purified ribosomal subunits.

Until recently, all attempts to study quantitatively the equilibrium of ribosomes (reaction rate and equilibrium parameters) stumbled over two main difficulties. The first was purely technical: the sedimentation techniques commonly used did not allow kinetic measurements. Moreover, the association equilibrium of the subunits could be strongly perturbed by the hydrostatic pressure in the ultracentrifugation cells during high speed sedimentations *(108, 109)*. The second is an artifact due to the structural heterogeneity of the ribosomes. Two types of 30 S + 50 S do in fact exist; they can be discriminated by their difference in sensitivity to hydrostatic pressure at high speed centrifugation in a sucrose gradient *(110–112, 288)*.

The first difficulty was overcome by using light-scattering techniques combined with stopped-flow measurements *(113–117)*. Light-scattering allows one to measure M_r's ranging from 1.4×10^6 for a mixture of subunits up to 2.6×10^6 for the 70 S ribosomes. Combination of both techniques brings about a minimum of perturbation in the system; moreover, they are fast enough to be used for dynamic studies.

As for the second difficulty, if the ribosomes are prepared by centrifugation in 5 mM Mg^{2+}, the "tight" couples (or "type-A" ribosomes), which are the most active and most resistant to hydrostatic pressure, can be selected *(110–112)*. This preparation procedure leads to physiologically active ribosomes. When special precautions are taken in growing the bacteria (harvesting at one-third of the exponential phase) and when the ribosomes are prepared immediately after harvesting, thus avoiding freezing the bacteria, a homogeneous preparation of type-A ribosomes is obtained *(112)*.

Quantitative analysis (at limiting ribosome concentrations) of optimal conditions required for converting the ribosomes into an initiation complex (using R17 RNA as messenger) indicates that type-A ribosomes are the ones forming the initiation complex with an efficiency of over 50% *(111)*.

Unfortunately, in most studies reported so far, type-B ribosomes sensitive to hydrostatic pressure have been used, a situation that considerably complicates the interpretation of results when it does not lead to wrong conclusions.

Debey et al. (112) showed that the subunit association curves, as a function of [Mg^{2+}], differ widely according to the type of ribosomes used: at a given concentration of monovalent cations, the interval of [Mg^{2+}] between 10% and 90% association is narrower, and the half-saturation lower, for type-A ribosomes. Furthermore, in contrast to type-B ribosomes, type-A exposed to 1 mM Mg^{2+} show no hysteresis effect in their reassociation curve (Fig. 5).

The structural differences between the two types of ribosomes are not clear; in most cases, the conformational difference appears to lie in the 50 S subunit. Type-B ribosomes are more flexible and undergo conformational changes that result in their being unfit for reassociation; in certain cases, these conformational changes can be reversed by thermal treatment.

The kinetics of association and dissociation of E. coli 30 S and 50 S ribosomal subunits of type A appear to fit the simple scheme (117)

$$30\text{ S} + 50\text{ S} \underset{k_1}{\overset{k_2}{\rightleftarrows}} 70\text{ S}$$

over a wide range of Mg^{2+} and ribosome concentrations. Therefore, the association of free 50 S and 30 S is not a bimolecular reaction limited by complex preassociation conformational changes of the subunits, but a second-order reaction governed by the rate of collision of these particles. Both rate constants strongly depend on [Mg^{2+}], monovalent ions, and the preparation of particles. When particles are washed with high concentrations of salt (50 mM NH_4Cl at 25°C), k_2 ranges from 0.04×10^6 to 21×10^6 $M^{-1}s^{-1}$ as Mg goes from 1.5 mM to 8 mM;[7] k_1 ranges from 150 to 2 s^{-1} in the interval of 1.0 to 3.0 mM Mg^{2+}. The kinetics are very similar in the presence of other divalent metals, such as Ca, Sr, Ba. Mn favors the rate of association (k_2 is increased while k_1 is lowered (0.78) (Wishnia, unpublished experiments). The highest rate may not be far from the diffusion-controlled limit. For comparison the association rate of hemoglobin dimers into tetramers is about 1/30 that of 30 S and 50 S subunits (at 8 mM Mg^{2+}) (118), while the association rate of succinate to aspartate transcarbamoylase is about 1/100 of it (119a). In contrast, this rate is quite similar to that for the association and dissociation of a triplet codon to the anticodon of tRNA (119b). This suggests that the ribosome particles are especially designed to associate without important energy barriers.

[7] If ribosomes are not washed with high salt (1.5 M), the results are generally similar, but the k_1 and K_d are lower (288).

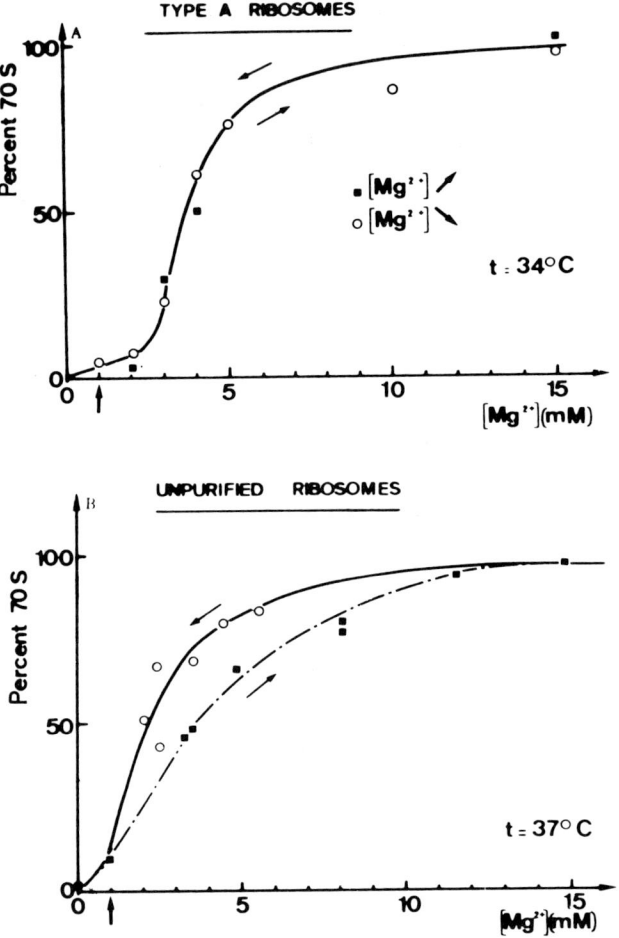

FIG. 5. Reversibility of subunit association. Association equilibrium curves were obtained either by direct association of the ribosomes in buffer containing the specified [Mg^{2+}] (○), or reassociation after incubation in buffer containing 1 mM Mg^{2+} (■). (A) Type-A ribosomes (final concentration 5 A_{260} units/ml); prior incubation at 1 mM Mg^{2+} was for 1 minute at 37°C. (B) Unpurified ribosomes (final concentration 5 A_{260} units/ml); prior incubation time at 1 mM Mg^{2+} was between 20 seconds and 1 minute at 37°C. From Debey et al. (112).

The equilibrium of ribosomal particles is temperature-dependent (119c). The process is exothermic with an enthalpy change of $\Delta H = -18$ kcal mol^{-1}; the entropy change is also negative. The enthalpy term is predominant for the reaction. These thermodynamic data for the overall equilibrium association are quite different from those obtained for reactions involving hydrophobic interactions, which are in general endothermic processes. This seems to eliminate any preponderant contribution of hydrophobic binding between the subunits. On the other hand, the equilibrium enthalpy change for ribosomal subparticles is close to that for oligonucleotide interaction, but the difference between these two types of interactions is shown in the entropy change. In accordance with the discussion in Section V, C, 3, the thermodynamic data suggest that interaction between ribosomal subunits is mainly due to interaction between RNA chains *(119c)*. The effect of Mg^{2+}, although strong enough in terms of rate to have important physiological consequences (500-fold increase for k_2 and 30-fold increase for k_1 between 1.5 and 8 mM Mg^{2+}), is relatively weak in terms of free energy (3.6 to 2.1 kcal); the system is well balanced and other factors may easily tip it one way or the other. The primary effect of Mg^{2+} (as calculated from the rather large changes in binding to the ribosome particles as a function of concentration) could be explained by decreasing the contribution of electrostatic repulsion to the free energy of activation. However, whether in addition specific interactions of Mg^{2+} with some RNA components or proteins contribute to the Mg-dependence is an unsettled question, but there is some indication that this is indeed the case *(119d)*.

At physiological Mg concentrations (4–5 mM), type-A ribosomes are nearly all associated. Since each initiation is preceded by a dissociation, this explains the need for protein factors. These factors, taking part in the initiation of protein synthesis in bacterial systems, act by changing the equilibrium between the 70 S ribosomes and their subunits, as explained below. The dissociation of 70 S ribosomes by factor IF3 was already well documented *(41, 120–122)*, but its mechanism had remained unclear.

The dissociating effect of IF3 and its enhancement by IF1 have recently been reinvestigated in parallel by light-scattering as well as by sucrose gradient centrifugation techniques *(116, 123)*. Two possible mechanisms could account for the IF3 "dissociating effect": (i) an entirely passive process in which IF3 (and eventually IF1) bind solely to the free 30 S subunit, resulting in a shifting of the equilibrium without affecting the dissociation rate; (ii) an active process in which initiation factors first attach to the 70 S ribosome, thus increasing the dissociation rate constant, k_1. It must be noted here that IF1 and IF3 do not attach to every 30 S molecule (this probably reflects the heterogeneity of isolated 30 S particles) but when they do, it is in the proportion of one molecule per subunit *(70, 71, 124)*. Most evidence favors the first hypothesis. IF3 fails to stimulate 50 S subunit exchange between free ^{3}H-labeled 50 S and those contained in the 70 S ribosomes *(121, 125, 126)*; moreover, IF3 prevents reassociation of the subunits *(125)*. But although IF3 has a much higher affinity for the 30 S than for the 70 S particle *(48, 70, 71, 125)* or even for the 30 S dimer *(127)*, a 70 S · IF3 complex may exist under certain conditions *(124)*, although such an intermediate complex has not been found. Furthermore, the stimulation of polyphenylalanine synthesis by purified IF3 in the presence of poly(U) at high [Mg^{2+}] *(128)* is also suggestive of a 70 S · IF3 complex.

The role of IF1 in dissociation has been more controversial. IF1 has never been observed to promote the dissociation of 70 S ribosomes in the absence of IF3, but its direct action on the association equilibrium has been inferred from its stimulation of 50 S subunit exchange *(121)*. However, the hypothetical 70 S · IF1 or 70 S · IF1 · IF2 complex, necessary to explain this, has never been isolated, whereas the 30 S · IF1 complex is stable enough to resist centrifugation to a certain extent in the presence of all the components of the initiation complex *(69)*.

It has been possible to show, without any ambiguity whatsoever, the part played by each factor by studying their influence on ribosomal equilibrium as well as on dissociation kinetics, as analyzed by scattered light *(116)*. With type-A ribosomes (in contrast to type B), IF3 does not bind significantly to the 70 S couple. Dissociation goes to completion at high concentrations of IF3, as it should; the fact that ribosome dissociation is complete leads to the conclusion that the association constant of IF3 with the 70 S is negligible as compared to that with the 30 S; with type-B ribosomes, the affinity for the 70 S is significant and cannot be neglected *(116)*.

The affinity of IF3 for the 30 S subunit was determined under various conditions (25°C to 37°C; 2 to 5 mM Mg^{2+}) and K_3^* was found to be 2.5 to 4 × 10^7 M^{-1} (Fig. 6). Furthermore, the rate constant k_2 is affected by the presence of IF3, which is not the case for k_1 (Table IV). The IF3 dissociating action is therefore purely passive: the factor binds to the 30 S subunit, which then becomes unfit for reassociation. This has recently been confirmed *(129)*.

In contrast, IF1 alone has but little effect on the proportion of associated ribosomes, whereas it drastically increases the two rate constants k_1 and k_2 (the ratio $k_1/k_2 = K_d$ varies only by a factor of 2 in the presence of excess IF1).[8] Moreover, IF1 does not affect the affinity of the 30 S particles for IF3 (Table IV).

[8] K and K' are the affinities of the initiation factor under study for the 30 S subunit and for the 70 S couple, respectively; the subscripts indicate the factor involved. K_d is the dissociation constant of the 70 S ribosome into subunits. The rates of reaction toward dissociation and association are k_1 and k_2, respectively.

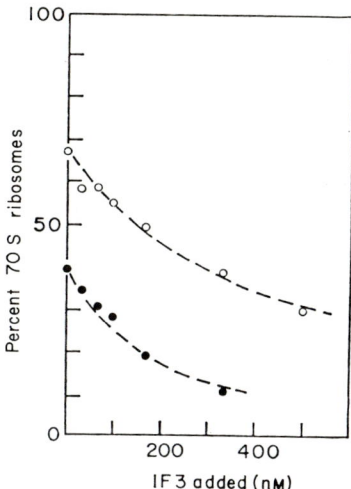

FIG. 6. Light-scattering titration of type-A ribosomes with IF3 at two Mg^{2+} concentrations. The equilibrium proportion of associated ribosomes was determined at 25°C on 150-μl mixtures of ribosomes and IF3 in the following buffers: 10mM TrisCl (pH 7.5); 100 mM NH_4Cl; 7 mM β-mercaptoethanol; magnesium acetate: (●) 2.2 mM; (○) 2.5 mM. Ribosome concentration was 96 nM. The points are experimental and the curves were calculated from the equation

$$K_{d,app} = [(50\ S)(30\ S)]/(70\ S) = K_d[(1 + K(IF))/(1 + K'(IF))],$$

using the following equilibrium constants: $K_3 = 2.5 \times 10^7 M^{-1}$; $K_3' = 0$; $K_d = 15$ nM at 2.5 mM Mg^{2+}; $K_d = 90$ nM at 2.2 mM Mg^{2+}. From Godefroy-Colburn et al. (116).

The results can therefore be interpreted by the following scheme, under conditions where the factors bind independently (no excess of IF1 or IF3):

$$\begin{array}{ccc}
70\ S & \overset{K_1'}{\rightleftharpoons} & 70\ S \cdot IF1 \\
k_2 \updownarrow k_1 & & k_2' \updownarrow k_1' \\
50\ S + 30\ S & \overset{K_1}{\rightleftharpoons} & 30\ S \cdot IF1 + 50\ S \\
\updownarrow K_3 & & \updownarrow K_3 \\
30\ S \cdot IF3 & \overset{K_1}{\rightleftharpoons} & 30\ S \cdot \begin{matrix}IF3 \\ IF1\end{matrix}
\end{array}$$

These results explain the need for both factors: one for increasing the dissociation of the 70 S particles, and the other for preventing their reassociation.

Insofar as IF2 is concerned, previous detailed studies of its interaction with the 30 S and 50 S subunits and with the 70 S ribosomes had shown that GTP [or its nonhydrolyzable analog, GMP-$P(CH_2)P$]

TABLE IV
EFFECT OF IF1 AND IF3 ON THE KINETIC CONSTANTS
OF RIBOSOME DISSOCIATION[a,b]

Experimental procedure	Concentrations (nM)			$k_{1,app}$ ($s^{-1} \times 10^3$)	$k_{2,app}$ ($\mu M^{-1} s^{-1}$)	$K_{d,app}$ (nM)
	Ribosomes	IF1	IF3			
1	39.6	—	—	10	0.24	42
1	50.0	—	—	10.9	0.21	52
1	50.1	—	—	8.0	0.19	42
1	51.9	30	—	17	0.21	81
1	51.3	31	—	15	0.21	72
1	51.0	59	—	20	0.21	96
1	51.3	65	—	21	0.27	79
1	51.7	88	—	24	0.28	88
1	41.8	116	—	21	0.32	65
1	51.0	144	—	33	0.36	90
1	41.8	—	117	10.7	0.077	139
2	38.7	—	117	7.5	0.053	143
2	63.5	—	146	7.0	0.040	173
2	51.2	87	58	36	0.19	188
2	41.0	114	114	19.5	0.060	325

[a] From Godefroy-Colburn et al. (116).
[b] The constants were determined at 2.5 mM Mg, 100 mM NH₄Cl by second-order analysis of recordings. Procedure 1 refers to dissociation of ribosomes by dilution in buffer already containing the initiation factor. Procedure 2 refers to dissociation by addition of IF3 into a pre-equilibrated solution of ribosomes and (eventually) IF1. Procedure 2 results in a much smaller change of the scattered light than procedure 1, and hence yields the constants with lower accuracy.

stabilize the complexes formed on the 30 S subunit (73). The 70 S ribosomes, on the contrary, were reported to bind IF2 only in the presence of the other components of the initiation complex. But the fact that IF2-dependent GTPase activity involves both ribosomal subunits leads one to believe that it requires formation of the intermediate complex, 70 S · IF2 · GTP. This has been documented by showing that IF2 has an association effect (116). In the presence of GTP [or of GMP-P(CH₂)P], this effect is specific, and the factor has a stronger affinity for the 70 S ribosome couple than for the 30 S subunit: the binding constant, K_2, for the 30 S subunit is about 5×10^5 M^{-1}, and K_2', for the 70 S particles, is about 1.0 to 1.5×10^7 M^{-1}. In the absence of the nucleotide, the association effect was much less pronounced and, furthermore, aggregates are formed. These results might indicate that GTP is necessary for a correct positioning of IF2. The light-scattering technique gave thus direct evidence that IF2 has a specific effect on the level of ribosome association. Perhaps this could not be shown earlier because of the zonal centrifugation method used, which displaced the equilibrium toward dissociation.

V. Role of Ribosomes and Initiation Factors in Recognizing the Initiation Signals

A. Requirements for IF3 with Synthetic Messengers

1. WITH POLYMERS CONTAINING THE INITIATION CODONS

There is a general contention that IF3 is required only for ribosomal binding of naturally occurring mRNAs, not when synthetic messengers containing the AUG or GUG initiation codons are used. This point is of particular relevance because some messengers, for instance phage RNA, contain numerous AUG triplets that do not operate as initiation codons. It has been suggested that IF3 is involved in the recognition of some specific starting signals on phage mRNA, the sequence (and possibly also the structure) of which is more complex than that of the initiator codon (130–132). In accordance with this hypothesis, it was found that binding of phage MS2 RNA to ribosomes requires the presence of IF3, whereas binding of synthetic messenger does not (133). The reported isolation of two messenger-discriminating species of IF3 also supports the role proposed for IF3 in messenger recognition (46).

Conflicting with this view, other results suggest that IF3 is essential for initiation complex formation, not only when phage RNAs are involved, but also with synthetic messengers (41, 134–139). However, most experiments had been performed with 70 S ribosomes, and IF3-stimulating activity could result from its effect on the dimer-subunit equilibrium rather than from a specific influence on initiation-complex formation. This interpretation appeared, at first glance, to be strengthened by the finding that IF3 stimulates IF1- and IF2-dependent fMet-tRNA binding in the presence of 70 S ribosomes, but not in the presence of 30 S + 50 S particles (under conditions where these subunits, probably type-B, do not spontaneously associate), nor in the presence of 30 S alone (140). However, other reports indicate that, in the presence of 30 S subunits, and synthetic messengers, IF2 alone does not suffice, whatever its concentration, to promote maximum binding of initiator tRNA: the addition of either IF3 or IF1 considerably stimulates binding (37). When the two ribosomal subunits, 50 S and 30 S, are added either separately or as 70 S couples, under conditions where their dissociation is not a limiting factor (50% equilibrium), IF3 still promotes considerable increase in initiation-complex formation. Thus, only when all three initiation factors are present can maximum binding be observed. Furthermore, with the 30 S + 50 S mixture, IF3 permits the recycling of IF2, even in the absence of IF1. It can therefore be concluded that IF3 stimulates initiation-complex formation with synthetic polymers such as poly(A,U,G), that it plays an active part in the basic process of AUG-dependent initiation-complex formation, and that its function is not restricted to ribosome "dissociation." However, it should be noted that poly(A,U,G) is rich in polypurine sequences, and this might be important for the recognition by IF3 (see Section V, C).

2. WITH HOMOPOLYMERS

IF3 stimulates poly(U)-directed polyphenylalanine synthesis at 18 mM Mg^{2+}, i.e., under conditions where ribosomes are 100% associated (47). This effect can in fact constitute the basis for a specific test of IF3 activity, for it does not depend on the presence of other initiation factors; IF1 and IF2 have no effect on this reaction by themselves (although the presence of IF1, but not IF2, somewhat increases the IF3-mediated stimulation). However, we have seen (Section IV) that IF3 exhibits no affinity for type-A 70 S couples. What, then, is the explanation for the stimulation of polyphenylalanine synthesis? The situation remains unclear. Factor IF3 does stimulate, but very slightly, Phe-tRNA attachment to the 30 S (141) and not to the 70 S ribosomes (J. Dondon and M. Grunberg-Manago, unpublished); on the contrary, at low or high [Mg^{2+}] this factor causes the release of aminoacyl-tRNA previously bound to the 30 S subunit, with the exception of initiator tRNA (141). It therefore appears that IF3-dependent stimulation of poly(U) translation cannot be accounted for by an effect on the positioning of the first phenylalanyl-tRNA but has to be sought at a further step. We rather believe that all these observations, taken together, indicate that, during elongation (presumably during the translocation step), the ribosomal subunits are not as tightly bound as during formation of the 70 S complex. In favor of this, it has been shown that type-B ribosomes, which are not as tightly bound, exhibit a true affinity for IF3 (116). Spirin suggested, in 1969 (142), that the 70 S couple could exist in two different states: one, which he called the "locked" state, where the subunits are tightly bound, and which corresponds to our type-A ribosomes; and another, designated as the "unlocked" state, where the particles remain slightly apart. He suggested that periodical "locking" and "unlocking" of the ribosomal subunits was the only operating mechanism whereby, during the translation process, tRNA, mRNA, and peptidyl-tRNA become displaced. The possibility therefore exists that during translocation the ribosomes are in the type-B conformation and that IF3 binds to the 30 S subunit of the 70 S particles.

In fact, if the 30 S + 50 S are cross-linked with suberimidate and thus become compactly "frozen," they can still synthesize polyphenylalanine but lose their capacity to be stimulated by IF3 (143). Experiments with antibodies against the protein located at the interface of the region involved in the association of ribosomal subunits (102) reinforce the theory of dual ribosome conformation during translation; as expected with these antibodies, no inhibition of partial reaction is revealed, but these antibodies, particularly those directed against L19, display the strongest inhibitory effect on the polymerization of phenylalanine at 18 mM Mg^{2+} (a condition where ribosome couples are associated). This implies that, during the elongation process, the antigenic determinants of interface proteins become available for antibody binding, and therefore that the ribosomal subunits open up during a step of protein synthesis, perhaps during translocation.

The mechanism by which IF3 stimulates polyphenylalanine synthesis is still an open question; it might act by favoring the release of discharged tRNA from the ribosomes, as it does with the aminoacyl-tRNAs (141).

B. Structure of Initiation Signals

From the preceding discussion, initiation factors appear to behave either as proteins catalyzing the recognition and transfer of initiator tRNA, or as elements modifying the general conformation of ribosomes, thus preventing association of the subunits and enabling mRNA recognition.

Some fundamental questions still have to be answered:

How are the different initiation "signals," which we know cannot be solely the AUG triplet, selected *in vivo*?

Can the process of signal selection be a way for the cell to regulate gene expression at the translation level? That such a fine regulation could exist is suggested by the fact that different cistrons of phage RNAs (such as MS2, Qβ, or R17) are translated *in vivo* with different frequencies, even when their molar concentration is the same (144).

Does the ribosome play a positive role in recognizing the initial signals on the messenger, and if so, what elements of this organelle are involved?

We have seen (Section V, A) that initiation with synthetic messengers is optimal in the presence of all three initiation factors, but does not absolutely require the presence of IF3, this factor being replaceable by IF1. With natural messengers, the situation is quite different, since the requirement for IF3 is absolute. This suggests that IF3 is the factor enabling the discrimination of the particular signals proper to natural messenger RNAs.

This point is clearly illustrated by the work of Revel et al. (145), out of which one important conclusion emerges: synthesis of a biologically active protein from a natural messenger is possible only when IF3 is present, while the simple formation of a complex binding initiator tRNA to the P site shows no absolute requirement for this factor. This is evidenced by the following experiments in which the synthesis of the lysozyme coded for by T4 messenger was compared to the binding of fMet-tRNA or to its AUG-directed transfer to puromycin (Table V). IF1 and IF2 are sufficient to catalyze binding of initiator tRNA opposite its AUG triplet, and the initiator is correctly positioned at site P. However, when a natural messenger is used, the stimulation by IF3 is stronger, as shown either by measurement of the lysozyme synthesis or by incorporation of fMet in a complex with puromycin (Table V).

The problem that arises is to know whether the capacity to recognize the signals is entirely bestowed on the ribosome through IF3, this factor having intrinsic recognition properties for specific sequences of the messenger (as is the case, for example, of a methylase), or whether the ribosome is the true recognition element, IF3 having only an amplifying role, but not any specificity per se.

1. RIBOSOMAL COMPONENT SPECIFICITY

Lodish first clearly demonstrated that the ribosomes play an active role in initiation signal recognition. Phage f2 RNA contains three cistrons corresponding, respectively, to protein "A" (or maturation protein), the coat protein, and the replicase (146). E. coli ribosomes recognize the beginning of these three regions, as was established by protection experiments of specific initiating sequences against ribonuclease digestion (147, 148) (see Section V, B, 2). On the contrary, the ribosomes from another bacterium, the gram-positive *Bacillus stearothermophilus*, translate only cistron A at their physiological temperatures, 65° or 47°C (149, 150). By crossing the various elements

TABLE V
ROLE OF PROKARYOTE INITIATION FACTORS IF1, IF2, AND
IF3 IN PROTEIN CHAIN INITIATION[a,b]

Factors added	T4 mRNA dependent		AUG-dependent	
	Lysozyme synthesis (units)	fMet-puromycin formation (pmol)	fMet-puromycin formation (30 S + 50 S) (pmol)	fMet-tRNA binding 30 S + 50 S (pmol)
IF1	1.2	—	0.28	0.61
IF3	1.3	—	0.25	—
IF2	1.4	—	1.20	0.65
IF1 + IF2	3.4	0.38	4.60	3.0
IF1 + IF3	6.5	—	0.40	—
IF3 + IF2	3.5	—	0.95	—
IF1 + IF3 + IF2	12.8	1.06	4.90	—
None	1.5	0.24	0.39	0.49

[a] From Revel et al. (145).
[b] Protein synthesis was carried out in a 0.125 ml reaction mixture containing: 40 mM TrisCl, pH 7.5; 120 mM NH_4Cl; 10 mM $MgCl_2$; 8 mM β-mercaptoethanol; 2 mM ATP; 4.8 mM phosphoenolpyruvate; 0.4 mM GTP; 0.2 mM CTP, UTP; 0.2 mM (20) amino acids; 5 μg pyruvate kinase; 50 μg of tRNA; 30 μl of high-speed supernatant fraction; 120 μg of salt-washed ribosomes; and 25 μg of factor IF1, 62 μg of factor IF3 and 75 μg of factor IF2 as indicated. T4 mRNA, 100 μg (extracted from 20-minute infected culture) was added; incubation was for 30 minutes at 37°C. A 100-μl aliquot was taken to measure lysozyme formed. Binding of fMet-tRNA to ribosomes was performed in a 50-μl incubation medium containing: 0.1 M TrisCl, pH 7.5, 0.08 M NH_4Cl; 7.5 mM $MgCl_2$; 8 mM dithiothreitol; 1 mM GTP; 45 μg of 30 S ribosomes and 100 μg of 50 S ribosomes (derived at 1 mM Mg^{2+} from high-salt-washed ribosomes), as indicated; 16 μg of poly(A,U,G) or 6 μg of A-U-G; and 10 μg of [^{3}H]fMet-tRNA (4000 cpm). Incubation was at 25°C for 5 minutes. Radioactivity retained on nitrocellulose filters was measured. Formation of fMet-puromycin was studied after adding puromycin (20 mM) by measuring ethylacetate-extractable radioactivity.

from the translation machinery at 47°C, Lodish showed that the specificity of action of E. coli and B. stearothermophilus ribosomes is not altered by the addition of heterologous initiation factors, tRNA, or supernatant enzymes, but that cistron selectivity of either bacterial species did reside in their 30 S particles. Recently, it was confirmed, with highly purified IF3 from E. coli or B. stearothermophilus, that IF3 and IF2 effects are not species-specific (52, 80).

Another example indicating that the small ribosomal subunit rather than initiation factors is involved in mRNA selection stems from results of Leffler and Szer (151). In the presence of purified initiation factors from either *E. coli* or *Caulobacter crescentis* (gram-positive), MS2 RNA is always translated by *E. coli* ribosomes, and Cb5 phage RNA by *C. crescentis* ribosomes. In no case was it possible to obtain translation of MS2 RNA in the presence of *C. crescentis* ribosomes, nor that of Cb5 phage RNA in the presence of *E. coli* ribosomes, although these RNAs are similar in both their physical properties and their genetic content.

In more recent work, the components of the 30 S ribosomal particles responsible for the specific recognition of initiation signals on messengers have been investigated *(152, 153)*, using heterologous 30 S particle reconstitution techniques *(154)*. For example, Held *et al. (152)* observed that *E. coli* 30 S at 37°C could translate the coat protein 10 to 20 times better than *B. stearothermophilus* 30 S at the same temperature. When hybrid ribosomes were reconstituted by mixing *E. coli* 16 S RNA with *B. stearothermophilus* proteins, translation of the coat cistron was very low; conversely, reconstituting 30 S particles from *B. stearothermophilus* 16 S RNA and *E. coli* proteins also resulted in a decrease of translation, but this decrease was less pronounced than with the previous hybrid.

Goldberg and Steitz *(153)* reached similar conclusions and analyzed the specificity of recognition in a more detailed way: their conclusion was that cistron selectivity is conferred primarily by the protein moiety of the 30 S particle. Thus, at 49°C, *E. coli* 30 S (+ factors) recognizes the sites for "A," coat, and replicase in the ratio of 1 : 2.3 : 1.4; when a hybrid 30 S particle is formed of *E. coli* proteins and *B. stearothermophilus* 16 S RNA, the ratio is 1 : 1.6 : 1.1, very close to that of *E. coli* 30 S. By contrast, when the symmetrical hybrid is prepared from *E. coli* 16 S RNA and *B. stearothermophilus* proteins, the ratio is quite different, 1 : 0.08 : 0.4, and more characteristic of *B. stearothermophilus* 30 S (since the latter discriminate with the ratio of 1 : 0.06 : 0.4).

Concerning the nature of the protein (or proteins) involved in signal discrimination, out of 17 proteins from *B. stearothermophilus*, singly substituted for *E. coli* proteins, only protein S12^5 entails a marked decrease in recognizing the signal for the cistron of the coat protein (about 50%) *(152)*.

These results thus suggest that the ability to discriminate among the initiation signals present on natural messengers (but not the recognition of the simple AUG triplet) is based upon the properties of protein S12, and that at least [according to Held *et al. (152)*] the 16 S RNA is also involved.

Another protein that appears to play an important role is S1. Addition of purified S1 from *E. coli*, to the *B. stearothermophilus* system at

39°C, resulted in a 10-fold stimulation of [^{14}C]valine incorporation (155); the stimulation has been reported to involve mainly the coat protein and replicase synthesis. S1 protein had already been shown to be an indispensable factor in protein synthesis (156), but the evidence discussed below does not involve that protein in cistron specificity.

As stated above, protein S1 is quite important for mRNA binding, as shown by the fact that poly(U)-directed polyphenylalanine synthesis is completely dependent on the presence of this protein (156). Washed ribosomes contain about 0.1–0.3 mol of S1 per molecule, but the protein is easily lost upon salt washing, and it is present in polysomes, as one copy per ribosome (156–159). The role of S1 is easily overlooked since this protein is present both on the ribosome and in the soluble fraction. As to the role of S1 in natural RNA translation, *E. coli* 30 S ribosomal subunits can be separated on agarose gel into two classes, one that can form an initiation complex with messenger RNAs in the absence of initiation factors and initiator tRNA at low temperature, and another that cannot (160–162). The differences between these two classes of 30 S are that only the one reacting with messenger RNA contains S1 (160, 161); as expected, S1 is essential for MS2 RNA translation (163), since it is involved in initiation complex formation. Unfolding of the messenger relieves the dependence on S1 of both translation and initiation. Moreover, protein S1 is in close contact with the messenger RNA on the ribosome (164). For these experiments the polynucleotide poly(s^{4}U), which can be photoactivated at 330 nm and react with proteins, was employed as the mRNA. Poly(s^{4}U) was bound to ribosomes in the presence of tRNA, and the resulting complex was irradiated. At 330 nm, the activity of the ribosome is not affected by irradiation. After isolation of the 30 S subunits and their digestion with RNases, the labeled proteins were separated by gel electrophoresis, and S1 was identified as the major component; two other proteins, S18 and S21, are also located at the messenger binding site near S1.

An excess of S1 inhibits initiation since the protein then binds to the messenger, even in the absence of ribosomes; this occurs at concentration exceeding one molecule of S1 per ribosome (165–167). It has been shown that under these conditions mRNA and S1 form a complex, and that the extent of inhibition correlates with the extent of complex formation (165, 166). Excess of mRNA overcomes this inhibition.

The effect of S1 depends on the structure and the number of initiation signals. This could result in an apparent selectivity of S1 on translation and might explain other reports (168, 169) of the existence of so called "interference factors," which were believed to inhibit selectively translation of some mRNAs such as MS2, but not that of T4 RNA. It is likely that interference factors are ribosomal proteins. One of these "interference factors," designated as "i factor" has been identified as protein S1 by various workers *(167–170);* S1 ("i") has also been shown to be a component of Qβ replicase *(171–173).*

Other ribosomal proteins also inhibit translation. For example, protein S21 inhibits fMet-tRNA binding *(174)* as well as MS2 translation *(141, 157, 175);* so do S3 and S14 *(163);* the inhibition by S3, S14, and S21 is relieved by increasing mRNA concentration.

In conclusion, from our present knowledge, it is not likely that these ribosomal proteins act by modulating the specificity of IF3, as was first suggested *(168);* they could, nevertheless, exert some important regulatory role in translation. However, this point must be reinvestigated with ribosome preparations whose protein equipment is well defined before one can draw conclusions about the *in vivo* functional meaning of the *in vitro* stimulation or inhibition. It is now realized that, during their isolation and purification by washing with high salt, ribosomes lose many proteins, which appear not only in the salt wash containing the initiation factors, but also in the supernatant. Thus, depending on the ribosome preparation, it is possible to isolate proteins that stimulate or inhibit when present in excess. This is the case not only for S1, but also for S2 and S3 *(175).* This situation also probably explains the report *(176)* that S21 is not inhibitory but is required for full activity of ribosomes; it is not surprising that, depending on the conditions of the test and on the ribosome preparation, contradictory results are found.

Two other approaches have been used to identify ribosomal proteins involved in initiation, the first being genetic. It is known that sensitivity to streptomycin is dependent on protein S12. In streptomycin-dependent mutants, Sm^D, protein S12 is modified *(177–179);* presumably in these mutants the antibiotic is required to ensure normal functioning of S12 in initiation. Therefore, when the Sm^D ribosomes are deprived of streptomycin, they should be capable of translating, for example, poly(U), but not natural messengers, and this is indeed what occurs. Further experiments have shown that the real defect in such mutants is that IF2 can no longer be recycled; thus fMet can be bound, but cannot form a peptide link *(180a).* Moreover, IF1 no longer stimulates the IF2-dependent positioning of initiator tRNA *(180b).* This, however, is observed only with strict Sm^D mutation, for streptomycin-dependence at other loci than StrA2 does not result in the same effect.

Another argument in favor of the role of S12 in initiation comes from experiments performed jointly by the group of Gros and Stöffler *(181, 182),* the principle of which was to analyze the effect of all the monovalent antibodies directed against each 30 S protein in *in vitro* translation reactions, including not only the translation of both synthetic and natural messengers, but also the positioning of the initiator and internal aminoacyl-tRNAs. Ribosomal proteins can tentatively be classified, according to their role at various steps of translation, on the basis of the effects exerted by the different antibodies on the various reactions. For example, anti-S6, -S12, -S14, -S19 and -S21 antibodies have a stronger inhibitory effect on translation of natural messengers than on poly(U); anti-S6 and -S12 specifically inhibit the binding of fMet-tRNA. This suggests that these proteins would be particularly involved in initiation, and is once more in agreement with what has been found about the role of S12. Also, S1, S2, S3, S10, S13, S19, S20 and S21 probably lie at an overlapping site, that is, a site between A and P, since the corresponding antibodies cause inhibition of both elongation and initiation reactions *(182)*. This is in agreement with what is known from other approaches, indicating that S2, S3 and S15 are part of both site A and site P *(183)*. Finally, there are some proteins that, curiously enough, do not, when blocked, affect the binding of tRNAs, but do inhibit translation of natural messengers (S4, S7, S15, S16).

To conclude, the immunochemical results agree rather well with the reconstitution and genetic data, and the important point is that, once again, S12 appears to be involved in initiation, as it plays a determinant part in the "recognition of initiation signals on natural messengers."

2. Nature of the Initiation Signals

The specificity of the mRNA initiation signals (and hence their discrimination by 30 S subunits) could possibly be related to a specific base sequence, or to the degree or type of the secondary structure in this region of the messenger, or to both. Most of our knowledge about initiation regions for protein synthesis comes from bacteriophage RNA; the sequences of viral mRNAs in eukaryotic systems are presently being determined.

The sequence of the initiation signals on a phage messenger, phage R17 RNA, was first established in 1969 *(147)*, and was based on the results of Takanami *et al.* *(184)* demonstrating that regions of bacteriophage RNA are protected from ribonuclease digestion by association with ribosomes. When a ^{32}P-labeled mRNA is mixed with washed *E. coli* ribosomes plus initiation factors and initiator tRNA, the ribosomes selectively bind at the initiation-sequence level and are unable to leave this site owing to the lack of internal aminoacyl-tRNAs and elongation factors. Upon digestion of the complex, either by pancreatic ribonuclease or by T1 ribonuclease, the whole mRNA is hydrolyzed, except for the portions associated with the ribosomes; the protected fragments can then be liberated upon treatment by a denaturing agent such as dodecyl sulfate, and the sequence determined by Sanger's procedure (185). The ease with which the single-stranded RNA genomes of these viruses can be labeled to a high specific activity and purified from the phage particles has made them an ideal material for RNA sequence analysis. Steitz was thus able to establish the sequence of initiation signals preceding the cistrons, which as already mentioned, code for only three proteins: (a) the coat protein (M_r 14,000) (cistron 2), which is synthesized in largest amounts both *in vivo* and *in vitro*; (b) the phage-specific subunit of the viral RNA replicase (M_r 60 to 65,000, cistron 3); and (c) the "A" or maturation protein (M_r about 40,000), which is normally translated in the lowest amount and is a minor constituent of the virus particle (cistron 1). A fourth protein (M_r 36,000) appears in Qβ virus and is the product of natural readthrough at the UGA termination signal of the Qβ coat cistron; it is therefore not considered to be a physiological protein.

In each case, it was shown that the amount of ^{32}P-labeled RNA remaining bound to ribosomes after fractionation from the bulk of the degraded RNA depends on the presence of N-formylmethionyl-tRNA in the original initiation reaction. This provided assurance that the interaction of the ribosome with the phage messenger was occurring at sites actually specifying the initiation of protein synthesis.

In general, the protected fragment is about 30 to 40 nucleotides long (186) (Fig. 7). Under normal conditions, *E. coli* ribosomes bind much more strongly to site 2 than to the others, which are, especially site 3, hidden by the messenger secondary structure.

Analysis of initiation signals 1, 2 and 3 showed for the first time that these sequences all include an AUG triplet preceding the first translated codons. The initiator triplet appears approximately in the middle of the protected sequence (Fig. 7). Indeed, knowing the N-terminal peptides from the maturation and coat proteins and from the replicase, it was easy to verify that the triplets to the right of AUG in the protected fragments were those predicted by the code.

What conclusions or comments can be drawn from an examination of the sequences of the various initiation signals?

1. Each of these signals bears an initiation codon AUG (underlined), except for one case (the MS2 maturation protein) where the

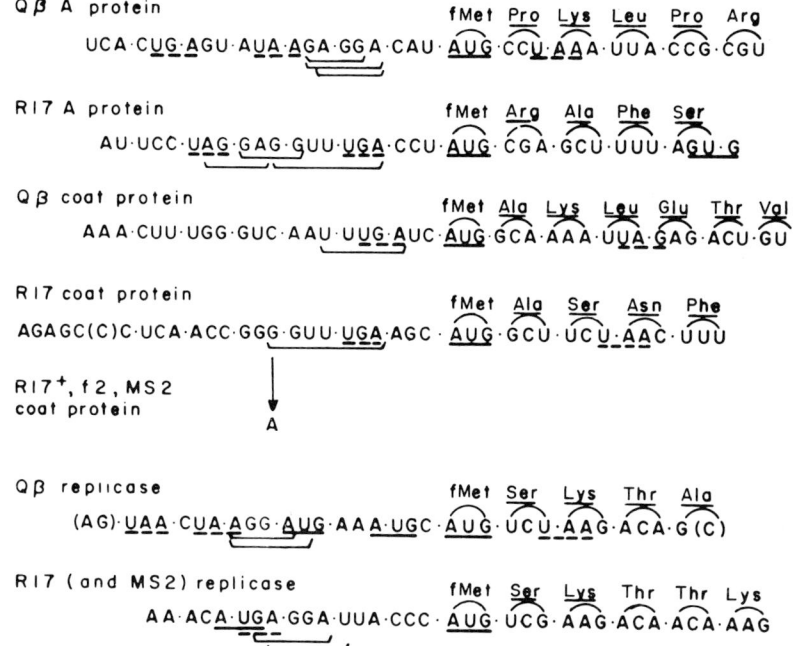

FIG. 7. RNA phage ribosome binding sites. The three ribosome-protected initiator regions from Qβ and R17 RNA are shown. A base change (G to A), in the coat-protein initiation site from another variant of R17 (denoted R17†) and from f2 and MS2 RNA, is indicated. The MS2 replicase initiator region is identical to that for R17. Initiation codons are indicated by ——, and termination triplets by ----. Base-sequence homologies of four nucleotides or more occurring at the 5' extremity of the initiator AUGs are indicated by ⌊——⌋. Experimentally determined amino-acid sequences (186–190) are underlined. From Steitz (186).

initiation codon is GUG (191). Of particular interest is the fact that these sequences sometimes contain more than one AUG triplet; this is especially clear, for example, for the initiation sequence of Qβ replicase, where one can distinguish two AUG in the reading phase, and one out of phase. The case is not unique: a similar situation is encountered, for instance, in the initiaton signal of phage R17 messenger. It strongly suggests that the true initiator AUG is probably the only one accessible to pairing with the anticodon CAU of fMet-tRNA.

2. A number of signals contain the sequence G-G-A-G-G-U, or at least (in seven cases out of twelve) A-G-G-A, preceding the initiator triplet at a certain distance. A large number of signals (in seven cases out of twelve) also contain the sequence R-U-U-U-U-R-R- (R for "a purine nucleoside").

3. The true initiation triplets are often preceded by termination triplets that stop the reading (they are generally found among the 15 bases preceding the initiator AUG). This is an expected consequence of the translation mechanism since ribosomes have to stop reading the cistron preceding the one at the level of which protein synthesis is reinitiated. However, none of the initiator AUG is directly preceded by an UAG, UAA, or UGA codon, the three known terminators in *E. coli*. Thus, intercistronic regions exist in polycistronic mRNA. Besides being implicated in ribosome recognition, such untranslated nucleotides presumably eliminate possible interference between ribosomes involved in termination and in initiation.

Particularly noteworthy is the fact that the region preceding AUG often contains more than one nonsense triplet. This is evidenced by the sequence preceding the initiation signal of Qβ replicase, which contains one stop signal, UAA, in phase, and two out of phase, UAA and UGA. Thus, even if the reading frame happens to be scrambled (that is to say, even if the ribosome no longer reads the triplets by groups of three), growth of the polypeptide chain will stop, and this will cause the system to reinitiate at the next AUG. The reason for this safety device is probably to prevent one ribosome from sliding and translating two adjacent cistrons into a single protein, a situation that would be lethal.

4. Although these initiation regions show a certain similarity, a single conformational principle cannot be drawn. A hairpinlike conformation might be constructed from some of these sequences, with triplet AUG at one end, as in the case for the coat-protein signal; but this is not always true, as shown by the sequences of Qβ replicase or β-galactosidase signals, where the codon for internal methionine residues appears to be masked by a secondary structure within the RNA. Moreover, the simple availability of an AUG or GUG codon is not sufficient to designate an actual initiation region in the phage RNA, since quite extensive fragmentation of the native RNA does not lead to spurious initiation or to ribosome binding *(192)*.

Triplet AUG is not even absolutely necessary for initiation *in vivo*. Thus, we have seen that in the cistron for MS2 maturation protein, the initiator is GUG. Better still, in studying the "reinitiation" process within the gene that codes for the repressor of the lac operon (gene i), Files *et al.*, after forcing termination by an amber mutation, were able to isolate C-terminal fragments and to show that they were indeed initiated by formylmethionine *(193)*. The "reinitiation" codons can be AUG, or the coding triplet for valine (GUG), or the codons for leucine (UUG or CUG). Thus, *in vivo*, these four codons can be used as initiators *(193)*.

According to Miller et al., the messenger for the lac operon repressor contains three codons in phase for reinitiation (among the first 70 codons, i.e., 210 nucleotides). When premature termination occurs, these reinitiation codons can be recognized; they are not, under normal circumstances, either because of the secondary structure, or because during normal translation they are constantly covered up by ribosomes. Premature termination, at a site preceding these codons, would allow their unmasking.

C. Mechanism of Cistron Recognition in Prokaryotes

1. Role of 16 S RNA in Cistron Recognition

From the results already discussed (Section V, B), it appears that 16 S RNA plays an important part in the specificity of translation. Shine and Dalgarno *(194, 195)* have put forward an attractive hypothesis to explain the specificity of this RNA. A sequence near the 3′ terminus of the 16 S RNA would directly participate in the initiation of protein biosynthesis by forming several Watson–Crick base-pairs with the messenger RNA in the vicinity of the initiator codon. It is significant that all coliphage RNA ribosome binding sites examined to date contain part, or the whole, of a purine-rich sequence (5′)-A-G-G-A-G-G-U-(3′) (see Table VI), at a relatively similar position on the 5′ side of the initiator triplet AUG, about 10 bases from this triplet (see preceding Section). Furthermore, the 3′-OH terminus of the small subunit RNA is very rich in pyrimidine. Shine and Dalgarno *(195)* determined the 3′ terminal sequences of 16 S RNA of different bacteria. In *E. coli* RNA, this sequence is Y-A-C-C-U-C-C-U-U-A-OH-(3′) (Y = "a pyrimidine nucleoside"), which has been confirmed by other laboratories *(196–198)*. Shine and Dalgarno postulate that this pyrimidine-rich sequence is involved in the binding to the polypurine stretch surrounding the AUG codon; seven base-pairs could possibly be involved in the interaction between the 16 S RNA and the initiation site of bacteriophage maturation protein (A), whereas four or five pairs could be used for binding to the other initiation sequences (Table VI). This is in agreement with the previous observation that ribosomes bind preferentially (about 11–40 times more efficiently) to the initiation sites of the maturation protein when added to a mixture of fragments of coat protein and replicase in the presence of initiation factors, plus GTP, and fMet-tRNA. The highly efficient binding of protein "A" initiator fragment contrasts with the relatively inefficient translation of intact coliphage RNA *in vitro*, as discussed in Section V, B, 2; during translation of intact RNA, the secondary structure of the molecule impedes binding of the 30 S subunits to their protein initiation re-

TABLE VI
INITIATION SEQUENCES RECOGNIZED BY *Escherichia coli* RIBOSOMES[a,b]

mRNA	Ribosome binding site
R17 A	GAU UCC UAG GAG GUU UGA CCU AUG CGA GCU UUU AGU G
Qβ A	UCA CUG AGU AUA AGA GGA CAU AUG CCU AAA UUA CCG CGU
R17 coat	CC UCA ACC GGG GUU UGA AGC AUG GCU UCU AAC UUU
Qβ coat	AAA CUU UGG GUC AAU UUG AUC AUG GCA AAA UUA GAG ACU
f2 coat	CC UCA ACCG(A,G)GUU UGA AGC AUG GCU UCC AAC UUU ACU
R17 replicase	AA ACA UGA GGA UUA CCC AUG UCG AAG ACA ACA AAG
Qβ replicase	AG UAA CUA AGG AUG AAA UGC AUG UCU AAG ACA G
f1 coat	UUU AAU GGA AAC UUC CUC AUG AAA AAG UCU UU
f1 gene 5	A AGG UAA UUC ACA AUG AUU AAA GUU GAA AU
f1 gene ?	A AAA AAG GUA AUU CAA AUG AAA UU
T7 *in vitro*	AAC AUG AGG UAA CAC CAA AUG AUU UUC ACU AAA GAG
T7 gene 0.3	ACG AGG UAA CAC AAG AUG GCU AUG
λ P_R	pppAUG UAC UAA GGA GGU UGU AUG GAA CAA CGC
ϕX174 spike (DNA)	TTT CTG CTT AGG AGT TTA ATC ATG TTT CAG ACT TTT ATT
trp leader	CAC GUA AAA AGG GUA UCG ACA AUG AAA GCA AUU UUC GUG
*trp*E	GAA CAA AAU UAG AGA AUA ACA AUG CAA ACA CAA AAA CCG
*trp*A	GAA AGC ACG AGG GGA AAU CUG AUG GAA CGC UAC GAA UCU
*lac*Z	AAU UUC ACA CAG GAA ACA GCU AUG ACC AUG AUU ACG GAU
*lac*i	pppG GAA GAG AGU CAA UUC AGG GUG AAU GUG GUG AAA CCA GUA ACG
*gal*E	AUA AGC CUA AUG GAG CGA AUU AUG AGA GUU CUG GUU ACC
16 S RNA 3' end	HOAUUCCUCCACUAG 5'

[a] From Steitz and Jakes (190) and Steitz *et al*. (214b).
[b] Underlining indicates contiguous bases complementary to the 3'-OH-terminal oligonucleotide of *E. coli* 16 S RNA. Dots indicate G·U base-pairs. Gaps appear at positions where a bulge in the rRNA strand is required to provide the indicated complementarity.

gions. Likewise, the 3' terminal sequence of 16 S RNA from other bacteria all have a high proportion of pyrimidine residues.

The postulate of Shine and Dalgarno seems attractive for several reasons. First, evidence has accumulated that an intact 3' terminus of 16 S RNA is necessary for protein synthesis, and more specifically for initiation, as shown by the inhibitory action of colicin E3, an antibiotic inducing a rapid shutoff of protein synthesis in sensitive *E. coli* cells (199). Colicin-E3 treatment of the sensitive cells in suspension results in the removal of about 50 nucleotides from the 3' terminus of the 16 S RNA by a single endonucleolytic cleavage (200–202). Second, cross-linking experiments and other chemical data suggest that the 3' terminus of 16 S RNA (the binding sites for initiation factors) and the ribosomal proteins S1, S7, and S12 (all three of which have been implicated in the initiation process) may all be neighbors in the 30 S ribosomal subunit (203–207). (This point is discussed in more detail in

Section C, 3.) Third, studies with the other initiation inhibitors, such as streptomycin *(208, 209)* and kasugamycin *(210)*, indicate that their ribosomal sites of action likewise lie in the vicinity of 16 S RNA 3' terminus. Fourth, the observation that random copolymers rich in A and G are the best competitive inhibitors for initiation on natural mRNAs *(211)* underscores the importance of polypurine stretches in ribosome binding to initiator regions. Finally, in the single well-documented instance of ribosome recognition of a noninitiating RNA sequence, the fragment of Qβ phage RNA bound by ribosomes from *B. stearothermophilus* at 65° was found to contain no initiator triplet *(150)*; however, it does possess a long polypurine stretch whose sequence is potentially complementary to the 3' terminus of *B. stearothermophilus* 16 S RNA.

More direct evidence of the validity of Shine and Dalgarno's hypothesis comes from experiments by Steitz and Jakes *(190)*, who treated initiation complexes formed by *E. coli* 70 S ribosomes with colicin E3 in the presence of the messenger fragment of the initiation cistron for the maturation protein (this fragment was ^{32}P-labeled). After disassembling the complex by exposure to 1% sodium dodecyl sulfate, they fractionated the components on polyacrylamide gel. An mRNA · rRNA hybrid containing approximately equimolar amounts of the 30-nucleotide mRNA fragment and of the 49- to 50-nucleotide colicin fragment was detected (Fig. 8). This hybrid, which exhibits electrophoretic mobility different from each of the above two fragments, appears only in the presence of all the components necessary for complex formation, in particular mRNA. Furthermore, it does not appear if the colicin treatment is omitted, since in that case the mRNA cosediments in sucrose gradient with intact 16 S RNA. Moreover, aurintricarboxylic acid, an inhibitor of mRNA binding to ribosomes *(214a)*, lowers the amount of complex formation. The hybrid can be dissociated by heat (55°C). With ^{32}P-labeled ribosomes in addition to the ^{32}P-labeled initiator region, chromatographic analysis revealed that the portion of the gel containing the complex consisted of oligonucleotides assigned to the colicin E3 fragment and to the R17 protein initiator fragment. These observations strongly support the hypothesis of Shine and Dalgarno that base-pairing between the mRNA and 3' terminus of 16 S RNA does occur during the formation of a functional initiation complex.

The diagram of Fig. 8 suggests that a specific secondary structure is assumed by the 3' terminal region of the 16 S RNA and that upon mRNA binding some of the intramolecular base-pairs are exchanged for intermolecular hydrogen bonds.

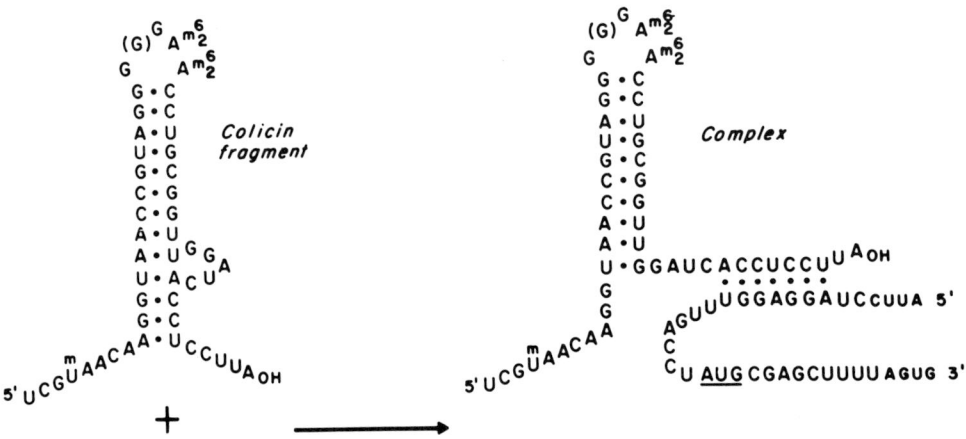

5' AUUCCUAGGAGGUUUGACCUAUGCGAGCUUUUAGUG 3'

R17 A protein initiator region

FIG. 8. Postulated hydrogen-bonding between the colicin fragment of 16 S RNA and the R17 A-protein initiator region. The secondary structure drawn, involving internal base-pairing, according to Shine and Dalgarno, is predicted to be stable under physiological conditions. From Steitz and Jakes *(190)*.

The temperature-jump relaxation methods are consistent with the structural conformation of the isolated colicin E3 fragment and the temperature-melting studies also support the existence of the intermolecular base-pairs in the complex containing the colicin fragment and the R17 A protein initiator region *(214b)*.

Therefore, it seems likely that mRNA–rRNA interactions play an important part in initiation-site selection. Fine control over the relative efficiency of ribosome binding to different cistrons could be achieved by varying the extent of complementarity of mRNA and rRNA. Although as few as three base-pairs may be formed by some initiation regions (it should be recalled that codon–anticodon interaction with this number of pairs is sufficient to ensure specificity), the ribosome, as a whole, will play a stabilizing part. In addition, the distance between the polypurine stretch and the initiator triplet is likely to influence the formation and stability of the mRNA · ribosome complex.

The hypothesis of Shine and Dalgarno *(195)* predicts a positive correlation between the translation of a particular bacterial mRNA cistron by bacterial ribosome and the degree of complementarity existing between the sequence of the ribosome binding site of that cistron and the 3'-terminal sequence of the 16 S ribosomal RNA. These workers therefore determined and compared the terminal sequence of 16 S RNA from *Pseudomonas aeruginosa, B. stearothermophilus,* and *C. crescentis* (Table VII), in relation to the translation specificity of the last two organisms, discussed in Section V, B, 1. The 3' terminus of *P. aeruginosa* 16 S RNA is identical to that of *E. coli* for seven nucleotides; it can be aligned in such a way that five, four, and three base-pairs are possible with the ribosome binding site sequences of the A protein, replicase, and coat protein cistron of R17 RNA, respectively (Table VIII). *P. aeruginosa* ribosomes translate all three cistrons of MS2 RNA in a manner analogous to that of *E. coli* ribosomes *(212).*

B. stearothermophilus and *B. subtilis* 16 S RNA have a 3'-terminal sequence significantly different from that of *E. coli* and *P. aeruginosa* 16 S RNA. Some complementarity of about four base-pairs nevertheless exist between this sequence and the ribosome binding site sequence of both the A protein and the replicase cistrons of R17 RNA. The degree of complementarity with the coat-protein binding site sequence is considerably less (one or two base-pairs). *B. stearothermophilus,* as discussed above, binds significantly only to the "A" protein site on native f2 or R17 RNA; while appreciable translation of both the replicase and "A" protein cistrons does occur when the RNA is unfolded by formaldehyde, no coat protein is synthesized under these conditions *(213a).*

The "abnormal" *B. stearothermophilus* ribosomes binding site on Qβ RNA, containing no initiation triplet and corresponding to none of

TABLE VII
3'-TERMINAL SEQUENCES OF 16 S rRNA[a,b]

Escherichia coli B	$GAUCACCUCCUUA_{OH}$
Pseudomonas aeruginosa	$G(X)_2YCUCUCCUU(A)_{OH}$[c]
Bacillus stearothermophilus	$G(X)_{\sim 5}YUCCUUUCU(A)_{OH}$[c,d]
B. stearothermophilus	$GAUCACCUCCUUUCUA_{OH}$[c]
B. subtilis	$G(X)_{\sim 7}YCUUUCU_{OH}$
Caulobacter crescentis (ATCC 15252)	$G(X)_3YUCCUUUCU_{OH}$

[a] From Shine and Dalgarno *(195).*
[b] X represents any nucleoside other than guanosine; Y = pyrimidine nucleoside.
[c] The variable presence of the 3'-terminal adenosine in 16 S RNA is found in a variety of bacteria and depends on the culture conditions.
[d] Sequence as determined by Sprague *et al. (213b).*

TABLE VIII
R17 Ribosome-Binding Site Sequences and Proposed Pairing with the 3′ Terminus of 16 S RNA[a,b]

Bacteria	R17 cistron	Possible pairing of 3' terminus with appropriate region of ribosome-binding site sequence	Number of base-pairs possible	Ribosome binding to unfolded R17, MS2 or f2 RNA
Escherichia coli B		ₒₕA U U C C U C C A C Y 16 S RNA		
	A-protein	(5')CU [AGGAGGU] UU(3')	7	+
	Replicase	ACAU [GAGG] AUU	4	+
	Coat protein	ACC [GG] G [G] UUUG → [A]	3(4)	+
Pseudomonas aeruginosa		ₒₕA U U C C U C U C Y 16 S RNA		
	A-protein	(5')CU [AGGAG] GUU(3')	5	+
	Replicase	UG [AGGA] UUAC	4	+
	Coat protein	ACC [GG] G [G] UUU → [A]	3(4)	+
Bacillus stearothermophilus		ₒₕA U C U U U C C U Y 16 S RNA		
	A-protein	(5')AUCCU [AGGA] G(3')	>4	+
	Replicase	ACAUG [AGGA] U	4	+
	Coat protein	AACCGG [GG] UU A	2(1)	−

Caulobacter crescentis

	$_{OH}$UCUUUCCUY	16 S RNA
A-protein	(5')UCCU [AGGA] G(3')	>4
Replicase	CAUC [AGGA] U	4
Coat protein	ACCGG [GG] UU	2(1)
	↓	
	A	

[a] From Shine and Dalgarno (*195*).

[b] The sequences shown represent all or part of the conserved sequence [5'-AGGAGGU-3'] from the A-protein, replicase, and coat-protein-initiator region of R17 RNA. The initiator AUG is located eight to nine bases to the 3' side of this sequence. Apart from the A-transition (see below), these sequences are identical to those available for the corresponding regions of f2 and MS2 RNA. Two sequences have been reported for this region of the coat-protein binding site in the two different R17 stocks used, the G → A transition represents a spontaneous mutation that occurred after the two stocks were separated. Such a transition may have some selective advantage, since it would increase the stability of interaction between the coat-protein initiation site and the 3'-end of 16 S RNA. The corresponding sequence from MS2 and f2 RNA contains the A substitution. The *in vitro* ribosome-binding data were obtained with R17 RNA containing the GGGG sequence and with f2 or MS2 RNA presumably containing the GGAG sequence. The apposition of a G and a U residue in the proposed helical region formed between the coat-protein binding site and 16 S RNA would cause little, if any, destabilization of the base-paired structure.

the three normal initiation regions, shows that mRNA is specifically recognized and bound by ribosomes independently of polypeptide chain initiation (150); however, the fragment contains a polypurine stretch [composed of G and A (A-A-A-G-A-G-G-G)] at a position similar to that of A-G-G-A-G- in the R17 "A" protein site. Therefore, it is significant that the complement of the 3' terminus of *B. stearothermophilus* RNA is A-G-A-A-A-G-G-A.

The 3' sequence of *C. crescentis* 16 S RNA is similar to that of *B. stearothermophilus;* at 37°C the initiation sites for "A" protein and for replicase are masked on MS2 RNA. *B. stearothermophilus* ribosomes bind to the "A" protein at 65°C.

It was therefore concluded that the extent of possible interaction between the 3' terminus of 16 S RNA from several different bacteria and the ribosome binding sites of coliphage RNA is consistent with the available data on translation specificity in these bacteria.

However, in order to correlate directly the mRNA · rRNA complementarity with the prokaryotes translation specificity, Sprague *et al.* (213b) determined the complete nucleotide sequences of the 3'-terminal T1 oligonucleotide of 16 S rRNA from *B. stearothermophilus* (Table VII) and the two polypurine stretches from Qβ bacteriophage RNA that are bound by ribosomes from this species at high temperature. Their results confirmed that specific recognition of the two Qβ regions can indeed be explained by an extensive mRNA · rRNA pairing. Masking of the two noninitiator Qβ regions by RNA secondary structure probably explains why these sites do not appear among the fragments protected by *B. stearothermophilus* ribosomes at lower temperatures. However, the translational specificity was more difficult to explain.When longer sequences than the extreme 3' termini of *B. stearothermophilus* rRNA are determined, a sequence homology is revealed between *E. coli* and *B. stearothermophilus* 16 S RNAs. The 3' termini of both RNAs is identical, except that the 3'-terminal adenosine of *E. coli* rRNA is replaced by U-C-U-A-OH in the thermophile. Thus, *B. stearothermophilus* ribosomes have the same potentiality to bind R17 of f2 coat protein initiator as do *E. coli* ribosomes, unless one postulates that the four extra nucleotides at the 3'-OH end prevents the complementary sequence from being available for mRNA recognition. We already discussed (Section V, B, 1) that proteins S12 and S1 are important selectivity determinants. One might postulate that they are involved in facilitating the mRNA–rRNA interaction by stabilizing the bonds of the RNA or by opening the intramolecular base-pairs of the colicin E3 fragment from the 16 S RNA ends. Nevertheless, the possibility exists that the proteins can create

2. Role of S1 in Cistron Recognition

The hypothesis of Shine and Dalgarno is further reinforced by the recent finding that the 3' end of 16 S RNA is the binding site for initiation factors and for certain ribosomal proteins, such as S1, S12 and S21, that are all involved in initiation (205–207). We already discussed (Section V, B, 1) the fact that S1 is involved in the binding of mRNA, S12 and S21 in initiation complex formation. Kurland et al. (206) found that protein S1 is located near the 3'-OH terminus of the 16 S RNA by showing that this protein is covalently attached to 16 S RNA after oxidation by periodate and reduction of the 30 S ribosomal subunits. This treatment oxidizes the 3'-ribosomal terminus of RNA, and the reduction of the product formed with the amino group of lysine creates a stable bond between the RNA and an appropriately positioned protein. When special precautions were taken to disrupt complexes of "nicked" RNA produced by periodate treatment, only S1 and S21 could be detected as covalently attached to the intact 16 S RNA. The same proteins can be recovered, cross-linked to 16 S RNA, when type-A 70 S ribosomes are similarly reduced and oxidized. Moreover, S1 was cross-linked to the 3'-OH of 23 S RNA; with type-B ribosomes, there is no detectable amount of S1 linked to 23 S RNA. This is additional indication of the very specific close arrangement of S1, S21 and the 3' end of 16 S and 23 S RNA in the tight 70 S ribosome. Furthermore, after treatment of the ribosomes with colicin E3, and centrifugation in sucrose gradient in the absence of Mg^{2+}, a hybrid containing S1 and the colicin nucleotide fragment can be isolated (207). At 0°C in the absence of ribosomes, S1 reversibly binds to the isolated colicin fragment; either aurintricarboxylic acid or mRNA disrupts the S1 · RNA complex.

The site of S1 interaction was examined more precisely by partial hydrolysis of the colicin fragment of 16 S RNA by RNase T1 in the presence and in the absence of S1. Since S1 did not give protection, the partial T1 digest was added to the protein and the mixture subjected to electrophoresis. An RNA fragment did bind to S1, and its structure was identified by total digestion with RNase T1 and pancreatic RNase, and chromatography. This fragment corresponds to the first 12 nucleotides from the 3' terminus of 16 S RNA, in the region that includes the sequence A-C-C-U-C-C, postulated to be the recognition site for mRNA cistrons. Dahlberg and Dahlberg propose that, in the presence of S1, the terminal 3' nucleotides of 16 S RNA are not paired; they would be stabilized in the "open" form, making this region accessible for hydrogen bonding with the ribosome binding sites in mRNA.

In this connection, it has been shown *(214b, 214c)* that homogeneous protein S1 unfolds a variety of stacked or helical single-stranded polynucleotides.[9] There is no sharp specificity in the binding of polynucleotide substrates to S1, e.g., single-stranded natural RNA and DNA are unfolded, but polypyrimidines appear to be a preferred target: poly(rA), for instance, is much less affected than are poly(rC) and poly(rU).

A reasonable mechanism for the observed conformation changes would be the displacement of an equilibrium between double- and single-stranded forms by the removal of single-stranded molecules from the equilibrium process. This mechanism has been suggested for several DNA denaturing proteins *(214d, 214e)*. However, it is also possible that the protein could remove the protons from the helix, causing some denaturation (see ref. *214e*); a possible interaction with double-stranded polynucleotides has also been detected *(214c)*. The specific binding by S1 to the resulting denatured regions could then account for its effect.

However, it is clear from the previous discussion that the presence of S1 on the ribosome is not sufficient for mRNA translation and that IF3 is required. One of the possibilities, discussed in the next section, is that IF3 stabilizes the "open" form, which, in the absence of IF3, is stable only at 0°C. In this connection one should recall (see Section V, B, 1) that, at 0°C, 30 S subunits containing S1 bind mRNA in the absence of initiation factors. It is also possible that S1 stabilizes mRNA · rRNA base-pairs and would thus be essential for stabilizing a relatively weak mRNA · rRNA complementation. In this connection, S1 stimulates differently ribosome recognition at the beginning of the three R17 cistrons: the highest stimulation was observed with the coat-protein cistron (Table IX) *(214b)*.

3. Role of Initiation Factors

To investigate whether or not IF3 also binds the ribosome in the immediate vicinity of the 16 S RNA 3' terminus, a cross-linking reagent, tartaryl diazide was used; it was found that the proteins from the 30 S subunits could indeed be cross-linked. They were compared in the presence and in the absence of IF3 *(215)*. An easily identified pair of cross-linked proteins is that of S7 and IF3. Further comparison of

---

[9] Three different proteins have recently been shown *(289–291)* to have properties similar to those of S1. The S1 we refer to in the text correspond to "i" factor (see Section V, B, 1).

TABLE IX
EFFECT OF S1 AND FACTORS ON RECOGNITION OF R17 INITIATOR REGIONS[a,b]

Expt. No.	Ribosomes	Factors	Ratio A : coat : replicase
I	30 S	IF2, IF3	1 : 2.0 : 0.5
	30 S	—	1 : 0.2 : 0.05
	30 S (−S1)	IF2, IF3	1 : 0.3 : 0.04
	30 S (−S1)	—	1 : 0.4 : 0.03
II	30 S	Crude	1 : 7.1 : 2.7
	30 S	IF2, IF3	1 : 3.6 : 0.8
	30 S	—	1 : 0.2 : 0.2
	30 S (−S1)	IF2, IF3	1 : 0.8 : 0.2
	30 S (−S1)	—	1 : 0.1 : 0.06

[a] From Steitz et al. (214b).
[b] Reaction mixtures contained, in 40 µl: 100 mM TrisCl, pH 7.5; 50 mM NH_4Cl; 9 mM Mg acetate; 0.25 mM GTP; 3.5 mM β-mercaptoethanol; 1.0 A_{260} unit of charged formylated mixed E. coli tRNA; 2.5 A_{260} units of 50 S ribosomal subunits; and 0.8 A_{260} unit of ^{32}P-labeled R17 RNA (specific activity = 1.5×10^6 cpm/µg). 30 S ribosomal subunits, 1.3 A_{260} units, either containing S1 at a level of 50% or depleted in S1 to a level of 2–3% by washing 2 times in 0.85 M NH_4Cl at a ribosome concentration of 5–15 A_{260}/ml, and 0.5 µg of IF2 plus 2.3 µg of IF3 or 70 µg of crude initiation factors were added as indicated. After formation of initiation complexes, the reactions were "trimmed" with nuclease and fractionated, and the protected sites were chromatographically identified. Parallel reactions in which purified S1 was added back to S1-depleted ribosomes showed that functional differences between the 30 S and 30 S (−S1) ribosomal subunits are due largely to their S1 content rather than to the presence or absence of some other component.

the cross-linked proteins, with and without IF3, reveals that, with the exception of the complexes containing the factor, the patterns are virtually identical and were unaffected by the addition of GTP, poly(U), MS2 RNA, or IF1. It was concluded (215) that the major features of the 30 S protein neighborhood are not dramatically altered by any of the above-mentioned components, including IF3. Furthermore, the S7-IF3 neighborhood was the major factor-containing complex produced when either 30 S particles or 70 S ribosomes were used.

Aurintricarboxylic acid depresses the binding of IF3 to ribosomes (141) and dramatically reduces the recovery of S7-IF3 complexes. From this it can be inferred that the interaction between IF3 and the 30 S subunit is functional and takes place at a site close to, or including, S7.

Other workers (216, 204) have stressed the association of IF3 with 30 S proteins other than S7 and, most important, have observed a cross-link between IF3 and S12. There are many potential sources for this discrepancy, but one is particularly relevant. Tartaryl diazide (6 Å) is much shorter than the reagents used by these other investigators. Since S7 is, by far, the protein most extensively cross-linked to IF3 by tartaryl diazide, we are inclined to believe that IF3 is positioned very close to this protein.

Since S7 is associated with the nucleotide sequences at the 3' terminus of 16 S RNA *(217–220)*, experiments were performed to determine whether or not ^{14}C-labeled IF3 could be cross-linked to the 3' terminus of 16 S RNA after oxidation and reduction of the ribosome with periodate *(221)*, as was done for S1. A significant fraction of the 16 S RNA molecules cross-linked to IF3 were recovered from 30 S subunits oxidized in the presence of [^{14}C]IF3. In the presence of aurintricarboxylic acid, no ^{14}C was recovered in the 16 S RNA; in addition, oxidation is required for the attachment of IF3 to the RNA.

The finding that IF3 bound to the 30 S ribosomal subunit can be cross-linked to both S7 and the 3' terminus of 16 S RNA indicates that IF3, S7, and the 3'-terminus of 16 S RNA are part of a ribosomal region that also includes S1, and that functions in the formation of a preinitiation complex with mRNA.

S1, S12, and IF3 yield cross-linked products in the 30 S subunit *(205)*, confirming that these proteins lie in the same neighborhood. When labeled IF3 was added to a mixture of 30 S and 50 S, or to 70 S ribosomes, and oxidized and reduced as above, not only was the 16 S RNA labeled, but also the 23 S RNA *(221)*. The site-specificity of the RNA-IF3 complexes formed in these experiments was tested in the same way as for the 30 S · IF3 complexes already described. In particular, aurintricarboxylic acid, which inhibits formation of the IF3 · 16 S RNA complex, also abolishes the formation of the complex with the 23 S ribosomal RNA. These results indicate that IF3 can also form a site-specific complex in the immediate vicinity of the 3' terminus of 23 S RNA.

This suggested that IF3 can bind to the 50 S subunits. Accordingly, when 50 S subunits were incubated with IF3 in the absence of 30 S particles, it was observed that IF · 23 S RNA complexes could be recovered after oxidation followed by reduction.

From an examination of the sequence of the 3'-OH end of 16 S and 23 S RNAs (Fig. 9), several observations can be made. As previously discussed, there is a self-complementary region at this terminus of 16 S RNA; in addition, according to Shine and Dalgarno, there are 4–7 bases that are complementary between this 16 S RNA 3'-OH terminus and the polypurine stretch near the initiator triplet. Moreover, there exist two regions complementary to the 23 S RNA, each containing at least five base-pairs. It is postulated *(221)* that these base-pairs are respona.

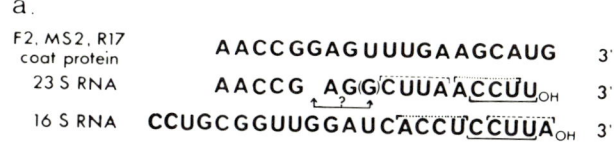

b.

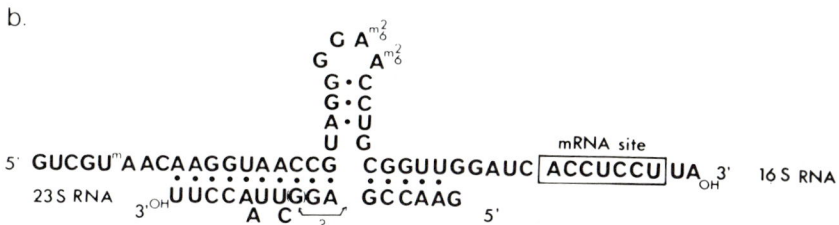

FIG. 9. Possible base-pairing between 3'-OH termini of 16 S RNA and 23 S or mRNA. (a) The nucleotide sequences of the 3' end of the 16 S RNA and 23 S RNA are compared with those of the bacteriophage RNA ribosomal binding site associated with the coat-protein cistron. The hexanucleotide of the 23 S RNA distal to the 3' end has been sequenced by C. Branlant et al. (223a). (b) Hydrogen-bonding schemes are depicted for the 16 S and 23 S RNA interaction. From Van Duin et al. (221).

sible for the association of the 30 S and 50 S subunits. It had in fact been suggested previously (222) that an interaction between the RNAs from each subunit could provide a convenient way of forming a 70 S couple (222).

Following the equilibrium scheme discussed in Section V, it is now postulated (221) that in the 70 S couple, 16 S RNA forms complementary base-pairs with the 23 S RNA at their 3'-OH ends. Upon dissociation these base-pairs are replaced by self-complementary base-pairs, existing in the 16 S RNA and 23 S RNA (223a). The resulting 30 S conformation is appropriate for the binding of mRNA, and the interaction is strengthened in the presence of IF3. The two reactions, 30 S mRNA and 30 S–50 S association, are thus mutually exclusive.

From the point of view of this model, the different in vitro effects of IF3 on the stability of 70 S couples and on the binding of mRNA can be seen as but two different aspects of a single function. In effect, IF3 could be either indirectly or directly (by binding to the sequence involved in base-pairing between 16 S and 23 S RNA) responsible for a series of transitions of complementary nucleotide interactions, from those between 16 S RNA and 23 S RNA to those between the self-complementary interactions of 16 S RNA, and finally to those between 16 S RNA and mRNA. As for S1, IF3 would be more essential for weak mRNA · rRNA base-pairing, and this is in fact observed (see Table IX).

Two tetranucleotide sequences at the 3' terminus of 16 S RNA are homologous with sequences at the 3' terminus of 23 S RNA (Fig. 9); it therefore seems possible that these sequences are also a part of the site for IF3.

There is additional evidence that IF3 may be involved in favoring one base-pair interaction over another. IF3 might per se have an effect on RNA secondary structure, in the absence of any other protein factor, and particularly of ribosomes *(223b)*. Thus, IF3 stimulates the renaturation of tryptophanyl-tRNA. This tRNA exists in two stable forms, native and denatured, which differ in their secondary structure [the denatured form cannot accept tryptophan from its specific synthetase whereas the native form can *(224)*]. Passage from one conformation to another is effected by incubating the tRNA in the presence of Mg^{2+} at a temperature of 37°C. IF3 facilitates the renaturation of a part of the tRNA and stabilizes both the renatured and denatured form *(223b);* IF3 is the only factor to exhibit this property, which is not shown by IF1, IF2, or S1. The mechanism of action of IF3 is probably similar, but not identical, to that of S1; this could already be inferred from the fact that that S1 does not prevent association with the 50 S subunit and that the two factors are necessary for optimum mRNA binding (see Table IX).

By binding to the 3' terminus of 16 S RNA, IF3 could modify not only the structure of the 3'-OH end of the RNA, but the overall conformation of the 30 S subunit. Indeed, important changes in the 30 S subunit upon IF3 binding have been reported by hydrodynamic methods and small-angle X-ray diffraction *(225)*. However, neutron small-angle scattering does not indicate important changes in this subunit, under conditions where it was checked (by using [^{14}C]IF3) that IF3 was stoichiometrically bound to every 30 S subunit *(226a)*. Cross-linking data *(215)* also indicate that there are no drastic changes in the neighborhood of the proteins. But there might be some rearrangements of amino-acid side chains, as suggested by circular dichroic changes of the 30 S ribosomal subunit induced by the binding of IF3 *(226b)*. Also, binding of IF3 to 30 S subunits significantly affects the rate and extent at which several 30 S ribosomal proteins react with radioactively labeled N-ethylmaleimide *(226c)*.

The binding of IF3 to *E. coli* 30 S subunits is inhibited by rRNA ligands, such as ethidium bromide, polyamines, and monovalent alkali metals. Furthermore, results obtained after modifying 30 S subunits chemically indicate that the IF3 binding site is preferentially lost when the rRNA becomes modified *(226d)*.

So far, only binary interactions involving 16 S RNA with mRNA or 23 S RNA in the presence of IF3 have been discussed, but physiologically, initiation-complex formation requires, in addition, the presence of initiator tRNA, IF2, and GTP, which would presumably modify the local interactions discussed above and lead to the 70 S initiation complex. In fact, cross-linking data indicate that IF2 is also located near the 3'-OH terminus of 16 S RNA *(216)*. IF2 binds to proteins S1 and S12. As already discussed, the existence of IF2 subunit-association activity has been shown by light-scattering techniques *(116)*. One could thus postulate that IF2 also binds to the 3'-OH terminal of 16 S RNA, but unlike IF3, IF2 exposes the sequences involved in the formation of complementary associations with 23 S RNA.

The site of action of IF1 could also lie at the 3'-OH end of the 16 S RNA. Like colicin E3, cloacin DF_{13} (a bacteriocin) cleaves the 16 S ribosomal RNA at a specific site, 49 nucleotides from its 3' terminus; after cleavage, the 3'-terminal fragment is retained by the ribosome *(227, 228)*. Polypeptide synthesis directed by MS2-RNA is inhibited as a result of cloacin action. However, under conditions where virtually all ribosomes contain a "nick" in their 16 S RNA and protein synthesis is completely abolished, the amount of initiation complex formed is unaffected while the rate at which it is formed is altered. This is due to the fact that ribosomes, rendered defective by cloacin, no longer respond to IF1, although they fully retain their ability to respond to IF3 and IF2. In particular the stimulation of dissociation by IF1 is not observed. This leads to the interpretation that the first consequence of 16 S RNA cleavage is a reduction in the rate at which the 30 S ribosomal particles are generated from 70 S couples. A second consequence is that defective ribosomes completely lose their capacity to bind fMet-tRNA in a second cycle of initiation. It is probable that, owing to improper functioning of IF1, the release of IF2 from the 70 S is abolished.

These results indicate that the 3'-OH end of 16 S RNA must be intact for the proper functioning of IF1,[10] and point to the important role of this factor in protein synthesis, as was previously indicated with the streptomycin-dependent mutants *(180b)*.

All these studies may be summarized by the statement that the cistron specificity of bacterial ribosomes, known to reside in the 30 S subunit, is likely to be localized in the long pyrimidine-rich decanucleotide located at the 3' terminus of 16 S RNA. The degree of complementarity of this region with the purine-rich segment within the ribosome-binding-site sequence on bacterial mRNA determines the

---

[10] IF1 and IF2 can be linked to the 3'-OH end of 16 S ribosomal RNA after periodate oxidation and reduction (S. Langberg and J. Hershey, personal communication), as was found for IF3 *(221)*.

intrinsic capacity of the ribosome to translate a particular cistron. While it appears that the 3'-OH terminus of 16 S RNA plays a special role, it is not excluded that other regions are involved in the recognition. Van Knippenberg *(229)* drew attention to a striking homology between the nucleotide sequence near the 5' end of 16 S RNA and ribosome binding sites of phage RNA. He speculated that both the 3' and the 5' termini of 16 S RNA could be functional in bringing the initiation codon AUG into a small loop, and that this might be required for initiation. Furthermore, since an A-U-G or G-U-G triplet can also direct the binding, other ribosomal components could contribute to the initiation of polypeptide chains. Moreover, as discussed in Section VI, D, the structure of initiator tRNA is responsible for its location at site P.

Since, in authentic messenger initiation sites, the length of the region complementary to rRNA can be as short as three nucleotides (see Table VIII) it is obvious that the potential for base-pairing plus the presence of the initiator codon is not sufficient to describe a true mRNA initiator region. In the nucleotide sequence of the genome of phage MS2, there are several internal and out-of-phase AUG triplets preceded by appropriate polypurine stretches *(191)*. These are generally involved in RNA secondary structures, which renders either the initiation triplet, or the polypurine sequence, or both, unavailable for base-pairing with the rRNA and initiator tRNA on the ribosome. It is likely that mRNA utilizes secondary and tertiary structures to prevent ribosome recognition of noninitiator AUG triplets.

Protein S1 and factor IF3 are both necessary to expose the 16 S RNA so that it can interact with the mRNA sequence. Depending on the degree of complementarity between 16 S RNA and the mRNA initiation site, the dependence on IF3 and S1 is more or less drastic (see Table IX). Base-pairing offers new possibilities as a selection mechanism for the regulation of protein synthesis; the extent of complementarity between ribosomes and their binding site may influence the frequency of initiation at certain sites. For instance, in the case of the three R17 initiation regions, ribosome recognition of the coat or replicase site is severalfold more dependent on the presence of initiation factors *(150, 214b)* and of fMet-tRNA than binding to the "A" protein initiator. Here, it seems likely that the comparatively higher degree of complementarity between the "A" site and the 16 S RNA 3' end substitutes to some extent for the usual requirements for the other components of the initiation reaction. The fact that site A can complement with a colicin fragment that has been removed from the ribosome *(190)* supports this idea.

In the case of poly(A,U,G) or poly(G,U) it is easy to propose pairing models; the purine sequences being present in larger amounts in these polymers, the action of IF3 would be less important.

It thus appears that there is no need to postulate different IF3 proteins to explain the specificity for cistron initiation. However, it should be mentioned that the presence in the cells of two species *(45, 46)* or more *(230)* of IF3 activity, which could discriminate between various mRNAs and between different initiation sites on the same mRNA, has been reported. In addition, after infection with bacteriophage T4, a new IF3 activity is formed that preferentially stimulates the binding of late T4 mRNA and discriminates against the translation of phage MS2 mRNA *(231–233)*. However, this is not universally accepted, since lysozyme and a large number of late T4 proteins can be synthesized *in vitro*, using late T4 RNA to direct synthesis on ribosomes taken from uninfected cells *(234–236)* and since no alteration occurs after T4 infection in the overall specificity of ribosomes or factors for the initiation of translation of several mRNA preparations *(237a)*. Furthermore, Leder *et al. (237b)* examined the effect of *E. coli* infection on the initiation system of this bacterium with a variety of phages (DNA phage T7, Qβ, and lysogenic cells induced for phage lambda). The most striking effect of this infection was a loss of initiation factor activity in all the infected cells. The defect induced by T7 infection did not appear to discriminate in favor of the translation of late T7 mRNA but, as with Qβ infection and lambda induction, resulted in a general loss of the ability to initiate late T7 and Qβ mRNAs as well as a loss in the ability to recognize the AUG codon. These results are not easily explained in terms of a model that assumes that phage infection results in altering the specificity of the initiation system. One explanation, suggested by previous discussions, is that one form of IF3 (IF3β) could result from proteolysis of IF3α and would thus be less active. Minimal amounts would stimulate binding of the initiation regions that require fewer factors, such as the cistron for the maturation protein or perhaps some phage cistron of early T4 mRNA, and would be insufficient when larger quantities of factor are required. There is indication that such a proteolysis occurs *(51b)*.

D. Mechanism of Cistron Recognition in Eukaryotes

Since the basic features of almost all fundamental biological processes are similar in prokaryotes and eukaryotes, it is tempting to assume that the 3' terminus of 18 S RNA in eukaryotic ribosomes likewise interacts with messenger RNA during polypeptide chain initiation. Indeed, it is striking that all species of 18 S RNA that have been analyzed (ranging from yeast to mammal) possess an identical 3'-terminal sequence: G-A-U-C-A-U-U-A$_{OH}$ *(238)*. Moreover, the first ribosome binding site to be sequenced from a eukaryotic mRNA (brome mosaic virus RNA) *(239)* does exhibit a four-base complementarity with this terminus. If the same mechanism operates in cells of higher organisms, there should be a compatibility between ribosomes and mRNAs from various eukaryotic species, at least with respect to the initiation of protein synthesis. By contrast, since prokaryotic messages have presumably been tailored to interact optimally with their particular 16 S RNA (here the 3'-terminal sequences do differ), the expression of most bacterial genes in eukaryotic cells (or vice versa) might be expected to encounter some difficulty at the point of polypeptide chain initiation. It is thus interesting that the Qβ and R17 bacteriophage genomes can be translated with some fidelity in eukaryotic protein-synthesizing systems *in vitro (240–244)*; however, their efficiency relative to eukaryotic mRNAs is much reduced.

On the other hand, we do not yet understand the part played in ribosome binding by the structures of eukaryotic mRNAs that differ significantly from prokaryotic mRNAs. In contrast to prokaryotic mRNAs, most eukaryotic cellular mRNAs and animal viral mRNAs have a polyadenylate sequence at their 3' terminus *(245)*.[5] This segment can be up to 200 nucleotides long and is added after transcription. Its function is not clear; it does not appear to be involved in the translation process since its removal does not inhibit mRNA activity, *in vitro (246)*. However, when the poly(A) segment was removed from native mRNA for rabbit globin and injected into *Xenopus laevis* oocytes, it appeared to have a shorter half-life than the corresponding poly(A)-containing mRNA *(247)*. This points to a connection between the presence of the poly(A) and the long half-lives of eukaryote mRNAs. The half-lives of poly(A)-containing mRNAs, however, vary considerably, from a few to 24 hours; this latter value is approximately the time of one cell generation *(248)*, and contrasts with prokaryote mRNAs, which have half-lives of about 2–3 minutes, much less than the time for one cell generation *(249)*.

Another striking and important feature of the eukaryotic mRNAs is that most, but not all, contain a 7-methylguanosine (m^{7}G) residue; this is added after transcription to their 5' terminus and linked through a 5'-5' pyrophosphate bond to the next residue on the mRNA *(250)*.[11] The presence of m^{7}G is required for translation of the mRNA *(251)* and more specifically for initiation *(252a)*, and there is some evidence that it is recognized by initiation factors *(252b)*.

Initiation of protein synthesis in eukaryotic cells was first believed to proceed from a single initiation site on the mRNA molecule *(244,253)*.

---

[11] See Vol. 19 of this series.

Eukaryotic proteins are either synthesized singly from monocistronic mRNA [such as alpha and beta chains of hemoglobin *(254)*] or cleaved from a large polypeptide precursor synthesized from one initiation site on a polycistronic mRNA (picornavirus protein) *(255a)*. Thus, poliovirus has one site *in vivo* or *in vitro* when translated in eukaryotic systems. This site is subsequently cleaved in a well-established series of events to yield the various mature proteins. Multiple initiation on a mRNA containing only one physiological initiation site occurs when poliovirus RNA is translated in an *E. coli* cell-free system *(244)*. Analyses of the cell-free product by peptide separation and by immunological methods indicate that protein synthesis is accurate; the genes for different proteins therefore have an initiation site recognized by *E. coli* ribosomes *(244)*. Similarly, *in vitro* translation of satellite tobacco necrosis virus RNA can occur accurately in an *E. coli* system. The *in vitro* synthesis carried in *E. coli* or wheat germ produces proteins with the amino-acid sequence of the coat protein of the virus *(255b)*. However, it should be pointed out that the amount of protein synthesized by any prokaryote, including gram-positive bacteria, is usually much less when mRNA from eukaryotic organism is used.

However, several mRNAs, when translated in a eukaryotic system, have at least two initiation sites, but one of them appears to be inaccessible in the large-precursor RNA and becomes active only after the precursor has been cleaved into a smaller mRNA *(244, 253, 256–259)*. Thus 42 S RNA from Semliki Forest virus virion has two initiation sites, but one of them, the coat-protein initiaton site, is partially inactive, while the smaller RNA species derived from the 42 S is highly active in coat-protein synthesis. The same thing occurs with polyoma virus 19 S and 16 S RNA, and with plant viral RNA. The alfalfa mosaic virus *(260, 261)* contains in its genome 4 species of RNA: 24 S, 20 S, 17 S, and 12 S. *In vitro*, using different eukaryotic cell-free systems, only the 12 S codes for coat protein, although the gene for this protein is present in the 17 S RNA. However, when translated in the *E. coli* system, the 17 S RNA does give rise to the coat protein (Hirth, personal communication). All these discrepancies may be resolved when the sequences of viral initiation sites, both recognized and hidden, become known. It is conceivable that for the hidden site there is not enough complementarity between the bases at the 3'-OH terminus of the 18 S RNA and the polypurine sequence at the 5' terminus of the initiation codon. After cleavage of the larger mRNA, there is addition of m^{7}G at the cleaved end, and we know that this modification is indeed compulsory for initiation of protein synthesis with many eukaryotic mRNAs. It is conceivable that this addition helps base-pairing between mRNA and the complementary stretch on 18 S RNA, or recognition by one of the initiation factors. Our understanding of eukaryotic initiation will probably make significant progress as soon as the sequence of initiation sites and of eukaryotic ribosomal 18 S RNA are known.

Another point that might be important for translation of eukaryotic mRNA, and that we still fail to understand, is the importance of proteins bound to eukaryotic mRNA as found in the cytoplasm (262).

Finally, at the time when, in prokaryotic systems, an "interference factor" was believed to change the specificity of initiation, a protein factor endowed with a similar property was reported to exist in the supernatant of Krebs II ascites cell (256, 263); this factor (IF-EMC) was absolutely necessary for initiation of EMC RNA translation, but not for that of hemoglobin mRNA translation (263). These mRNAs fall into two broad categories (Table X): one where the requirement for IF-EMC in translation is almost absolute (EMC RNA, Mengo RNA) and another where there is no significant dependence on this factor for translation (e.g., TMV RNA, hemoglobin RNA) (256). We have mentioned that a factor purified from ribosomal wash, eIF-4A, is essential in the initiation of hemoglobin (see Section II, B, 2). Because eIF-4A and IF-EMC show similarities in their chromatographic behaviors, Marcker *et al.* (256) investigated the requirement for both these factors in an (IF-EMC)-deficient cell-free protein-synthesizing system prepared from Krebs II ascites cells; they showed that IF-EMC and eIF-4A [purified by Staehelin *et al.* (55)] are functionally interchangeable. It was therefore concluded that IF-EMC is not a tissue-specific initiation factor, but a common factor required for the initiation of protein synthesis; however, the amount required for initiation of translation differs widely according to the various mRNAs tested. This recalls the situation for the IF3 requirement in prokaryotes.

A possible explanation is that eIF-4A is involved in the formation of base-pairing between the 3'-OH terminus of the 18 S ribosomal RNA and the complementary region of the initiation sites on mRNA, just as postulated for IF3 and S1.

The complex between 18 S RNA and EMC RNA may involve relatively few base-pairs, resulting in a high dependency on IF-EMC for translation; on the other hand, mRNAs such as TMV RNA or hemoglobin RNA may form more base-pairs with ribosomal RNA, resulting in a greatly reduced requirement for IF-EMC. The amount of this factor *in vivo* and the existence of a regulatory mechanism that would modulate this amount remain to be investigated.

Finally, small "translational control" RNAs have been isolated

TABLE X
THE EFFECT OF THE ADDITION OF IF-EMC ON THE TRANSLATION
OF VARIOUS mRNAs IN THE ASCITES CELL-FREE SYSTEM[a,b]

mRNA	Methionine incorporation (nmol) Control	Control + IF-EMC	Stimulation by IF-EMC over control (%)
Encephalomyocarditis virus (EMC)	7	55	686
Mengo virus	5	88	1660
Foot-and-mouth disease virus (FMDV)	18	62	233
Poliovirus	11	10	0
Rous sarcoma virus (70 S RNA)	3	4	33
Semliki Forest virus (42 S RNA)	15	12	0
Semliki Forest virus (26 S RNA)	12	18	50
Vaccinia virus *in vitro* mRNA	26	27	4
Tobacco mosaic virus (TMV)	42	44	5
Alfalfa mosaic virus	17	17	0
Brome mosaic virus (unfractionated RNA)	11	13	18
Satellite tobacco necrosis virus	34	48	37
Tobacco rattle virus (unfractionated RNA)	45	56	24
Hemoglobin 9 S RNA	10	14	40
Oviduct polysomal RNA	12	10	0

[a] From Marcker *et al.* (256).
[b] Each mRNA was assayed at saturing mRNA concentration and the effect of IF-EMC was tested. Incorporations in the presence of mRNA were corrected for background incorporation in the absence of added mRNA; this varied between 0.010 and 0.015 pmol of methionine (1 pmol is approximately equivalent to 25×10^4 cpm).

(264, 265); it is conceivable that small RNA molecules could form base-pairs with the 3' terminus of 18 S RNA and might thus inhibit polypeptide chain initiation. But it still has to be proved that these molecules do not represent degraded ribosomal RNA fragments. At any rate their physiological role is far from clear. This also applies to the inhibition of hemoglobin synthesis by double-stranded RNA *(266)*.

Recently, protein kinase activities, independent of cyclic nucleotide, have been isolated from postribosomal supernatant fractions of rabbit reticulocytes; they phosphorylate factor IF2, and it is suggested that protein synthesis in reticulocyte lysates is controlled by the reversible phosphorylation of eIF2 (P. Farrell, K. Balkow, T. Hunt, and C. R. Jackson; J. A. Traugh, S. M. Tahara, S. B. Sharp, B. Safer, and W. C. Merrick, personal communications).

VI. Pending Questions

A. Precise Operational Sequence of Initiation Complex Formation in Prokaryotes

It is generally agreed that, in eukaryote protein synthesis, the first step is the formation of an EF2 · GTP · (Met-tRNA) complex that attaches to the ribosome before mRNA is bound. In prokaryotes, there has been some controversy concerning the order of attachment of mRNA and fMet-tRNA to the 30 S subunit *(267, 268, 124)*. One possible sequence of events was that the ribosome, with the help of IF3, binds to a specific initiation sequence on the mRNA, the appropriate initiation codon, AUG or GUG, then specifying the IF2-dependent attachment of initiator tRNA and of GTP to this preformed complex. This sequence of events implies that binding of initiator tRNA, is a consequence of the selection of the messenger initiation sites, rather than a prerequisite to messenger binding. This concept is supported by the observation that the 30 S particle, MS2 RNA, and IF3 can interact stoichiometrically in the absence of fMet-tRNA to form a complex (unstable) that is strongly stabilized by the subsequent addition of IF1 + IF2 *(124, 228)*; it was further strengthened when it was found *(72)* that subsequent addition of initiator tRNA causes the liberation of IF3 and is indispensable for the translation of natural messenger.

Furthermore, using an *in vitro* coupled system where λ and φ80 DNA transcription occurs in the presence of washed ribosomes (plus or minus the various components of the initiation reaction), Crepin *et al. (269)* have shown that the interaction between mRNA, transcribed by RNA polymerase and ribosomes, is IF3-dependent, confirming an earlier report *(270)* that the 30 S subunits, when alone, stimulate *in vitro* transcription. This effect is enhanced 4- to 5-fold by the presence of IF3, but very little by the presence of IF1 + IF2; simultaneous addition of the three factors results in a 6-fold enhancement, optimum "coupling" being obtained when initiator tRNA is also present, although it is not absolutely required. Thus, the "nascent" messenger can easily bind the 30 S particle during a reaction that does not require initiator tRNA but is dependent on the factors.

From this information, there is no doubt that a natural messenger, such as MS2 RNA, can form a ternary complex with a 30 S subunit in the presence of IF3, but the question is whether this complex has a functional significance as intermediate of initiation. This is denied by Noll and Noll *(121)* as well as by Jay and Kaempfer *(267, 268)*. The latter obtained evidence for the existence of unstable intermediates between fMet-tRNA, the 30 S particle and initiation factors, in the absence of messenger RNA, and believe that binding of fMet-tRNA precedes that of the messenger; moreover, they claim that this prior binding is essential for the correct attachment of messenger to ribosomes. Indeed, they could demonstrate, on sucrose gradients, a 34 S complex that would correspond to the transient interaction between 30 S, fMet-tRNA, GTP and IF2. Formation of this complex required only the presence of IF2, not that of the messenger. By the same technique, they also showed a 46 S preinitiation complex composed of the 30 S particle, fMet-tRNA, IF2, IF3, and phage 17 RNA; subsequent junction of the 50 S subunit leads to the initiation complex. Formation of the 34 S complex is extremely fast, and the subsequent binding of R17 RNA in the presence of IF3 very slow. When the 30 S subunits are incubated beforehand with R17 RNA and IF3, not only is there no stimulation of initiation complex formation, but the latter is even slowed down (267). This can result from incorrect binding of R17 RNA to the 30 S particle in the absence of initiator tRNA.

The detection of 30 S · mRNA complexes could be taken as a suggestion that perhaps the sequence of events in initiation of protein synthesis might follow alternative paths with either fMet-tRNA or mRNA binding first to the 30 S subunit. Labelled fMet-tRNA from the 34 S complex preferentially enters the 46 S initiation complex containing R17 RNA, even in the presence of excess of free fMet-tRNA, 30 S, R17 and IF3 (267). On the other hand, incubation of 30 S particles with R17 RNA and all the initiation components, with the exception of fMet-tRNA, does not yield the 30 S · (R17 RNA) complexes that can subsequently bind fMet-tRNA, since in the presence of inhibitors preventing further binding of mRNA [such as edein (271a) or excess of S1], no 46 S complexes could be detected. These would probably occur if functional 30 S · (R17 RNA) complexes had accumulated during the prior incubation. In the absence of mRNA, the 34 S fMet-tRNA complex is very unstable; this might explain why, in earlier studies, binding of fMet-tRNA appeared to be dependent on the presence of mRNA.

In conclusion, it appears likely that fMet-tRNA would first have to bring about the right conformational changes on the 30 S particles before R17 RNA can bind and be correctly positioned. If this thesis is correct, it would have the advantage of correlating the initiation process in eukaryotes and prokaryotes.

Unfortunately, Kaempfer's experiments were done in the absence of IF1, which is essential for initiation. Furthermore, the binding of the subunits to MS2 RNA can take place in the absence of fMet-tRNA^{fMet}

(G. Van Diggelen, W. Van Prooyen, J. Van Duin, unpublished) as previously reported (124, 128). Finally, basing their opinion on kinetic results, Risuleo et al. (271b) consider that either mRNA or fMet-tRNA^{fMet} could first bind to the ribosomes, the preferential route depending on the mRNA structure. However, it is not known whether this is not an *in vitro* artifact and whether *in vivo* only one route exists.

In this connection, an initiation complex containing fMet-tRNA correctly attached at the P site (as shown by the puromycin reaction) can be obtained in the absence of messenger, provided the Mg^{2+} is increased to 35 mM (see Section V, D) (272, 273). On the other hand, it should be recalled that it has never been possible to show, in the absence of ribosomes, a GTP · (fMet-tRNA) · IF2 complex with a high enough association constant to be considered as a physiological complex. If such a complex exists, it can only be formed in the presence of ribosomes (see discussion in Section III, A).

B. Number of tRNA Binding Sites on 30 S Subunits

All the evidence is consistent with the existence of two sites on active 70 S ribosomes: a puromycin-reactive site for initiator tRNA (P site) as well as a site for EF-Tu-dependent internal aminoacyl-tRNA binding (A site). However, there is no evidence that a codon–anticodon interaction exists simultaneously for the two tRNAs on the ribosome; also it is not clear whether the 30 S particle has one or two different binding site(s) for tRNA; simultaneous binding of two tRNA molecules on the 30 S subunit has never been demonstrated. Nevertheless, both initiator and internal aminoacyl-tRNA can bind the 30 S particle, but the mRNA · AA-RNA · 30 S complex is much less stable than the one involving initiator tRNA (274). In an attempt to answer this question, Pongs et al. (275) covalently linked to ribosomes a labeled, chemically reacting triplet analogous to AUG. When *AUG was cross-linked to freshly prepared 30 S or 70 S particles, a defined pattern of labeled proteins was found; *AUG reacted mainly with protein S18 and slightly with S4. However, the amount of *AUG covalently bound was very low. After treating 70 S or 30 S particles under different conditions (cold, incubation at 37°C), the amount of covalently bound *AUG increased, the pattern was unchanged, but more label was found in S4. In contrast, when the 30 S subunit was frozen prior to cross-linking, the label was found in both S12 and S4. The amount of *AUG bound never exceeded 1 mole per mole of ribosome.

In summary, the following three preparations of 30 S subunits containing the label in different proteins were obtained: *AUG · 30 S (· S18), *AUG · 30 S (· S4 · S18) and *AUG · 30 S (· S4 · S12). This again points to the variability of conformation taken by ribosomes and explains the varying results found by different authors with crosslinking reagents. All the 30 S ribosomes crossed-linked with *AUG can bind fMet-tRNA^{fMet}, and this binding is dependent on IF2. Initiation complexes thus formed and incubated with 50 S were assayed for puromycin reaction and GTP hydrolysis. The initiator tRNA bound to *AUG · 30 S (· S18) or to *AUG · 30 S (S14 · S18) was resistant to puromycin. However, with the preparation where *AUG was bound not only to S4 but also to S12, fMet-tRNA^{fMet} was partly displaced by puromycin. The authors did not check, however, whether IF2 was absent from the preparations resistant to puromycin. This point is particularly important, since in these preparations the IF2-dependent GTPase was inactive and the possibility exists that IF2 was not recycled and inhibited puromycin binding.

Except for this reservation, the labeling data are compatible with two independent codon-binding sites, the accessibility of which would strongly depend on the 30 S subunit conformation. However, Pongs et al. were not able to bind simultaneously two labeled AUGs per ribosome. Therefore, their data are also compatible with a model containing a preferential codon-binding site, where, according to its conformation, the tRNA can be either puromycin resistant or puromycin sensitive, as proposed previously for streptomycin reactivity (85, 123, 183, 276, 277).

C. ATP Requirement

Another unanswered question is that of the ATP requirement. We have already discussed this for the synthesis of proteins in eukaryotes, both in reticulocytes and in wheat germ (see Section III, B). In reticulocytes, ATP appears to be involved in mRNA binding; a similar requirement was first reported with the wheat germ system. It was also postulated that this nucleotide is used for the synthesis of ppGpp, and this may be the actual role of ATP in initiation (278). Met-tRNA^{fMet} does not react with puromycin unless ATP is present, which might indicate that the correct initiation complex can be formed only in the presence of ATP.

No requirement for ATP in E. coli has been established, and initiation complex formation is not stimulated by this nucleotide. However, when attempts were made to synthesize proteins in vitro in the presence of natural messenger, or poly(A,U,G) or poly(U,G), using precharged tRNAs, a very weak synthesis was observed unless ATP was present; no such requirement was observed with the other synthetic homopolymers (Springer and Grunberg-Manago, unpublished). The role of ATP is not a trivial one. Changing [Mg^{2+}] or recharging GDP to GTP does not eliminate the ATP requirement. However, one cannot fully eliminate the possibility that ATP might be necessary to recharge some minor species of unstable tRNA by some synthetase contaminating one of the factors of the translation apparatus, since no ATP requirement was found for initiation complex formation.

In contrast, in the *B. stearothermophilus* system, an ATP stimulation of initiation complex formation was observed *(279)*. This stimulation appears to involve *B. stearothermophilus* ribosomes and not the factors, as shown by cross experiments with *E. coli*. Obvious artifacts, such as magnesium complexes or phosphorylation of GDP as the site of the ATP stimulation, have been eliminated. This role of ATP is not yet understood; the nucleotide could be used for the formation of high-energy guanine derivatives, as suggested by Marcus *(278)*. However, the fMet-tRNA bound in its absence reacts with puromycin; the situation is therefore not similar to what has been observed in the wheat-germ system. Whether the ATP requirement is a property of gram-positive organisms should be interesting to investigate; we already know that there are differences in the initiation process between gram-positive and gram-negative organisms. The natural mRNA of gram-positive bacteria is not translated in systems from gram-negative bacteria and vice versa (J. Rabinowitz, private communication).

However, it should be noted that the 5 S RNA · L18 · L12 complex exhibits both GTPase and ATPase activity in *B. stearothermophilus* as well as in *E. coli (280a)*. ATP could be required not for the formation of the initiation complex, but as an energy source for correct selection of the translation frame as a generalization of a scheme such as that proposed by Hopfield *(280b)*.

D. Role of the Formyl Group in Prokaryote Initiator tRNA

In prokaryotes, in contrast to eukaryotes, polypeptide chain initiation proceeds via a mechanism involving a formylated species of initiator tRNA (tRNA^{fMet}), and several exceptions to this rule are discussed in Section II, A. Moreover, it has been shown that *E. coli* can grow to a certain extent in a medium where the level of formylation is strongly depressed *(281, 282)*.

In the *E. coli* system, when nonformylated initiator tRNA is used, formation of the 30 S · tRNA initiation complex is strongly stimulated by all three initiation factors and is messenger-dependent, as in the case of fMet-tRNA *(272)*; it is also optimal at higher Mg^{2+} (Table XI). Therefore, formylation does not affect formation of the preinitiation

TABLE XI
Effect of GTP and Magnesium Concentration on the Binding of fMet-tRNA and Met-tRNA^{fMet} to 30 S Ribosomal Subunits in the Absence and in the Presence of Initiation Factors[a,b]

	Mg^{2+} concentration (mM)		fMet-tRNA^{fMet} (pmol)	Met-tRNA^{fMet} (pmol)
	5	−IF	0.12	0.18
		+IF	1.12	0.68
+GTP	5	−IF	0.21	0.17
		+IF	1.83	0.35
	15	−IF	0.46	0.45
		+IF	1.27	1.07
	35	−IF	0.49	0.52
		+IF	1.17	0.95

[a] From Petersen et al. (273).
[b] Each assay tube (50 µl) contained: 15 pmol of 30 S ribosomal subunits; 0.11 A_{260} unit of poly(A,U,G); 50 mM TrisCl (pH 7.4); 50 mM NH$_4$Cl; 26 µg of crude IF; 2.1 pmol of fMet-tRNA^{fMet} (4100 cpm/pmol) or 3.0 pmol of Met-tRNA^{fMet} (4100 cpm/pmol); magnesium acetate as indicated. When GTP is present, its concentration was 1 mM.

complex, but its formation does not require GTP and is even somewhat inhibited by this nucleotide. When unformylated tRNA is first bound to the 30 S particles and the 50 S subunits are subsequently added, most of the Met-tRNA^{fMet} bound reacts with puromycin, indicating that it is correctly positioned. The situation is quite different with the 70 S ribosomes (273): only the attachment of the formylated species is stimulated by the initiation factors, which strongly inhibit binding of the unformylated Met-tRNA^{fMet}. In the absence of initiation factors, both species are bound to the 70 S ribosomes, but only the formylated tRNA reacts completely with puromycin. Experiments with tetracycline support the idea that, although bound in a different manner than formylated tRNA, the unformylated species is not located preferentially at the A site under these conditions.

It appears, therefore, that 70 S ribosomes are able to distinguish not only between initiator tRNA^{fMet} and noninitiator tRNA^{Met}, but also between formylated and unformylated acylated initiator tRNA (tRNA^{fMet}), and this even in the absence of initiation factors. Since no energy is required in this recognition step (although the presence of IF2 and GTP enhance the ability to discriminate between the two initiator tRNA species), one may presume the existence of at least either two sites or two 70 S ribosome conformations for the binding of initiator tRNA. Binding experiments, and experiments on the formation of aminoacyl puromycin, as a function of magnesium, potassium, or initiation factors, led Danchin et al. (273) to propose a two-state equilibrium for 70 S particles, involving a minor, active conformation, and a major one that is not readily active. The formyl group would act as a specific trigger to select the active conformation. From this model, equilibrium parameters and kinetic constants of the peptidyl-transferase activity have been tentatively derived.

Finally, this leads to proposing a general function for the regulatory action of the formyl group. Initiation of polypeptide synthesis generally proceeds via a complex involving initiator tRNA, messenger RNA, and the 30 S ribosomal subunit. Since this subunit is unable to discriminate between the formylated and unformylated initiator tRNA species, one may wonder whether, in special cases, the 70 S particle might not be used *as such* in initiation complexes. In particular, in the case of polycistronic messengers, it may be that the first cistron is translated via dissociated ribosomes, but that when passing to the next cistron, an appreciable proportion of the ribosomes remains as 70 S particles, requiring formylation for the initiation step. Formylation might, therefore, be used as a regulatory device in the translation of polycistronic messengers, since lowering the formylation level would increase the polarity effect.

This model has been tested *in vivo*. Studies on the polarity of the expression of the lactose operon genes showed that the polarity strongly increased when the level of formylation decreased (U. Petersen, A. Ullmann, and A. Danchin, manuscript in preparation). The observation that a termination factor releases the polypeptide chain without releasing the ribosome from mRNA also suggests that, after reaching the termination codon, the ribosome can slide across the intercistronic region until it hits the AUG triplet at the beginning of the next cistron *(121)*.

VII. Conclusion

Despite the fact that many questions about initiation remain unanswered, we begin to have a comprehensive picture of this step and particularly of the role of the different components involved. In fact, the initiation of protein synthesis is actually better understood than the initiation of replication or of transcription, these latter two processes appearing more complicated than had previously been realized. However, it should be emphasized that most *in vitro* studies make use of either bacteriophage or viral mRNAs, which are not typical messengers. Hence the mechanisms, and particularly the control mechanisms deduced from these experiments are not necessarily the same as those involved in translation of cellular mRNAs. Another important point is that while the role of initiation factors appears to be well documented from *in vitro* studies, we have no proof of their role *in vivo* for lack of appropriate mutants: only one attempt has been made to select such mutants *(51);* this should be developed.

It is obvious that ribosomes can no longer be thought of as static machinery. This had been suggested *(85, 123, 183, 276, 277)* but now appears much more documented. In addition to providing a structure on which amino acids are polymerized, ribosomes play a dynamic part in selecting which mRNA is to be translated. It also begins to be clear that the secondary structure of ribosomal RNAs not only plays a structural role in ribosome morphopoiesis, but is directly involved in mRNA · rRNA interactions, as has already been underlined. Other RNA · RNA interactions are also involved in various steps of ribosome functioning [codon–anticodon and 5 S RNA with the G-T-Ψ-C region of tRNA *(283–285)*]. This could support the idea that the primitive translation machinery may have consisted solely of RNA molecules *(286).* But many possible complementarities exist in the different RNAs; the ribosomal RNAs, for example, can assume different intramolecular as well as intermolecular structures *(223a).* This could be reflected in different ribosome conformations, such as type-A or type-B ribosomes, and on the reversible transitions that occur during various steps of protein synthesis, such as translocation, initiation or termination. An important aspect of ribosome function is the efficient control of these transitions by a series of extraribosomal proteins. Ribosomal proteins and initiation factors could recognize some feature of the ribosomal RNA and help it to adopt the right configuration for the different base-pairing. The role of protein interactions taking place on the ribosome should, therefore, not be minimized. The various interactions that take place in ribosomes make these organelles one of the most interesting models for the study of RNA · RNA, RNA · protein, and protein · protein interactions. Some ribosomal proteins are now available in reasonable amounts and in a high degree of purity, which results in the ribosome becoming a choice material for physical chemists. In the not-too-distant future, the general principles governing all these interactions should be established.

Acknowledgments

We wish to express our gratitude to Drs. Steitz, Staehelin, and Bosch for making available their manuscripts before publication. We also want to thank Mme Costinesco for her help in preparing the manuscript.

REFERENCES

1. H. F. Lodish, *Nature* **251**, 385 (1974).
2. J. P. Waller, *JMB* **7**, 483 (1963).
3. K. A. Marcker and F. Sanger, *JMB* **8**, 835 (1964).
4. B. F. C. Clark and K. A. Marcker, *JMB* **17**, 394 (1966).
5. K. A. Marcker, *JMB* **14**, 63 (1965).
6. J. M. Adams and M. R. Capecchi, *PNAS* **55**, 147 (1966).
7. R. E. Webster, D. L. Engelhardt and N. D. Zinder, *PNAS* **55**, 155 (1966).
8. H. F. Lodish, *Nature* **220**, 345 (1968).
9. M. Takeda and R. E. Webster, *PNAS* **60**, 1487 (1968).
10. D. M. Livingston and P. Leder, *Bchem* **8**, 435 (1969).
11. J. M. Adams, *JMB* **33**, 571 (1968).
12. A. T. Matheson and A. J. Dick, *FEBS Lett.* **6**, 235 (1970).
13. N. K. Gupta, *JBC* **243**, 4959 (1968).
14. U. L. RajBhandary and H. P. Ghosh, *JBC* **244**, 1104 (1969).
15. C. T. Caskey, B. Redfield and H. Weissbach, *ABB* **120**, 119 (1967).
16. K. Takeishi, T. Ukita and S. Nishimura, *JBC* **243**, 5761 (1968).
17. A. E. Smith and K. A. Marcker, *Nature* **226**, 607 (1970).
18. J. C. Brown and E. A. Smith, *Nature* **226**, 610 (1970).
19. D. Housman, M. Jacobs-Lorena, U. L. RajBhandary and H. F. Lodish, *Nature* **227**, 913 (1970).
20. A. E. Smith and K. A. Marcker, *JMB* **38**, 241 (1968).
21. F. Berthelot, D. Bogdanovsky, G. Shapira and F. Gros, *Mol. Cell. Biochem.* **1**, 63 (1973).
22. R. S. Ranu and I. G. Wool, *Nature* **257**, 616 (1975).
23. S. K. Dube, K. A. Marcker, B. F. C. Clark and S. Cory, *Nature* **218**, 232 (1968); *EJB* **8**, 244 (1969); S. K. Dube and K. A. Marcker, *ibid.* **8**, 256 (1969).
24. S. Cory, K. A. Marcker, S. K. Dube and B. F. C. Clark, *Nature* **220**, 1039 (1968); S. Cory and K. A. Marcker, *EJB* **12**, 177 (1970); Z. Ohasi, K. Murao, T. Yahagi, D. L. von Minden, J. A. McCloskey and S. Nishimura, *BBA* **262**, 209 (1972); H. Ishikura, Y. Yamada, K. Murao, M. Saneyoshi and S. Nishimura, *BBRC* **37**, 990 (1969).
25. M. Simsek and U. L. RajBhandary, *BBRC* **49**, 508 (1972).
26. P. W. Piper and B. F. C. Clark, *Nature* **247**, 516 (1974); M. Simsek, U. L. RajBhandary, M. Boisnard and G. Petrissant, *ibid.* **247**, 518 (1974).
27. Y. Ono, A. Skoultchi, A. Klein and P. Lengyel, *Nature* **220**, 1304 (1968).
28. A. S. Delk and J. C. Rabinowitz, *Nature* **252**, 106 (1974).
29. A. S. Delk and J. C. Rabinowitz, *PNAS* **72**, 528 (1975).
30. H. H. Arnold, W. Schmidt and H. Kersten *FEBS Proc. Meet., 10th, Abstr.* No. 388 (1975).
31. R. O. R. Kaempfer, M. Meselson and H. J. Raskas, *JMB* **31**, 277 (1968).
32. M. Nomura and C. V. Lowry, *PNAS* **58**, 946 (1967); M. Nomura, C. V. Lowry and C. Guthrie, *ibid.* **58**, 1487 (1967); C. Guthrie and M. Nomura, *Nature* **219**, 232 (1968).
33. W. M. Stanley, Jr., M. Salas, A. J. Wahba and S. Ochoa, *PNAS* **56**, 290 (1966).
34. M. Revel and F. Gros, *BBRC* **25**, 124 (1966).
35. J. M. Eisenstadt and G. Brawerman, *Bchem* **5**, 2777 (1966).
36a. J. W. B. Hershey, E. Remold O'Donnell, D. Kolafoksky, K. F. Dewey and R. E. Thach, *in* "Methods of Enzymology" (Moldave and Grossman, eds.), Vol. XX, p. 235. Academic Press, New York, 1971.

36b. J. L. Fakunding, J. A. Traugh, R. R. Traut, J. W. Hershey, *in* "Methods of Enzymology" (Moldave and Grossman, eds.), Vol. XXX, p. 24. Academic Press, New York,1974; S. Lee-Huang and S. Ochoa, *ibid.*, p. 31.
37. J. Dondon, Th. Godefroy-Colburn, M. Graffe and M. Grunberg-Manager, *FEBS Lett.* **45**, 82 (1974).
38. S. Lee-Huang, M. A. G. Sillero and S. Ochoa, *EJB* **18**, 536 (1971).
39. F. Gros, J. C. Lelong, F. Berthelot, J. Dondon and M. Grunberg-Manager, *Int. Congr. Bchem., 8th, Abstr.* p. 197 (1970).
40. J. C. Lelong, Thèse Doctorat d'Etat, Paris VII (1975).
41. M. J. Miller and A. J. Wahba, *JBC* **248**, 1084 (1973).
42. J. C. Lelong, M. Grunberg-Manager, J. Dondon, D. Gros and F. Gros, *Nature* **226**, 505 (1970).
43. J. L. Fakunding, J. A. Traugh, R. R. Traut and J. W. B. Hershey, *JBC* **247**, 6365 (1972).
44. S. L. Hedrick and A. J. Smith, *ABB* **126**, 155 (1968).
45. S. Lee-Huang and S. Ochoa, *Nature NB* **234**, 236 (1971).
46. S. Lee-Huang and S. Ochoa, *ABB* **156**, 84 (1973).
47. N. Schiff, M. J. Miller and A. J. Wahba, *JBC* **249**, 3797 (1974).
48. S. Sabol, M. A. G. Sillero, K. Iwasaki and S. Ochoa, *Nature* **228**, 1269 (1970).
49. C. Gualerzi, C. L. Pon and A. Kaji, *BBRC* **45**, 1312 (1971).
50. J. S. Dubnoff and U. Maitra, *PNAS* **68**, 318 (1971).
51a. M. Springer and M. Grunberg-Manager, *FEBS Proc. Meet., 10th, Abstr. No. 457* (1975).
51b. M. Springer, M. Graffe and J. Hennecke, *PNAS*, in press (1977).
51c. M. Springer, M. Graffe and M. Grunberg-Manager, *Mol. Gen. Genet.* **151**, 17 (1977).
52. A. C. Kay, M. Graffe, and M. Grunberg-Manager, *Biochimie* **58**, 183 (1976).
53. S. Leffler and W. Szer, *JBC* **249**, 1458 (1974).
54. S. Leffler and W. Szer, *JBC* **249**, 1465 (1974).
55. T. Staehelin, H. Trachsel, B. Erni, A. Boschetti and M. H. Schreier, *FEBS Proc. Meet., 10th*, **39**, 309 (1975); M. H. Schreier, B. Erni, and T. Staehelin, *JMB* in press (1977).
56. M. H. Schreier and T. Staehelin, *JMB* **73**, 329 (1973); *in* "Regulation of Transcription and Translation in Eukaryotes" (E. F. K. Bautz, P. Karlson and H. Kersten, eds.), 24th Mosbach Colloq. p. 335. Springer-Verlag, Berlin and New York, 1973.
57a. P. M. Prichard, J. M. Gilbert, D. A. Shafritz and W. F. Anderson, *Nature* **226**, 511 (1970); D. A. Shafritz, P. M. Prichard, J. M. Gilbert, W. C. Merrick and W. F. Anderson, *PNAS* **69**, 983 (1972).
57b. B. Safer, S. L. Adams, W. M. Kemper, W. F. Anderson and W. C. Merrick, *Int. Congr. Biochem., 10th, Abstr.* p. 117 (1976).
58. Y. C. Chen, C. L. Woodley, K. K. Bose and N. K. Gupta, *BBRC* **48**, 1 (1972); N. K. Gupta and R. J. Aerni, *ibid.* **51**, 907 (1973).
59. D. H. Levin, D. Kyner and G. Acs, *JBC* **248**,6416 (1973).
60. S. M. Heywood, *PNAS* **67**, 1782 (1970); S. M. Heywood, D. S. Kennedy and A. J. Bester, *ibid.* **71**, 2428 (1974).
61. G. L. Dettman and W. M. Stanley, Jr., *BBA* **287**, 124 (1972); G. L. Dettman and W. M. Stanley, Jr., *ibid.* **299**, 142 (1973).
62. M. Zasloff and S. Ochoa, *PNAS* **69**, 1796 (1972); M. Zasloff and S. Ochoa, *ibid.* **68**, 3059 (1971).
63. P. H. van Knippenberg, J. van Duin and H. Lentz, *FEBS Lett.* **34**,95 (1973).
64. R. Mazumder, Y. B. Chae and S. Ochoa, *FP* **28**, 597, No. 1883 (1969).
65. P. S. Rudland, W. A. Whybrow and B. F. C. Clark, *Nature NB* **231**, 76 (1971).
66. A. H. Lockwood, P. R. Chakraborty and U. Maitra, *PNAS* **68**, 3122 (1971).

67a. Y. Groner and M. Revel, *EJB* **22**, 144 (1971).
67b. Y. Groner and M. Revel, *JMB* **74**, 407 (1973).
68. A. Majumdar, K. K. Bose, N. K. Gupta and A. J. Wahba, *JBC* in press (1977).
69. J. W. B. Hershey, K. F. Dewey and R. E. Thach, *Nature* **222**, 944 (1969).
70. S. Sabol and S. Ochoa, *Nature NB* **234**, 233 (1971).
71. J. Thibault, A. Chestier, D. Vidal and F. Gros, *Biochimie* **54**, 829 (1972).
72. C. Vermeer, R. J. de Kievit, W. J. van Alphen and L. Bosch, *FEBS Lett.* **31**, 273 (1973).
73. J. L. Fakunding and J. W. B. Hershey, *JBC* **248**, 4206 (1973).
74. D. Kolakofsky, K. F. Dewey, J. W. B. Hershey and R. E. Thach, *PNAS* **61**, 1066 (1968).
75. J. S. Dubnoff and U. Maitra, *JBC* **247**, 2876 (1972).
76. J. S. Dubnoff, A. H. Lockwood and U. Maitra, *JBC* **247**, 2884 (1972).
77. R. Benne and H. O. Voorma, *FEBS Lett.* **20**, 347 (1972).
78. R. Benne, R. Arentzen and H. O. Voorma, *BBA* **269**, 304 (1972).
79. A. C. Kay and M. Grunberg-Manago, *BBA* **277**, 225 (1972).
80. A. C. Kay, Thèse de Doctorat d'Etat, Paris VII (1975).
81. J. D. Watson, *Bull. Soc. Chim. Biol.* **46**, 1399 (1964).
82. R. Hauptmann, A. P. Czernilofsky, H. O. Voorma, G. Stöffler and E. Kuechler, *BBRC* **56**, 331 (1974).
83. H. Oen, M. Pellegrini, D. Eilat and C. R. Cantor, *PNAS* **70**, 2799 (1973).
84. A. P. Czernilofsky, E. E. Collatz, G. Stöffler and E. Kuechler, *PNAS* **71**, 230 (1974).
85. M. Springer, J. Dondon, M. Graffe, M. Grunberg-Manago, J. C. Lelong and F. Gros, *Biochimie* **53**, 1047 (1971).
86a. M. Grunberg-Manago, J. Dondon and M. Graffe, *FEBS Lett.* **22**, 217 (1972).
86b. D. Kolakofsky, T. Ohta and R. E. Thach, *Nature* **220**, 244 (1968).
87. R. Mazumder, *PNAS* **70**, 1939 (1973).
88. A. Kay, G. Sander and M. Grunberg-Manago, *BBRC* **51**, 979 (1973).
89. J. L. Fakunding, R. R. Traut and J. W. B. Hershey, *JBC* **248**, 8555 (1973).
90a. A. H. Lockwood, U. Maitra, N. Brot and H. Weissbach, *JBC* **249**, 1213 (1974).
90b. R. L. Heimark, J. W. B. Hershey and R. R. Traut, *JBC* in press.
91. E. Hamel, M. Koka and T. Nakamoto, *JBC* **247**, 805 (1972); E. Hamel and T. Nakamoto, *ibid.* **247**, 6810 (1972).
92. J. H. Highland, J. W. Bodley, J. Gordon, R. Hasenbank and G. Stöffler, *PNAS* **70**, 147 (1973).
93. G. Sander, R. C. Marsh and A. Parmeggiani, *BBRC* **47**, 866 (1972).
94. H. Yokosawa, N. Inoue-Yokosawa, K. I. Arai, M. Kawakita and Y. Kaziro, *JBC* **248**, 375 (1973).
95. N. Inoue-Yokosawa, C. Ishikawa and Y. Kaziro, *JBC* **249**, 4321 (1974).
96. N. V. Belitsina, M. A. Glukhova and A. S. Spirin, *FEBS Lett.* **54**, 35 (1975).
97. J. R. Horne and V. A. Erdmann, *PNAS* **70**, 2870 (1973).
98. M. G. Klöpfer and V. A. Erdmann, *BBA* **390**, 226 (1975).
99. J. H. Highland, G. A. Howard, E. Ochsner, G. Stöffler, R. Hasenbank and J. Gordon, *JBC* **250**, 1141 (1975).
100. J. Modolell, B. Cabrer, A. Parmeggiani and D. Vasquez, *PNAS* **68**, 1796 (1971).
101. J. Modolell, D. Vazquez and R. E. Monro, *Nature NB* **230**, 109 (1971).
102. G. Stöffler, *in* "Ribosomes" (M. Nomura, A. Tissières and Lengyel, eds.), p. 615. Cold Spring Harbor Laboratory, Cold Spring Harbor, New York, 1974.
103a. R. C. Marsh and A. Parmeggiani, *PNAS* **70**, 151 (1973).
103b. J. Voigt, G. Sander, K. Nagel and A. Parmeggiani, *BBRC* **57**, 1279 (1974).
104. G. Sander, R. C. Marsh, J. Voigt and A. Parmeggiani, *Bchem* **14**, 1805 (1975).

105a. H. Wolf, G. Chinali and A. Parmeggiani, *PNAS* **71**, 4910 (1974).
105b. G. Edwin Wilson, Jr. and M. Cohn, in press (1976).
106. C. Darnbrough, S. Legon, T. Hunt and R. J. Jackson, *JMB* **76**, 379 (1973).
107. A. Marcus, *JBC* **245**, 955, 962 (1970).
108. A. A. Infante and R. Baierlein, *PNAS* **68**, 1780 (1971).
109. A. S. Spirin, *FEBS Lett.* **14**, 399 (1971).
110. O. P. van Diggelen and L. Bosch, *EJB* **39**, 499 (1973).
111. H. Noll, M. Noll, B. Hapke and G. van Deijen, *in* "Regulation of Transcription and Translation in Eukaryotes" 24th Mosbach Colloq., (E. F. K. Bautz, P. Karlson and H. Kersten, eds.), p. 257. Springer-Verlag, Berlin and New York, 1973.
112. P. Debey, G. Hui Bon Hoa, P. Douzou, Th. Godefroy-Colburn, M. Graffe and M. Grunberg-Manago, *Bchem* **14**, 1553 (1975).
113. J. Reale, A. Scafati, M. R. Stornainolo and P. Novaro, *Biophys. J.* **11**, 370 (1971).
114. Y. Igarashi, T. Imamura, M. Suzuki and Y. Miyazawa, *Biochem. J. (Tokyo)* **73**, 683 (1973).
115. R. S. Zitomer and J. G. Flaks, *JMB* **71**, 263 (1972).
116. Th. Godefroy-Colburn, A. D. Wolfe, J. Dondon, M. Grunberg-Manago, Ph. Dessen and D. Pantaloni, *JMB* **94**, 461 (1975).
117. A. Wishnia, A. Boussert, M. Graffe, Ph. Dessen and M. Grunberg-Manago, *JMB* **93**, 499 (1975).
118. M. E. Andersen, J. K. Moffat and Q. H. Gibson, *JBC* **246**, 2796 (1971).
119a. B. D. Sykes, P. G. Schmidt and G. R. Stark, *JBC* **245**, 1180 (1970).
119b. K. Yoon, D. H. Turner and I. Tinoco, Jr., *JMB* **99**, 507 (1975).
119c. G. Hui Bon Hoa, M. Graffe and M. Grunberg-Manago, *Bchem*, in press (1977).
119d. J. S. Kliber, G. Hui Bon Hoa, P. Douzou, M. Graffe and M. Grunberg-Manago, *NARes* **3**, 3423 (1976).
120. J. S. Dubnoff and U. Maitra, *PNAS* **68**, 318 (1971).
121. M. Noll and H. Noll, *Nature NB* **238**, 225 (1972).
122. A. R. Subramanian and B. D. Davis, *Nature* **228**, 1273 (1970).
123. M. Grunberg-Manago, Th. Godefroy-Colburn, A. D. Wolfe, Ph. Dessen, D. Pantaloni, M. Springer, M. Graffe, J. Dondon and A. Kay, *in* "Regulation of Transcription and Translation in Eukaryotes" (E. F. K. Bautz, P. Karlson and H. Kersten, eds.), 24th Mosbach Colloq., p. 213. Springer-Verlag, Berlin and New York, 1973.
124. C. Vermeer, J. Boon, A. Talens and L. Bosch, *EJB* **40**, 283 (1973).
125. R. Kaempfer, *JMB* **71**, 583 (1972).
126. S. Sabol, D. Meier and S. Ochoa, *EJB* **33**, 332 (1973).
127. C. Gualerzi, M. R. Wabl and C. L. Pon, *FEBS Lett.* **35**, 313 (1973).
128. N. Schiff, M. J. Miller and A. J. Wahba, *JBC* **249**, 3797 (1974).
129. M. Gottlieb, B. D. Davis and R. C. Thompson, *PNAS* **72**, 4238 (1975).
130. K. Iwasaki, S. Sabol, A. J. Wahba and S. Ochoa, *ABB* **125**, 542 (1968).
131. M. Revel, J. C. Lelong, G. Brawerman and F. Gros, *Nature* **219**, 1016 (1968).
132. M. Revel, H. Aviv (Greenshpan), Y. Groner and Y. Pollack, *FEBS Lett.* **9**, 213 (1970).
133. C. Vermeer, W. van Alphen, P. van Knippenberg and L. Bosch, *EJB* **40**, 295 (1973).
134. A. J. Wahba, K. Iwasaki, M. J. Miller, S. Sabol, M. A. G. Sillero and C. Vasquez, *CSHSQB* **34**, 291 (1969).
135. M. Grunberg-Manago, J. C. Rabinowitz, J. Dondon, J. C. Lelong and F. Gros, *FEBS Lett.* **19**, 193 (1971).
136. M. Yoshida and P. S. Rudland, *JMB* **68**, 465 (1972).
137. J. S. Dubnoff, A. H. Lockwood and U. Maitra, *ABB* **149**, 528 (1972).
138. D. P. Suttle, M. A. Haralson and J. M. Ravel, *BBRC* **51**, 376 (1973).

139. S. D. Bernal, B. M. Blumberg and T. Nakamoto, *PNAS* **71**, 774 (1974).
140. D. Meier, S. Lee Huang and S. Ochoa, *JBC* **248**, 8613 (1973).
141. C. L. Pon and C. Gualerzi *PNAS* **71**, 4950 (1974).
142. A. S. Spirin, *CSHSQB* **34**, 197 (1969).
143. D. A. Hawley, M. J. Miller, L. I. Slobin and A. J. Wahba, *BBRC* **61**, 329 (1974).
144. M. Kozak and D. Nathan, *Bact. Rev.* **36**, 109 (1972).
145. M. Revel, M. Herzberg and H. Greenshpan, *CSHSQB* **34**, 261 (1969).
146. H. E. Lodish and H. D. Robertson, *CSHSQB* **34**, 655 (1969).
147. J. A. Steitz, *Nature* **224**, 957 (1969).
148. H. F. Lodish and H. D. Robertson, *JMB* **45**, 9 (1969).
149. H. E. Lodish, *Nature* **226**, 705 (1970).
150. J. A. Steitz, *JMB* **73**, 1 (1973).
151. S. Leffler and W. Szer, *PNAS* **70**, 2364 (1973).
152. W. A. Held, W. R. Gette and M. Nomura, *Bchem* **13**, 2115 (1974).
153. M. L. Goldberg and J. A. Steitz, *Bchem* **13**, 2123 (1974).
154. K. Higo, W. Held, L. Kahan and M. Nomura, *PNAS* **70**, 944 (1973).
155. S. Isono and K. Isono, *EJB* **56**, 15 (1975).
156. J. van Duin and P. H. van Knippenberg, *JMB* **84**, 185 (1974).
157. H. F. Noller, C. Chang, G. Thomas and J. Aldridge, *JMB* **61**, 669 (1971).
158. M. Tal, M. Aviram, A. Kanarek and A. Weiss, *BBA* **218**, 381 (1973).
159. P. H. van Knippenberg, P. J. J. Hooykaas and J. van Duin, *FEBS Lett.* **41**, 323 (1974).
160. W. Szer and S. Leffler, *PNAS* **71**, 3611 (1974).
161. J. M. Hermoso and W. Szer, *PNAS* **71**, 4708 (1974).
162. A. E. Dahlberg, *JBC* **249**, 7673 (1974).
163. G. van Dieijen, C. J. van der Laken, P. H. van Knippenberg and J. van Duin, *JMB* **93**, 351 (1975).
164. I. Fiser, K. H. Scheit, G. Stöffler and E. Kuechler, *FEBS Lett.* **56**, 226 (1975).
165. G. Jay and R. Kaempfer, *JMB* **82**, 193 (1974).
166. G. Jay and R. Kaempfer, *JBC* **250**, 5749 (1975).
167. A. J. Wahba, M. J. Miller, A. Niveleau, T. A. Landers, G. G. Carmichael, K. Weber, D. A. Hawley and L. I. Slobin, *JBC* **249**, 3314 (1974).
168. M. Revel, Y. Pollack, Y. Groner, R. Schebs, H. Inouye, H. Banssi and H. Zeller, *FEBS Proc. Meet., 8th* **27**, 261 (1972).
169. S. Lee-Huang and S. Ochoa, *BBRC* **49**, 371 (1972).
170. H. Inouye, Y. Pollack and J. Petre, *EJB* **45**, 109 (1974).
171. Y. Groner, R. Scheps, R. Kamen, D. Kolakofsky and M. Revel, *Nature NB* **239**, 19 (1972).
172. R. Kamen, M. Kondo, W. Römer and C. Weissman, *EJB* **31**, 44 (1972).
173. T. Blumenthal, T. A. Landers and K. Weber, *PNAS* **69**, 1313 (1972).
174. J. van Duin, P. H. van Knippenberg, M. Dieben and C. G. Kurland, *Mol. Gen. Genet.* **116**, 181 (1972).
175. L. L. Randall-Hazelbauer and C. G. Kurland, *Mol. Gen. Genet.* **115**, 234 (1972).
176. W. A. Held, M. Nomura and J. W. B. Hershey, *Mol. Gen. Genet.* **128**, 11 (1974).
177. M. Ozaki, S. Mizushima and M. Nomura, *Nature* **222**, 333 (1969).
178. E. A. Birge and C. G. Kurland, *Science* **166**, 1282 (1969).
179. G. Funatsu and H. G. Wittmann, *JMB* **68**, 547 (1972).
180a. M. Lazar and F. Gros, *Biochimie* **55**, 171 (1973).
180b. M. Lazar, Thèse Doctorat ès Sciences, Paris (1975).

181. J. C. Lelong, D. Gros, F. Gros, A. Bollen, R. Maschler and G. Stöffler, *PNAS* **71**, 248 (1974).
182. J. C. Lelong, C. Jeantet, M. Crepin, R. Maschler, G. Stöffler and F. Gros, in preparation.
183. C. G. Kurland, D. Donner, J. van Duin, M. Green, L. Lutter, L. Randall, H. W. Hazelbauer, H. W. Schaup and H. Zeichhardt, *FEBS Proc. Meet., 8th* **27**, 225 (1972).
184. M. Takanami, Y. Yan and T. H. Jukes, *JMB* **12**, 761 (1965).
185. B. G. Barrell (G. L. Cantoni and D. R. Davies, Eds.), Proc. in *NARes*, Harper and Row Publishers, **59**, 275 (1971).
186. J. A. Steitz, in "RNA Phages" (N. D. Zinder, ed.), p. 319. Cold Spring Harbor Laboratory, Cold Spring Harbor, New York, 1975.
187. S. L. Gupta, J. Chen, L. Schaefer, P. Lengyel and S. M. Weissman, *BBRC* **39**, 883 (1970).
188. W. Min Jou, G. Haegeman, M. Ysebaert and W. Fiers, *Nature* **237**, 82 (1972).
189. H. F. Lodish, *JMB* **32**, 681 (1968).
190. J. A. Steitz and K. Jakes, *PNAS* **72**, 4734 (1975).
191. G. Volckaert and W. Fiers, *FEBS Lett.* **35**, 91 (1973); R. de Wachter, J. Merregaert, A. Vandenberghe, R. Contrevas and W. Fiers, *EJB* **22**, 400 (1971); W. Fiers, R. Contrevas, F. Duerinck, G. Haegeman, D. Iserentant, J. Merregaert, W. Min Jou, F. Molemans, A. Raeymaekers, A. van der Berghe, G. Volckaert and M. Ysebaert, *Nature* **260**, 500 (1976).
192. J. A. Steitz, *PNAS* **70**, 2605 (1973).
193. J. G. Files, K. Weber and J. H. Miller, *PNAS* **71**, 667 (1974).
194. J. Shine and L. Dalgarno, *PNAS* **71**, 1342.
195. J. Shine and L. Dalgarno, *Nature* **254**, 34 (1975).
196. H. F. Noller and W. Hen, *Mol. Biol. Rep.* **8**, 437 (1974).
197. C. Ehresmann, P. Stiegler and J. P. Ebel, *FEBS Lett.* **49**, 47 (1974).
198. K. U. Sprague and J. A. Steitz, *NARes.* **2**, 787 (1975).
199. J. Konisky and M. Nomura, *JMB* **26**, 181 (1967).
200. C. M. Bowman, J. E. Dahlberg, T. Ikemura, J. Konisky and M. Nomura, *PNAS* **68**, 964 (1971).
201. B. W. Senior and I. B. Holland, *PNAS* **68**, 959 (1971).
202. T. Boon, *PNAS* **68**, 2421 (1971).
203. R. A. Kenner, *BBRC* **51**, 932 (1973).
204. D. A. Hawley, L. I. Slobin and A. J. Wahba, *BBRC* **61**, 544 (1974).
205. A. Bollen, R. L. Heimark, A. Cozzone, R. R. Traut, J. W. B. Hershey and L. Kahan, *JBC* **250**, 4310 (1975).
206. A. P. Czernilofsky, O. G. Kurland and G. Stöffler, *FEBS Lett.* **58**, 281 (1975).
207. A. E. Dahlberg and J. E. Dahlberg, *PNAS* **72**, 2940 (1975).
208. A. E. Dahlberg, E. Lund, N. O. Kjeldgaard, C. M. Bowman and M. Nomura, *Bchem* **12**, 948 (1973).
209. M. Crepin, J. C. Lelong and F. Gros, *Karolinska Symp. Res. Methods Reprod. Endocrinol., 6th* p. 33 (1973).
210. T. L. Melser, J. E. Davies and J. E. Dahlberg, *Nature NB* **233**, 12 (1971).
211. M. Revel and H. Greenshpan, *EJB* **16**, 117 (1970).
212. W. Szer and J. Brenowitz, *BBRC* **38**, 1154 (1970).
213a. H. F. Lodish, *JMB* **56**, 627 (1971).
213b. K. U. Sprague, J. A. Steitz, R. M. Grenley and C. E. Stocking, *Nature* in press (1977).

214a. A. P. Grollman and M. L. Stewart, *PNAS* **61**, 719 (1968).
214b. J. A. Steitz, K. U. Sprague, D. A. Steege, R. C. Yuan, M. Laughrea, P. B. Moore and A. J. Wahba, in "Nucleic Acid-Protein Recognition" (H. J. Vogel, ed.), p. 491. Academic Press, New York, 1977.
214c. D. G. Bear, R. Ng, D. Van Derveer, N. P. Johnson, G. Thomas, T. Schleich and H. Noller, *Bchem*, in press (1976).
214d. W. Szer, J. M. Hermoso and M. Boublik, *BBRC* **70**, 957 (1976).
214e. P. H. von Hippel and J. D. McGhee, *ARB* **41**, 231 (1972).
215. J. Van Duin; C. G. Kurland, J. Dondon and M. Grunberg-Manago, *FEBS Lett.* **59**, 287 (1975).
216. A. Bollen, R. L. Heimark, A. Cozzone, R. R. Traut and J. W. B. Hershey, *JBC* **250**, 4310 (1975); R. L. Heimark, L. Kahan, K. Johnston, J. W. B. Hershey and R. R. Traut, *JMB* **105**, 219 (1976).
217. R. A. Zimmermann, A. Muto, P. Fellner, C. Ehresmann and C. Branlant, *PNAS* **69**, 1282 (1972).
218. J. P. Ebel, P. Fellner, C. Ehresmann, P. Stiegler, J. L. Fischel, C. Branlant, A. Krol and J. Sriwidada, *FEBS Proc. Meet.*, 9th 109 (1973).
219. R. A. Zimmermann, A. Muto and G. A. Mackie, *JMB* **86**, 433 (1974).
220. A. Yuki and R. Brimacombe, *EJB* **56**, 23 (1975).
221. J. Van Duin, C. G. Kurland, J. Dondon, M. Grunberg-Manago, C. Branlant and J. P. Ebel, *FEBS Lett.* **62**, 111 (1976).
222. C. G. Kurland, in "Ribosomes" (M. Nomura, A. Tissières and P. Lengyel, eds.), p. 369. Cold Spring Harbor Laboratory, Cold Spring Harbor, New York, 1974.
223a. C. Branlant, J. Sriwidada, A. Krol and J. P. Ebel, *NARes* **3**, 1671 (1976).
223b. R. H. Buckingham and M. Grunberg-Manago, in preparation.
224. R. H. Buckingham, A. Danchin and M. Grunberg-Manago, *BBRC* **56**, 1 (1974).
225. H. H. Paradies, A. Franz, C. L. Pon and C. Gualerzi, *BBRC* **59**, 600 (1974).
226a. P. Baudry, U. Petersen, M. Grunberg-Manago and B. Jacrot, *BBRC* **72**, 391 (1976).
226b. C. Gualerzi, M. Grandolfo, H. H. Paradies and C. Pon, *JMB* **95**, 569 (1975).
226c. R. Ewald, C. Pon and C. Gualerzi, *Bchem* **15**, 4786 (1976).
226d. C. L. Pon and C. Gualerzi, *Bchem* **15**, 804 (1976).
227. R. A. Baan, J. J. Duijfjes, E. van Leerdam, P. H. van Knippenberg and L. Bosch, *PNAS* **73**, 702 (1976).
228. F. K. de Graaf, H. G. D. Niekus and J. Klootwijk, *FEBS Lett.* **35**, 161 (1973); L. Bosch, *FEBS Proc. Meet.*, 10th **39**, 275 (1975).
229. P. H. van Knippenberg, *NARes.* **2**, 79 (1975).
230. M. Yoshida and P. S. Rudland, *JMB* **68**, 465 (1972).
231. W. T. Hsu and S. B. Weiss, *PNAS* **64**, 345 (1969).
232. E. B. Klem, W. T. Hsu and S. B. Weiss, *PNAS* **67**, 696 (1970).
233. J. A. Steitz, S. K. Dube and P. S. Rudland, *Nature* **226**, 824 (1970).
234. W. Salser, R. F. Gesteland and B. Ricard, *CSHSQB* **34**, 771 (1969).
235. E. Goldman and H. F. Lodish, *JMB* **67**, 35 (1972).
236. J. M. Wilhelm and R. Haselkorn, *Virology* **43**, 198 (1971).
237a. E. Goldman and H. F. Lodish, *J. Virol.* **8**, 417 (1971); *JMB* **67**, 35 (1972).
237b. P. Leder, L. S. Skogerson and R. Callahan, *ABB* **153**, 814 (1972).
238. J. Shine and L. Dalgarno, *BJ* **141**, 608 (1974).
239. R. Das Gupta, D. S. Shih, C. Saris and P. Kaesberg, *Nature* **256**, 624 (1975).
240. H. Aviv, I. Boime, B. Loyd and P. Leder, *Science* **178**, 1293 (1972).
241. T. G. Morrisson and H. F. Lodish, *PNAS* **70**, 315 (1973).
242. M. H. Schreier, T. Staehelin, R. F. Gesteland and P. F. Spahr, *JMB* **75**, 575 (1973).
243. J. W. Davies and P. Kaesberg, *J. Virol.* **12**, 1434 (1973).

244. A. E. Smith, *Symp. Soc. Gen. Microbiol.*, **25th**, 183 (1975).
245. J. E. Darnell, W. R. Jelmek and G. R. Molloy, *Science* **181**, 1215 (1973).
246. A. E. Sippel, J. G. Stavrianopoulos, G. Schutz and P. Feigelson, *PNAS* **71**, 4635 (1974).
247. G. Huez, G. Marbaix, E. Hubert, M. Leclercq, U. Nudel, H. Soreq, R. Salomon, B. Lebleu, M. Revel and U. Z. Littauer, *PNAS* **71**, 3143 (1974).
248. R. H. Singer and S. Penman, *JMB* **78**, 321 (1973).
249. D. Schlessinger, in "The Mechanism of Protein Synthesis and Its Regulation" (L. Bosch, ed.), p. 44. North-Holland Publ. Amsterdam, 1972.
250. K. Miura, Y. Furuichi, K. Schimotohno, T. Urushibara and M. Sugiura, *FEBS Proc. Meet.*, **10th**, **39**, 95 (1975).
251. S. Muthukrishan, G. W. Both, Y. Furuichi and A. J. Shatkin, *Nature* **255**, 33 (1975).
252a. G. W. Both, A. R. Banerjee and A. J. Shatkin, *PNAS* **72**, 1189 (1975).
252b. D. A. Shafritz, J. A. Weinstein, B. Safer, W. C. Merrick, L. A. Weber, E. D. Hickey and C. Baglioni, *Nature* **261**, 291 (1976).
253. A. E. Smith, S. T. Bayley, W. F. Mangel, H. Shure, T. Wheeler and R. I. Kamen, *FEBS Proc. Meet.*, *10th* **39**, 151 (1975).
254. M. R. Morrison, S. A. Brinkley, J. Gorski and J. B. Lingrel, *JMB* **249**, 5290 (1974).
255a. M. F. Jacobson, J. Asso and D. Baltimore, *JMB* **49**, 657 (1970).
255b. W. H. Klein, C. Nolan, J. M. Lazar and J. M. Clark, Jr., *Bchem* **11**, 2009 (1972).
256. K. A. Marcker, G. E. Blair, H. H. M. Dahl and J. C. Lelong, *FEBS Proc. Meet.*, *10th* **39**, 297 (1975).
257. C. Clegg and I. Kennedy, *EJB* **53**, 175 (1975).
258. A. E. Smith, T. Wheeler, N. Glanville and L. Kaariainen, *EJB* **49**, 101 (1974).
259. D. T. Simons and J. H. Strauss, *JMB* **86**, 397 (1974).
260. E. Mohier, L. Hirth, M. A. Le Meur and P. Gerlinger, *FEBS Proc. Meet.*, *10th* **39**, 171 (1975).
261. M. N. Thang, L. Dondon, D. C. Thang, E. Mohier, L. Hirth, M. A. LeMeur and P. Gerlinger, in "Colloques INSERM" (A. L. Haenni and G. Beaud, eds.), 225 (1975).
262. R. Williamson, *FEBS Lett.* **37**, 1 (1973).
263. D. T. Wigle and A. E. Smith, *Nature NB* **242**, 136 (1973).
264. D. Bogdanovsky, W. Hermann and G. Schapira, *BBRC* **54**, 25 (1973); A. Berus, M. Salden, D. Bogdanovsky, J. M. Raymond, G. Schapira and H. Bloemendal, *PNAS* **72**, 714 (1973).
265. D. S. Kennedy, A. J. Bester and S. M. Heywood, *BBRC* **61**, 415 (1974).
266. R. Kaempfer, *PNAS* **70**, 1222 (1973).
267. G. Jay and R. Kaempfer, *PNAS* **71**, 3199 (1974).
268. G. Jay and R. Kaempfer, *JBC* **250**, 5742 (1975).
269. M. Crepin, J. C. Lelong and F. Gros, *Karolinska Symp. Res. Methods Reprod. Endocrinol.*, *6th*, p. 33 (1973).
270. D. Shine and K. Moldave, *JMB* **21**, 231 (1966).
271a. W. Szer and L. K. Porowska, in "Molecular Mechanism of Antibiotic Action on Protein Biosynthesis and Membranes" (E. Monos, F. G. Fernandez and D. Vasquez, eds.), p. 57. Elsevier, Amsterdam, 1972.
271b. C. Gualerzi, G. Risuleo, and C. L. Pon, *Biochem.* in press (1977).
272. H. U. Petersen, A. Danchin and M. Grunberg-Manago, *Bchem* **15**, 1357 (1976).
273. H. U. Petersen, A. Danchin and M. Grunberg-Manago, *Bchem* **15**, 1362 (1976).
274. M. Grunberg-Manago, B. F. C. Clark, M. Revel, P. S. Rudland and J. Dondon, *JMB* **40**, 33 (1969).
275. O. Pongs, G. Stöffler and E. Lanka, *JMB* **99**, 301 (1975).
276. C. G. Kurland, *ARB* **41**, 377 (1972).

277. L. Zagorska, J. Dondon, J. C. Lelong, F. Gros and M. Grunberg-Manago, *Biochimie* **53**, 63 (1971).
278. W. H. Anderson, L. Bosch, F. Gros, M. Grunberg-Manago, S. Ochoa, A. Reich and T. Staehelin, *FEBS Lett.* **48**, 1 (1974).
279. A. Kay, M. Graffe and M. Grunberg-Manago, *FEBS Lett.* **58**, 112 (1975).
280a. J. R. Horne and V. A. Erdmann, *PNAS* **70**, 2870 (1973); M. Gaunt Klopfer and V. A. Erdmann, *BBA* **390**, 226 (1975).
280b. J. J. Hopfield, *PNAS* **71**, 4135 (1974).
281. A. Danchin, *FEBS Lett.* **34**, 327 (1973).
282. R. J. Harvey, *J. Bact.* **114**, 309 (1973).
283. D. Richter, V. A. Erdmann and M. Sprinzl, *Nature NB* **246**, 132 (1973).
284. V. A. Erdmann, H. Sprinzl and O. Pongs, *BBRC* **54**, 942 (1975).
285. C. G. Kurland, R. Rigler, M. Ehrenberg and C. Blomberg, *PNAS* **72**, 4248 (1975).
286. F. H. C. Crick, *JMB* **38**, 367 (1969).
287. J. W. Hershey, J. Yanov, K. Johnston and J. L. Fakunding, *ABB*, in press (1977).
288. M. Noll and H. Noll, *JMB* **105**, 111 (1976).
289. L. Visentine, F. Hasnain, W. Gallin, K. Johnson, D. Griffith and A. Wabbha, *FEBS Lett.* in press.
290. A. R. Subramanian, C. Haase and M. Giesen, *EJB* **67**, 591 (1976).
291. D. Bonnet, H. U. Petersen, A. Danchin and M. Grunberg-Manago, in preparation.
292. D. H. Levin, R. S. Ranu, V. Ernst and I. M. London, *PNAS* **73**, 3112 (1976).
293. J. A. Traugh, S. M. Tahara, S. B. Sharp, B. Safer and W. C. Merrick, *Nature* **263**, 163 (1976).
294. O. G. Issinger, R. Benne, J. W. B. Hershey and R. R. Trent, *JBC* **251**, 6471 (1976).

Subject Index

A

Adenosine triphosphate, requirement, initiation complex and, 271–272
Adenovirus, human, endonuclease, 107–108
Antiviral activity, bleomycin and, 51
Apurinic sites, endonuclease and, 116–118

B

Base pairs, non-Watson-Crick, 8–9
Bleomycin
  chromatin effects, 44–46
  deoxyribonucleases and, 42–43
  DNA-dependent DNA polymerases and, 39–41
  DNA-dependent RNA polymerases and, 42
  DNA ligase and, 42
  DNA repair and, 48, 53–54
  DNA synthesis and, 46–48, 53
  effect on DNA in vivo, 43–44
  future directions
    DNA repair mechanisms, 53–54
    DNA synthesis, 53
    nucleotide sequences in DNA, 51–53
  gene expression and, 48–49
  medical implications, 49
    antiviral activity, 51
    cell kinetics, 50
    tumor specificity, 50–51
  protein synthesizing system and, 42
  reaction with native DNA in vitro, 23
    base specificity, 33–34
    production of strand scissions, 24–30
    reaction conditions, 30–33
    reaction mechanism, 34–37
  reaction with polynucleotides in vitro
    DNA · RNA hybrid, 38
    poly (ADP-ribose), 38
    RNA, 38
    single-stranded DNA, 37–38
  ribonucleases and phosphodiesterases, 43
  RNA-dependent DNA polymerases and, 42
  RNA effects in vivo, 46

C

Cell kinetics, bleomycin and, 50
Chromatin, bleomycin effects on, 44–46
2′ : 3′-Cyclic nucleotide 3′-phosphodiesterase, mammalian, 102

D

Deoxyribonuclease(s), bleomycin and 42–43
Deoxyribonuclease I, mammalian, 82–85
Deoxyribonuclease II, mammalian, 87–89
Deoxyribonuclease III, mammalian, 93–94
Deoxyribonucleic acid
  bleomycin effect in vivo, 43–44
  degradation in plants, 161–163, 203
    biochemical evidence, 163–188
    biological evidence, 188–196
    facts and fancy, 196–198
    missing links and further speculation, 198–203
  nucleotide sequences, bleomycin and, 51–53
  reaction with bleomycin in vitro, 23
    base specificity, 33–34
    production of strand scissions, 24–30
    reaction conditions, 30–33
    reaction mechanism, 34–37
  repair, bleomycin and, 48, 53–54
  single-stranded, bleomycin and, 37–38
  synthesis, bleomycin and, 46–48, 53
Deoxyribonucleic acid ligase, bleomycin and, 42
Deoxyribonucleic acid polymerases
  DNA-dependent, bleomycin and, 39–41
  RNA-dependent, bleomycin and, 42
Deoxyribonucleic acid · ribonucleic acid hybrids, bleomycin and, 38

SUBJECT INDEX

E

Endodeoxyribonucleases
  Ca, Mg-dependent, 85–87
  deoxyribonuclease I, 82–85
  deoxyribonuclease II, 87–89
  "nicking-closing" enzymes, 91–93
  "nicking" enzymes, 89–91
Endo-exodeoxyribonuclease, with preference for poly (dT), 95
Endonuclease, single-stranded-nucleate, sugar nonspecific, 95–97
Endoribonucleases
  hybrid nuclease, 78–80
  nucleolar ribonuclease, 72–73
  "processing" ribonucleases, 76–78
  ribonuclease I, 61–72
  ribonuclease II, 73–75
  ribonucleases attacking double-stranded RNA, 75–76
Eukaryotes
  initiation signals, mechanism of cistron recognition, 263–267
  protein synthesis in, 226–227
Exodeoxyribonucleases
  deoxyribonuclease III, 93–94
  exodeoxyribonuclease IV, 94–95
Exonuclease(s), sugar nonspecific
  phosphodiesterase I, 97–100
  phosphodiesterase II, 100–102
Exoribonucleases
  5'-exoribonuclease, 80–81
  polynucleotide phosphorylase, 81–82

F

Formyl groups, prokaryote initiator tRNA and, 272–274

G

Gene expression, bleomycin and, 48–49

I

Initiation factors, protein synthesis and, 215–220
Initiation signals
  role of ribosomes and initiation factors
    mechanism of cistron recognition in eukaryotes, 263–267
    mechanism of cistron recognition in prokaryotes, 247–263
    requirement for IF3 with synthetic messengers, 235–237
    structure of initiation signals, 237–247

N

Nicking enzymes, mammalian, 89–93
Nucleases
  cellular repair processes, 114–115
    endonuclease specific for apurinic sites, 116–118
    excision exonucleases, 118–119
    mammalian "incision" endonuclease, 115–116
    deficiency in xeroderma pigmentosa, 119–120
  localization and function, 102–106
  virus-associated, 106–107
    deoxyribonucleases in other viruses, 108–110
    endodeoxyribonuclease in polyoma virus, 107
    human adenovirus endonuclease, 107–108
    ribonuclease H, 110–114
    ribonucleases, 110

P

Phosphodiesterase(s), bleomycin and, 43
Phosphodiesterase I, mammalian, 97–100
Phosphodiesterase II, mammalian, 100–102
Plants, exogenous DNA degradation in, 161–163, 203
  biochemical evidence, 163–188
  biological evidence, 188–196
  facts and fancy, 196–198
  missing links and further speculation, 198–203
Poly(adenosine diphosphate ribose), bleomycin and, 38
Polynucleotide phosphorylase, mammalian, 81–82
Polyoma virus, endodeoxyribonuclease, 107
Prokaryotes
  initiation signals, mechanism of cistron recognition, 247–263
  initiator tRNA, formyl groups and, 268–270

SUBJECT INDEX

operational sequence of initiation complex formation in, 268–270
protein synthesis in, 220–226
Protein, synthesizing system, bleomycin and, 42
Protein synthesis
  components involved in initiation step
    initiation factors, 215–220
    initiator tRNA, 210–214
  initiation mechanisms
    ATP requirement, 271–272
    number of tRNA binding sites on 30 S subunit, 270–271
    operational sequence of initiation complex formation in prokaryotes, 268–270
    role of formyl groups in prokaryote initiator tRNA, 272–274
  reversible association of ribosomal subunits, 227–235
  total reaction sequence
    eukaryotes, 226–227
    prokaryotes, 220–226
Pseudouridine, position of, 9–10

R

Ribonuclease(s)
  bleomycin and, 43
  double-stranded RNA and, 75–76
  hybrid, 78–80
  nucleolar, 72–73
  "processing," 76–78
  viral, 110
Ribonuclease H, viral, 110–114
Ribonuclease I, mammalian, 61–72
Ribonuclease II, mammalian, 73–75
Ribonucleic acid
  bleomycin and, 38, 46
  tumor virus
    4 S to 28 S, 136–138
    35 S and 70 S, 134–136
Ribonucleic acid polymerases, DNA-dependent, bleomycin and, 42
Ribosomes
  subunits, reversible association, 227–235
  30 S subunits, number of tRNA binding sites, 270–271
Ribosomes and initiation factors, recognition of initiation signals and

cistron recognition in eukaryotes, 263–267
cistron recognition in prokaryotes, 247–263
requirement for IF3, 235–237
structure of signals, 237–247

S

Synthetic messengers, requirement of IF3 and, 235–237

T

Transfer ribonucleic acid
  biological activities, 1–4
  chemical reactivity and three-dimensional structure, 17–18
Transfer ribonucleic acid
  exceptions to generalized structure, 10–11
  initiator, protein synthesis and, 210–214
  phenylalanine, three-dimensional structure, 11–15
  psuedouridine position and, 9–10
  significance of G · U base pairs, 8–9
  structural correlations, 15–17
  subclassification and generalized primary structure, 4–8
  tumor virus,
    4 S, 138–140
    70 S, 141–146
    function of, 146–155
Tumor specificity, bleomycin and, 50–51
Tumor viruses
  host tRNA and, 131–133
  nucleic acid components of RNA
    35 S and 70 S, 134–136
    4 S to 28 S, 136–138
  transfer RNA
    4 S, 138–140
    70 S, 141–146
    function of, 146–155

V

Viruses, nucleases of, 106–114

X

Xeroderma pigmentosa, repair deficiency in, 119–120

Contents of Previous Volumes

Volume 1

"Primer" in DNA Polymerase Reactions
F. J. Bollum

The Biosynthesis of Ribonucleic Acid in Animal Systems
R. M. S. Smellie

The Role of DNA in RNA Synthesis
Jerard Hurwitz and J. T. August

Polynucleotide Phosphorylase
M. Grunberg-Manago

Messenger Ribonucleic Acid
Fritz Lipmann

The Recent Excitement in the Coding Problem
F. H. C. Crick

Some Thoughts on the Double-Stranded Model of Deoxyribonucleic Acid
Aaron Bendich and Herbert S. Rosenkranz

Denaturation and Renaturation of Deoxyribonucleic Acid
J. Marmur, R. Rownd, and C. L. Schildkraut

Some Problems Concerning the Macromolecular Structure of Ribonucleic Acids
A. S. Spirin

The Structure of DNA as Determined by X-Ray Scattering Techniques
Vittorio Luzzati

Molecular Mechanisms of Radiation Effects
A. Wacker

Author Index—Subject Index

Volume 2

Nucleic Acids and Information Transfer
Liebe F. Cavalieri and Barbara H. Rosenberg

Nuclear Ribonucleic Acid
Henry Harris

Plant Virus Nucleic Acids
Roy Markham

The Nucleases of *Escherichia coli*
I. R. Lehman

Specificity of Chemical Mutagenesis
David R. Krieg

Column Chromatography of Oligonucleotides and Polynucleotides
Matthys Staehelin

Mechanism of Action and Application of Azapyrimidines
J. Skoda

The Function of the Pyrimidine Base in the Ribonuclease Reaction
Herbert Witzel

Preparation, Fractionation, and Properties of sRNA
G. L. Brown

Author Index—Subject Index

Volume 3

Isolation and Fractionation of Nucleic Acids
K. S. Kirby

Cellular Sites of RNA Synthesis
David M. Prescott

Ribonucleases in Taka-Diastase: Properties, Chemical Nature, and Applications
Fujio Egami, Kenji Takahashi, and Tsuneko Uchida

Chemical Effects of Ionizing Radiations on Nucleic Acids and Related Compounds
Joseph J. Weiss

The Regulation of RNA Synthesis in Bacteria
  FREDERICK C. NEIDHARDT

Actinomycin and Nucleic Acid Function
  E. REICH AND I. H. GOLDBERG

De Novo Protein Synthesis in Vitro
  B. NISMAN AND J. PELMONT

Free Nucleotides in Animal Tissues
  P. MANDEL

AUTHOR INDEX—SUBJECT INDEX

Volume 4

Fluorinated Pyrimidines
  CHARLES HEIDELBERGER

Genetic Recombination in Bacteriophage
  E. VOLKIN

DNA Polymerases from Mammalian Cells
  H. M. KEIR

The Evolution of Base Sequences in Polynucleotides
  B. J. MCCARTHY

Biosynthesis of Ribosomes in Bacterial Cells
  SYOZO OSAWA

5-Hydroxymethylpyrimidines and Their Derivatives
  T. L. V. ULBRIGHT

Amino Acid Esters of RNA, Nucleotides, and Related Compounds
  H. G. ZACHAU AND H. FELDMANN

Uptake of DNA by Living Cells
  L. LEDOUX

AUTHOR INDEX—SUBJECT INDEX

Volume 5

Introduction to the Biochemistry of 4-Arabinosyl Nucleosides
  SEYMOUR S. COHEN

Effects of Some Chemical Mutagens and Carcinogens on Nucleic Acids
  P. D. LAWLEY

Nucleic Acids in Chloroplasts and Metabolic DNA
  TATSUICHI IWAMURA

Enzymatic Alteration of Macromolecular Structure
  P. R. SRINIVASAN AND ERNEST BOREK

Hormones and the Synthesis and Utilization of Ribonucleic Acids
  J. R. TATA

Nucleoside Antibiotics
  JACK J. FOX, KYOICHI A. WATANABE, AND ALEXANDER BLOCH

Recombination of DNA Molecules
  CHARLES A. THOMAS, JR.

  Appendix I. Recombination of a Pool of DNA Fragments with Complementary Single-Chain Ends
    G. S. WATSON, W. K. SMITH, AND CHARLES A. THOMAS, JR.

  Appendix II. Proof that Sequences of A, C, G, and T Can Be Assembled to Produce Chains of Ultimate Length, Avoiding Repetitions Everywhere
    A. S. FRAENKEL AND J. GILLIS

The Chemistry of Pseudouridine
  ROBERT WARNER CHAMBERS

The Biochemistry of Pseudouridine
  EUGENE GOLDWASSER AND ROBERT L. HEINRIKSON

AUTHOR INDEX—SUBJECT INDEX

Volume 6

Nucleic Acids and Mutability
  STEPHEN ZAMENHOF

Specificity in the Structure of Transfer RNA
  KIN-ICHIRO MIURA

Synthetic Polynucleotides
   A. M. MICHELSON, J. MASSOULIÉ,
   AND W. GUSCHBAUER

The DNA of Chloroplasts, Mitochondria, and Centrioles
   S. GRANICK AND AHARON GIBOR

Behavior, Neural Function, and RNA
   H. HYDÉN

The Nucleolus and the Synthesis of Ribosomes
   ROBERT P. PERRY

The Nature and Biosynthesis of Nuclear Ribonucleic Acids
   G. P. GEORGIEV

Replication of Phage RNA
   CHARLES WEISSMANN AND
   SEVERO OCHOA

AUTHOR INDEX—SUBJECT INDEX

Volume 7

Autoradiographic Studies on DNA Replication in Normal and Leukemic Human Chromosomes
   FELICE GAVOSTO

Proteins of the Cell Nucleus
   LUBOMIR S. HNILICA

The Present Status of the Genetic Code
   CARL R. WOESE

The Search for the Messenger RNA of Hemoglobin
   H. CHANTRENNE, A. BURNY, AND
   G. MARBAIX

Ribonucleic Acids and Information Transfer in Animal Cells
   A. A. HADJIOLOV

Transfer of Genetic Information during Embryogenesis
   MARTIN NEMER

Enzymatic Reduction of Ribonucleotides
   AGNE LARSSON AND PETER REICHARD

The Mutagenic Action of Hydroxylamine
   J. H. PHILLIPS AND D. M. BROWN

Mammalian Nucleolytic Enzymes and Their Localization
   DAVID SHUGAR AND
   HALINA SIERAKOWSKA

AUTHOR INDEX—SUBJECT INDEX

Volume 8

Nucleic Acids—The First Hundred Years
   J. N. DAVIDSON

Nucleic Acids and Protamine in Salmon Testes
   GORDON H. DIXON AND
   MICHAEL SMITH

Experimental Approaches to the Determination of the Nucleotide Sequences of Large Oligonucleotides and Small Nucleic Acids
   ROBERT W. HOLLEY

Alterations of DNA Base Composition in Bacteria
   G. F. GAUSE

Chemistry of Guanine and Its Biologically Significant Derivatives
   ROBERT SHAPIRO

Bacteriophage φX174 and Related Viruses
   ROBERT L. SINSHEIMER

The Preparation and Characterization of Large Oligonucleotides
   GEORGE W. RUSHIZKY AND
   HERBERT A. SOBER

Purine N-Oxides and Cancer
   GEORGE BOSWORTH BROWN

The Photochemistry, Photobiology, and Repair of Polynucleotides
   R. B. SETLOW

What Really Is DNA? Remarks on the Changing Aspects of a Scientific Concept
ERWIN CHARGAFF

Recent Nucleic Acid Research in China
TIEN-HSI CHENG AND ROY H. DOI

AUTHOR INDEX—SUBJECT INDEX

Volume 9

The Role of Conformation in Chemical Mutagenesis
B. SINGER AND H. FRAENKEL-CONRAT

Polarographic Techniques in Nucleic Acid Research
E. PALEČEK

RNA Polymerase and the Control of RNA Synthesis
JOHN P. RICHARDSON

Radiation-Induced Alterations in the Structure of Deoxyribonucleic Acid and Their Biological Consequences
D. T. KANAZIR

Optical Rotatory Dispersion and Circular Dichroism of Nucleic Acids
JEN TSI YANG AND TATSUYA SAMEJIMA

The Specificity of Molecular Hybridization in Relation to Studies on Higher Organisms
P. M. B. WALKER

Quantum-Mechanical Investigations of the Electronic Structure of Nucleic Acids and Their Constituents
BERNARD PULLMAN AND ALBERTE PULLMAN

The Chemical Modification of Nucleic Acids
N. K. KOCHETKOV AND E. I. BUDOWSKY

AUTHOR INDEX—SUBJECT INDEX

Volume 10

Induced Activation of Amino Acid Activating Enzymes by Amino Acids and tRNA
ALAN H. MEHLER

Transfer RNA and Cell Differentiation
NOBORU SUEOKA AND TAMIKO KANO-SUEOKA

N^6-(Δ^2-Isopentenyl)adenosine: Chemical Reactions, Biosynthesis, Metabolism, and Significance to the Structure and Function of tRNA
ROSS H. HALL

Nucleotide Biosynthesis from Preformed Purines in Mammalian Cells: Regulatory Mechanisms and Biological Significance
A. W. MURRAY, DAPHNE C. ELLIOTT, AND M. R. ATKINSON

Ribosome Specificity of Protein Synthesis in Vitro
ORIO CIFERRI AND BRUNO PARISI

Synthetic Nucleotide-peptides
ZOE A. SHABAROVA

The Crystal Structures of Purines, Pyrimidines and Their Intermolecular Complexes
DONALD VOET AND ALEXANDER RICH

AUTHOR INDEX—SUBJECT INDEX

Volume 11

The Induction of Interferon by Natural and Synthetic Polynucleotides
CLARENCE COLBY, JR.

Ribonucleic Acid Maturation in Animal Cells
R. H. BURDON

Liporibonucleoprotein as an Integral Part of Animal Cell Membranes
V. S. SHAPOT AND S. YA. DAVIDOVA

Uptake of Nonviral Nucleic Acids by Mammalian Cells
  Pushpa M. Bhargava and G. Shanmugam

The Relaxed Control Phenomenon
  Ann M. Ryan and Ernest Borek

Molecular Aspects of Genetic Recombination
  Cedric I. Davern

Principles and Practices of Nucleic Acid Hybridization
  David E. Kennell

Recent Studies Concerning the Coding Mechanism
  Thomas H. Jukes and Lila Gatlin

The Ribosomal RNA Cistrons
  M. L. Birnstiel, M. Chipchase, and J. Speirs

Three-Dimensional Structure of tRNA
  Friedrich Cramer

Current Thoughts on the Replication of DNA
  Andrew Becker and Jerard Hurwitz

Reaction of Aminoacyl-tRNA Synthetases with Heterologous tRNA's
  K. Bruce Jacobson

On the Recognition of tRNA by Its Aminoacyl-tRNA Ligase
  Robert W. Chambers

Author Index—Subject Index

Volume 12

Ultraviolet Photochemistry as a Probe of Polyribonucleotide Conformation
  A. J. Lomant and Jacques R. Fresco

Some Recent Developments in DNA Enzymology
  Mehran Goulian

Minor Components in Transfer RNA: Their Characterization, Location, and Function
  Susumu Nishimura

The Mechanism of Aminoacylation of Transfer RNA
  Robert B. Loftfield

Regulation of RNA Synthesis
  Ekkehard K. F. Bautz

The Poly(dA-dT) of Crab
  M. Laskowski, Sr.

The Chemical Synthesis and the Biochemical Properties of Peptidyl-tRNA
  Yehuda Lapidot and Nathan de Groot

Subject Index

Volume 13

Reactions of Nucleic Acids and Nucleoproteins with Formaldehyde
  M. Ya. Feldman

Synthesis and Functions of the -C-C-A Terminus of Transfer RNA
  Murray P. Deutscher

Mammalian RNA Polymerases
  Samson T. Jacob

Poly(adenosine diphosphate ribose)
  Takashi Sugimura

The Stereochemistry of Actinomycin Binding to DNA and Its Implications in Molecular Biology
  Henry M. Sobell

Resistance Factors and Their Ecological Importance to Bacteria and to Man
  M. H. Richmond

Lysogenic Induction
  Ernest Borek and Ann Ryan

Recognition in Nucleic Acids and the Anticodon Families
JACQUES NINIO

Translation and Transcription of the Tryptophan Operon
FUMIO IMAMOTO

Lymphoid Cell RNA's and Immunity
A. ARTHUR GOTTLIEB

SUBJECT INDEX

Volume 14

DNA Modification and Restriction
WERNER ARBER

Mechanism of Bacterial Transformation and Transfection
NIHAL K. NOTANI AND JANE K. SETLOW

DNA Polymerases II and III of *Escherichia coli*
MALCOLM L. GEFTER

The Primary Structure of DNA
KENNETH MURRAY AND ROBERT W. OLD

RNA-Directed DNA Polymerase—Properties and Functions in Oncogenic RNA Viruses and Cells
MAURICE GREEN AND GRAY F. GERARD

SUBJECT INDEX

Volume 15

Information Transfer in Cells Infected by RNA Tumor Viruses and Extension to Human Neoplasia
D. GILLESPIE, W. C. SAXINGER, AND R. C. GALLO

Mammalian DNA Polymerases
F. J. BOLLUM

Eukaryotic RNA Polymerases and the Factors That Control Them
B. B. BISWAS, A. GANGULY, AND D. DAS

Structural and Energetic Consequences of Non-complementary Base Oppositions in Nucleic Acid Helices
A. J. LOMANT AND JACQUES R. FRESCO

The Chemical Effects of Nucleic Acid Alkylation and Their Relation to Mutagenesis and Carcinogenesis
B. SINGER

Effects of the Antibiotics Netropsin and Distamycin A on the Structure and Function of Nucleic Acids
CHRISTOPH ZIMMER

SUBJECT INDEX

Volume 16

Initiation of Enzymic Synthesis of Deoxyribonucleic Acid by Ribonucleic Acid Primers
ERWIN CHARGAFF

Transcription and Processing of Transfer RNA Precursors
JOHN D. SMITH

Bisulfite Modification of Nucleic Acids and Their Constituents
HIKOYA HAYATSU

The Mechanism of the Mutagenic Action of Hydroxylamines
E. I. BUDOWSKY

Diethyl Pyrocarbonate in Nucleic Acid Research
L. EHRENBERG, I. FEDORCSÁK, AND F. SOLYMOSY

SUBJECT INDEX

CONTENTS OF PREVIOUS VOLUMES

Volume 17

The Enzymic Mechanism of Guanosine 5',3'-Polyphosphate Synthesis
  FRITZ LIPMANN AND JOSE SY

Effects of Polyamines on the Structure and Reactivity of tRNA
  TED T. SAKAI AND SEYMOUR S. COHEN

Information Transfer and Sperm Uptake by Mammalian Somatic Cells
  AARON BENDICH, ELLEN BORENFREUND, STEVEN S. WITKINS, DELIA BEJU, AND PAUL J. HIGGINS

Studies on the Ribosome and Its Components
  PNINA SPITNIK-ELSON AND DAVID ELSON

Classical and Postclassical Modes of Regulation of the Synthesis of Degradative Bacterial Enzymes
  BORIS MAGASANIK

Characteristics and Significance of the Polyadenylate Sequence in Mammalian Messenger RNA
  GEORGE BRAWERMAN

Polyadenylate Polymerases
  MARY EDMONDS AND MARY ANN WINTERS

Three-Dimensional Structure of Transfer RNA
  SUNG-HOU KIM

Insights into Protein Biosynthesis and Ribosome Function through Inhibitors
  SIDNEY PESTKA

Interaction with Nucleic Acids of Carcinogenic and Mutagenic N-Nitroso Compounds
  W. LIJINSKY

Biochemistry and Physiology of Bacterial Ribonuclease
  ALOK K. DATTA AND SALIL K. NIYOGI

SUBJECT INDEX

Volume 18

The Ribosome of Escherichia coli
  R. BRIMACOMBE, K. H. NIERHAUS, R. A. GARRETT AND H. G. WITTMANN

Structure and Function of 5 S and 5.8 S RNA
  VOLKER A. ERDMANN

High-Resolution Nuclear Magnetic Resonance Investigations of the Structure of tRNA in Solution
  DAVID R. KEARNS

Premelting Changes in DNA Conformation
  E. PALEČEK

Quantum-Mechanical Studies on the Conformation of Nucleic Acids and Their Constituents
  BERNARD PULLMAN AND ANIL SARAN

SUBJECT INDEX

Volume 19

I. The 5'-Terminal Sequence ("Cap") of mRNAs

Caps in Eukaryotic mRNAs: Mechanism of Formation of Reovirus mRNA 5'-Terminal $m^7GpppGm$-C
  Y. FURUICHI, S. MUTHUKRISHNAN, J. TOMASZ AND A. J. SHATKIN

Nucleotide Methylation Patterns in Eukaryotic mRNA
  FRITZ M. ROTTMAN, RONALD C. DESROSIERS AND KAREN FRIDERICI

Structural and Functional Studies on the "5'-Cap": A Survey Method for mRNA
  HARRIS BUSCH, FRIEDRICH HIRSCH, KAUSHAL KUMAR GUPTA, MANCHANAHALLI RAO, WILLIAM SPOHN AND BENJAMIN C. WU

Modification of the 5'-Terminals of mRNAs by Viral and Cellular Enzymes
  BERNARD MOSS, SCOTT A. MARTIN, MARCIA J. ENSINGER, ROBERT F. BOONE AND CHA-MER WEI

Blocked and Unblocked 5' Termini in Vesicular Stomatitis Virus Product RNA *in Vitro:* Their Possible Role in mRNA Biosynthesis

RICHARD J. COLONNO, GORDON ABRAHAM AND AMIYA K. BANERJEE

The Genome of Poliovirus Is an Exceptional Eukaryotic mRNA

YUAN FON LEE, AKIO NOMOTO AND ECKARD WIMMER

II. Sequences and Conformations of mRNAs

Transcribed Oligonucleotide Sequences in Hela Cell hnRNA and mRNA

MARY EDMONDS, HIROSHI NAKAZATO, E. L. KORWEK AND S. VENKATESAN

Polyadenylylation of Stored mRNA in Cotton Seed Germination

BARRY HARRIS AND LEON DURE III

mRNAs Containing and Lacking Poly(A) Function as Separate and Distinct Classes during Embryonic Development

MARTIN NEMER AND SAUL SURREY

Sequence Analysis of Eukaryotic mRNA

N. J. PROUDFOOT, C. C. CHENG AND G. G. BROWNLEE

The Structure and Function of Protamine mRNA from Developing Trout Testis

P. L. DAVIES, G. H. DIXON, L. N. FERRIER, L. GEDAMU AND K. IATROU

The Primary Structure of Regions of SV40 DNA Encoding the Ends of mRNA

KIRANUR N. SUBRAMANIAN, PRABHAT K. GHOSH, RAVI DHAR, BAYAR THIMMAPPAYA, SAYEEDA B. ZAIN, JULIAN PAN AND SHERMAN M. WEISSMAN

Nucleotide Sequence Analysis of Coding and Noncoding Regions of Human β-Globin mRNA

CHARLES A. MAROTTA, BERNARD G. FORGET, MICHAEL COHEN/SOLAL AND SHERMAN M. WEISSMAN

Determination of Globin mRNA Sequences and Their Insertion into Bacterial Plasmids

WINSTON SALSER, JEFF BROWNE, PAT CLARKE, HOWARD HEINDELL, RUSSELL HIGUCHI, GARY PADDOCK, JOHN ROBERTS, GARY STUDNICKA AND PAUL ZAKAR

The Chromosomal Arrangement of Coding Sequences in a Family of Repeated Genes

G. M. RUBIN, D. J. FINNEGAN AND D. S. HOGNESS

Mutation Rates in Globin Genes: The Genetic Load and Haldane's Dilemma

WINSTON SALSER AND JUDITH STROMMER ISAACSON

Heterogeneity of the 3' Portion of Sequences Related to Immunoglobulin κ-Chain mRNA

URSULA STORB

Structural Studies on Intact and Deadenylylated Rabbit Globin mRNA

JOHN N. VOURNAKIS, MARCIA S. FLASHNER, MARYANN KATOPES, GARY A. KITOS, NIKOS C. VAMVAKOPOULOS, MATTHEW S. SELL AND REGINA M. WURST

Molecular Weight Distribution of RNA Fractionated on Aqueous and 70% Formamide Sucrose Gradients

HELGA BOEDTKER AND HANS LEHRACH

III. Processing of mRNAs

Bacteriophages T7 and T3 as Model Systems for RNA Synthesis and Processing

J. J. DUNN, C. W. ANDERSON, J. F. ATKINS, D. C. BARTELT AND W. C. CROCKETT

The Relationship between hnRNA and mRNA

ROBERT P. PERRY, ENZO BARD, B. DAVID HAMES, DAWN E. KELLEY AND UELI SCHIBLER

A Comparison of Nuclear and Cytoplasmic Viral RNAs Synthesized Early in Productive Infection with Adenovirus 2

HESCHEL J. RASKAS AND ELIZABETH A. CRAIG

CONTENTS OF PREVIOUS VOLUMES

Biogenesis of Silk Fibroin mRNA: An Example of Very Rapid Processing?
  PAUL M. LIZARDI

Visualization of the Silk Fibroin Transcription Unit and Nascent Silk Fibroin Molecules on Polyribosomes of Bombyx mori
  STEVEN L. MCKNIGHT, NELDA L. SULLIVAN AND OSCAR L. MILLER, JR.

Production and Fate of Balbiani Ring Products
  B. DANEHOLT, S. T. CASE, J. HYDE, L. NELSON AND L. WIESLANDER

Distribution of hnRNA and mRNA Sequences in Nuclear Ribonucleoprotein Complexes
  ALAN J. KINNIBURGH, PETER B. BILLINGS, THOMAS J. QUINLAN AND TERENCE E. MARTIN

IV. Chromatin Structure and Template Activity

The Structure of Specific Genes in Chromatin
  RICHARD AXEL

The Structure of DNA in Native Chromatin as Determined by Ethidium Bromide Binding
  J. PAOLETTI, B. B. MAGEE AND P. T. MAGEE

Cellular Skeletons and RNA Messages
  RONALD HERMAN, GARY ZIEVE, JEFFREY WILLIAMS, ROBERT LENK AND SHELDON PENMAN

The Mechanism of Steroid-Hormone Regulation of Transcription of Specific Eukaryotic Genes
  BERT W. O'MALLEY AND ANTHONY R. MEANS

Nonhistone Chromosomal Proteins and Histone Gene Transcription
  GARY STEIN, JANET STEIN, LEWIS KLEINSMITH, WILLIAM PARK, ROBERT JANSING AND JUDITH THOMSON

Selective Transcription of DNA Mediated by Nonhistone Proteins
  TUNG Y. WANG, NINA C. KOSTRABA AND RUTH S. NEWMAN

V. Control of Translation

Structure and Function of the RNAs of Brome Mosaic Virus
  PAUL KAESBERG

Effect of 5'-Terminal Structures on the Binding of Ribopolymers to Eukaryotic Ribosomes
  S. MUTHUKRISHNAN, Y. FURUICHI, G. W. BOTH AND A. J. SHATKIN

Translational Control in Embryonic Muscle
  STUART M. HEYWOOD AND DORIS S. KENNEDY

Protein and mRNA Synthesis in Cultured Muscle Cells
  R. G. WHALEN, M. E. BUCKINGHAM AND F. GROS

VI. Summary

mRNA Structure and Function
  JAMES E. DARNELL

SUBJECT INDEX

QP
551
P695
v.20
1977

SEP 26 1977